T0275581

LONDON MATHEMATICAL SOCIETY LECTURE NOTE SERIES

Managing Editor: Professor J.W.S. Cassels, Department of Pure Mathematics and Mathematical Statistics, University of Cambridge, 16 Mill Lane, Cambridge CB2 1SB, England

The titles below are available from booksellers, or, in case of difficulty, from Cambridge University Press.

34   Representation theory of Lie groups, M.F. ATIYAH *et al*
46   p-adic analysis: a short course on recent work, N. KOBLITZ
50   Commutator calculus and groups of homotopy classes, H.J. BAUES
59   Applicable differential geometry, M. CRAMPIN & F.A.E. PIRANI
66   Several complex variables and complex manifolds II, M.J. FIELD
69   Representation theory, I.M. GELFAND *et al*
76   Spectral theory of linear differential operators and comparison algebras, H.O. CORDES
77   Isolated singular points on complete intersections, E.J.N. LOOIJENGA
83   Homogeneous structures on Riemannian manifolds, F. TRICERRI & L. VANHECKE
86   Topological topics, I.M. JAMES (ed)
87   Surveys in set theory, A.R.D. MATHIAS (ed)
88   FPF ring theory, C. FAITH & S. PAGE
89   An F-space sampler, N.J. KALTON, N.T. PECK & J.W. ROBERTS
90   Polytopes and symmetry, S.A. ROBERTSON
92   Representation of rings over skew fields, A.H. SCHOFIELD
93   Aspects of topology, I.M. JAMES & E.H. KRONHEIMER (eds)
94   Representations of general linear groups, G.D. JAMES
95   Low-dimensional topology 1982, R.A. FENN (ed)
96   Diophantine equations over function fields, R.C. MASON
97   Varieties of constructive mathematics, D.S. BRIDGES & F. RICHMAN
98   Localization in Noetherian rings, A.V. JATEGAONKAR
99   Methods of differential geometry in algebraic topology, M. KAROUBI & C. LERUSTE
100 Stopping time techniques for analysts and probabilists, L. EGGHE
104 Elliptic structures on 3-manifolds, C.B. THOMAS
105 A local spectral theory for closed operators, I. ERDELYI & WANG SHENGWANG
107 Compactification of Siegel moduli schemes, C-L. CHAI
108 Some topics in graph theory, H.P. YAP
109 Diophantine analysis, J. LOXTON & A. VAN DER POORTEN (eds)
110 An introduction to surreal numbers, H. GONSHOR
113 Lectures on the asymptotic theory of ideals, D. REES
114 Lectures on Bochner-Riesz means, K.M. DAVIS & Y-C. CHANG
115 An introduction to independence for analysts, H.G. DALES & W.H. WOODIN
116 Representations of algebras, P.J. WEBB (ed)
118 Skew linear groups, M. SHIRVANI & B. WEHRFRITZ
119 Triangulated categories in the representation theory of finite-dimensional algebras, D. HAPPEL
121 Proceedings of *Groups - St Andrews 1985*, E. ROBERTSON & C. CAMPBELL (eds)
122 Non-classical continuum mechanics, R.J. KNOPS & A.A. LACEY (eds)
125 Commutator theory for congruence modular varieties, R. FREESE & R. MCKENZIE
126 Van der Corput's method of exponential sums, S.W. GRAHAM & G. KOLESNIK
128 Descriptive set theory and the structure of sets of uniqueness, A.S. KECHRIS & A. LOUVEAU
129 The subgroup structure of the finite classical groups, P.B. KLEIDMAN & M.W. LIEBECK
130 Model theory and modules, M. PREST
131 Algebraic, extremal & metric combinatorics, M-M. DEZA, P. FRANKL & I.G. ROSENBERG (eds)
132 Whitehead groups of finite groups, ROBERT OLIVER
133 Linear algebraic monoids, MOHAN S. PUTCHA
134 Number theory and dynamical systems, M. DODSON & J. VICKERS (eds)
135 Operator algebras and applications, 1, D. EVANS & M. TAKESAKI (eds)
136 Operator algebras and applications, 2, D. EVANS & M. TAKESAKI (eds)
137 Analysis at Urbana, I, E. BERKSON, T. PECK, & J. UHL (eds)
138 Analysis at Urbana, II, E. BERKSON, T. PECK, & J. UHL (eds)
139 Advances in homotopy theory, S. SALAMON, B. STEER & W. SUTHERLAND (eds)
140 Geometric aspects of Banach spaces, E.M. PEINADOR & A. RODES (eds)
141 Surveys in combinatorics 1989, J. SIEMONS (ed)
144 Introduction to uniform spaces, I.M. JAMES
145 Homological questions in local algebra, JAN R. STROOKER
146 Cohen-Macaulay modules over Cohen-Macaulay rings, Y. YOSHINO
147 Continuous and discrete modules, S.H. MOHAMED & B.J. MÜLLER
148 Helices and vector bundles, A.N. RUDAKOV *et al*
149 Solitons, nonlinear evolution equations and inverse scattering, M. ABLOWITZ & P. CLARKSON
150 Geometry of low-dimensional manifolds 1, S. DONALDSON & C.B. THOMAS (eds)
151 Geometry of low-dimensional manifolds 2, S. DONALDSON & C.B. THOMAS (eds)
152 Oligomorphic permutation groups, P. CAMERON
153 L-functions and arithmetic, J. COATES & M.J. TAYLOR (eds)
154 Number theory and cryptography, J. LOXTON (ed
155 Classification theories of polarized varieties, TAKAO FUJITA
156 Twistors in mathematics and physics, T.N. BAILEY & R.J. BASTON (eds)

158 Geometry of Banach spaces, P.F.X. MÜLLER & W. SCHACHERMAYER (eds)
159 Groups St Andrews 1989 volume 1, C.M. CAMPBELL & E.F. ROBERTSON (eds)
160 Groups St Andrews 1989 volume 2, C.M. CAMPBELL & E.F. ROBERTSON (eds)
161 Lectures on block theory, BURKHARD KÜLSHAMMER
162 Harmonic analysis and representation theory, A. FIGA-TALAMANCA & C. NEBBIA
163 Topics in varieties of group representations, S.M. VOVSI
164 Quasi-symmetric designs, M.S. SHRIKANDE & S.S. SANE
165 Groups, combinatorics & geometry, M.W. LIEBECK & J. SAXL (eds)
166 Surveys in combinatorics, 1991, A.D. KEEDWELL (ed)
167 Stochastic analysis, M.T. BARLOW & N.H. BINGHAM (eds)
168 Representations of algebras, H. TACHIKAWA & S. BRENNER (eds)
169 Boolean function complexity, M.S. PATERSON (ed)
170 Manifolds with singularities and the Adams-Novikov spectral sequence, B. BOTVINNIK
171 Squares, A.R. RAJWADE
172 Algebraic varieties, GEORGE R. KEMPF
173 Discrete groups and geometry, W.J. HARVEY & C. MACLACHLAN (eds)
174 Lectures on mechanics, J.E. MARSDEN
175 Adams memorial symposium on algebraic topology 1, N. RAY & G. WALKER (eds)
176 Adams memorial symposium on algebraic topology 2, N. RAY & G. WALKER (eds)
177 Applications of categories in computer science, M. FOURMAN, P. JOHNSTONE, & A. PITTS (eds)
178 Lower K- and L-theory, A. RANICKI
179 Complex projective geometry, G. ELLINGSRUD et al
180 Lectures on ergodic theory and Pesin theory on compact manifolds, M. POLLICOTT
181 Geometric group theory I, G.A. NIBLO & M.A. ROLLER (eds)
182 Geometric group theory II, G.A. NIBLO & M.A. ROLLER (eds)
183 Shintani zeta functions, A. YUKIE
184 Arithmetical functions, W. SCHWARZ & J. SPILKER
185 Representations of solvable groups, O. MANZ & T.R. WOLF
186 Complexity: knots, colourings and counting, D.J.A. WELSH
187 Surveys in combinatorics, 1993, K. WALKER (ed)
188 Local analysis for the odd order theorem, H. BENDER & G. GLAUBERMAN
189 Locally presentable and accessible categories, J. ADAMEK & J. ROSICKY
190 Polynomial invariants of finite groups, D.J. BENSON
191 Finite geometry and combinatorics, F. DE CLERCK et al
192 Symplectic geometry, D. SALAMON (ed)
193 Computer algebra and differential equations, E. TOURNIER (ed)
194 Independent random variables and rearrangement invariant spaces, M. BRAVERMAN
195 Arithmetic of blowup algebras, WOLMER VASCONCELOS
196 Microlocal analysis for differential operators, A. GRIGIS & J. SJÖSTRAND
197 Two-dimensional homotopy and combinatorial group theory, C. HOG-ANGELONI,
    W. METZLER & A.J. SIERADSKI (eds)
198 The algebraic characterization of geometric 4-manifolds, J.A. HILLMAN
199 Invariant potential theory in the unit ball of C^n, MANFRED STOLL
200 The Grothendieck theory of dessins d'enfant, L. SCHNEPS (ed)
201 Singularities, JEAN-PAUL BRASSELET (ed)
202 The technique of pseudodifferential operators, H.O. CORDES
203 Hochschild cohomology of von Neumann algebras, A. SINCLAIR & R. SMITH
204 Combinatorial and geometric group theory, A.J. DUNCAN, N.D. GILBERT & J. HOWIE (eds)
205 Ergodic theory and its connections with harmonic analysis, K. PETERSEN & I. SALAMA (eds)
206 An introduction to noncommutative differential geometry and its physical applications, J. MADORE
207 Groups of Lie type and their geometries, W.M. KANTOR & L. DI MARTINO (eds)
208 Vector bundles in algebraic geometry, N.J. HITCHIN, P. NEWSTEAD & W.M. OXBURY (eds)
209 Arithmetic of diagonal hypersurfaces over finite fields, F.Q. GOUVÊA & N. YUI
210 Hilbert C*-modules, E.C. LANCE
211 Groups 93 Galway / St Andrews I, C.M. CAMPBELL et al
212 Groups 93 Galway / St Andrews II, C.M. CAMPBELL et al
214 Generalised Euler-Jacobi inversion formula and asymptotics beyond all orders, V. KOWALENKO,
    N.E. FRANKEL, M.L. GLASSER & T. TAUCHER
215 Number theory, S. DAVID (ed)
216 Stochastic partial differential equations, A. ETHERIDGE (ed)
217 Quadratic forms with applications to algebraic geometry and topology, A. PFISTER
218 Surveys in combinatorics, 1995, PETER ROWLINSON (ed)
220 Algebraic set theory, A. JOYAL & I. MOERDIJK
221 Harmonic approximation, S.J. GARDINER
222 Advances in linear logic, J.-Y. GIRARD, Y. LAFONT & L. REGNIER (eds)
223 Analytic semigroups and semilinear initial boundary value problems, KAZUAKI TAIRA
224 Computability, enumerability, unsolvability, S.B. COOPER, T.A. SLAMAN & S.S. WAINER (eds)
226 Novikov conjectures, index theorems and rigidity I, S. FERRY, A. RANICKI & J. ROSENBERG (eds)
227 Novikov conjectures, index theorems and rigidity II, S. FERRY, A. RANICKI & J. ROSENBERG (eds)
228 Ergodic theory of Z^d actions, M. POLLICOTT & K. SCHMIDT (eds)
229 Ergodicity for infinite dimensional systems, G. DA PRATO & J. ZABCZYK
230 Prolegomena to a middlebrow arithmetic of curves of genus 2, J.W.S. CASSELS & E.V. FLYNN
231 Semigroup theory and its applications, K.H. HOFMANN & M.W. MISLOVE (eds)

London Mathematical Society Lecture Note Series. 229

Ergodicity for Infinite Dimensional Systems

G. Da Prato
Scuola Normale Superiore, Pisa

J. Zabczyk
Polish Academy of Sciences, Warsaw

CAMBRIDGE
UNIVERSITY PRESS

Published by the Press Syndicate of the University of Cambridge
The Pitt Building, Trumpington Street, Cambridge CB2 1RP
40 West 20th Street, New York, NY 10011-4211, USA
10 Stamford Road, Oakleigh, Melbourne 3166, Australia

First published 1996

Library of Congress cataloging in publication data available

British Library cataloguing in publication data available

ISBN   0 521 579007 paperback

Transferred to digital printing 2003

Contents

Preface   ix

I   Markovian Dynamical Systems   1

1 General Dynamical Systems   3
   1.1   Basic concepts   3
   1.2   Ergodic Systems and the Koopman–von Neumann The-
       orem   5

2 Canonical Markovian Systems   11
   2.1   Markovian semigroups   11
   2.2   Canonical systems and their continuity   14

3 Ergodic and mixing measures   20
   3.1   The Krylov–Bogoliubov existence theorem   20
   3.2   Characterizations of ergodic measures   22
   3.3   The strong law of large numbers   30
   3.4   Mixing and recurrence   33
   3.5   Limit behaviour of P_t, $t \geq 0$ 39

4 Regular Markovian systems   41
   4.1   Regular, strong Feller and irreducible semigroups . . .   41
   4.2   Doob's theorem   43

**II   Invariant measures for stochastic evolution
    equations**   49

v

5 Stochastic Differential Equations   **51**

5.1   Introduction 51

5.2   Wiener and Ornstein–Uhlenbeck processes 52

    5.2.1   Stochastic integrals and convolutions 56

5.3   Stochastic evolution equations 65

5.4   Regular dependence on initial conditions and Kolmogorov equations 69

    5.4.1   Differentiable dependence on initial datum   . . 69

    5.4.2   Kolmogorov equation 70

5.5   Dissipative stochastic systems 71

    5.5.1   Generalities about dissipative mappings 72

    5.5.2   Existence of solutions for deterministic equations 75

    5.5.3   Existence of solutions for stochastic equations in Hilbert spaces 80

    5.5.4   Existence of solutions for stochastic equations in Banach spaces 87

6 Existence of invariant measures   **89**

6.1   Existence from boundedness 89

6.2   Linear systems 96

    6.2.1   A description of invariant measures 97

    6.2.2   Invariant measures and recurrence 102

6.3   Dissipative systems 104

    6.3.1   General noise 105

    6.3.2   Additive noise 108

6.4   Genuinely dissipative systems 114

6.5   Dissipative systems in Banach spaces 117

7 Uniqueness of invariant measures   **121**

7.1   Strong Feller property for non–degenerate diffusions . 121

7.2   Strong Feller property for degenerate diffusion 129

7.3   Irreducibility for non–degenerate diffusions 137

7.4   Irreducibility for equations with additive noise 140

8 Densities of invariant measures   **147**

8.1   Introduction 147

8.2   Sobolev spaces 149

8.3   Properties of the semigroup R_t, $t > 0$, on $L^2(H,\mu)$ . . 153

8.4   Existence and absolute continuity of the invariant mea-
      sure of P_t, $t > 0$, with respect to μ 156
8.5   Locally Lipschitz nonlinearities 159
8.6   Gradient systems 160
8.7   Regularity of the density when $\mathcal{L}$ is variational 165
8.8   Further regularity results in the diagonal case 168

III   Invariant measures for specific models   175

9   Ornstein–Uhlenbeck processes   177

9.1   Introduction 177
9.2   Ornstein–Uhlenbeck processes of wave type 178
      9.2.1   General properties 178
      9.2.2   Second order dissipative systems 181
      9.2.3   Comments on nonlinear equations 183
9.3   Ornstein–Uhlenbeck processes in finance 184
9.4   Ornstein–Uhlenbeck processes in chaotic environment   186
      9.4.1   Cylindrical noise 187
      9.4.2   Chaotic noise 194

10   Stochastic delay systems   199

10.1   Introduction 199
10.2   Linear case 200
10.3   Nonlinear equations 203

11   Reaction–Diffusion equations   211

11.1   Introduction 211
11.2   Finite interval. Lipschitz coefficients 213
       11.2.1   Existence and uniqueness of solutions 214
       11.2.2   Existence and uniqueness of invariant measures 215
11.3   Equations with non–Lipschitz coefficients 217
11.4   Reaction–diffusion equations on d dimensional spaces . 219

12   Spin systems   225

12.1   Introduction 225
12.2   Classical spin systems 227
12.3   Quantum lattice systems 235

13 Systems perturbed through the boundary   **241**
13.1 Introduction 241
13.2 Equations with non–homogeneous boundary conditions 243
13.3 Equations with Neumann boundary conditions 248
13.4 Ergodic solutions 254

14 Burgers equation   **257**
14.1 Introduction 257
14.2 Existence of solutions 260
14.3 Strong Feller property 265
14.4 Invariant measure 268
    14.4.1 Existence 268
    14.4.2 Uniqueness 276

15 Navier–Stokes equations   **281**
15.1 Preliminaries 281
15.2 Local existence and uniqueness results 284
15.3 A priori estimates and global existence 291
15.4 Existence of an invariant measure 295

IV   Appendices   **305**

A Smoothing properties of convolutions   **307**
  A.1 307

B An estimate on modulus of continuity   **311**
  B.1 311

C A result on implicit functions   **317**
  C.1 317

Bibliography   **321**

Preface

This book is devoted to asymptotic properties of solutions of stochastic evolution equations in infinite dimensional spaces. It is divided into three parts: Markovian Dynamical Systems, Invariant Measures for Stochastic Evolution Equations, and Invariant Measures for Specific Models.

In the first part of the book we recall basic concepts of the theory of dynamical systems and we link them with the theory of Markov processes. In this way such notions as ergodic, mixing, strongly mixing Markov processes will be special cases of well known concepts of a more general theory. We also give a proof of the Koopman–von Neumann ergodic theorem and, following Doob, we apply it in Chapter 4 to a class of regular Markov processes important in applications. We also include the Krylov–Bogoliubov theorem on existence of invariant measures and give a semigroup characterization of ergodic and mixing measures.

The second part of the book is concerned with invariant measures for important classes of stochastic evolution equations. The main aim is to formulate sufficient conditions for existence and uniqueness of invariant measures in terms of the coefficients of the equations.

We develop first two methods for establishing existence of invariant measures exploiting either compactness or dissipativity properties of the drift part of the equation. We also give necessary and sufficient conditions for existence of invariant measures for general linear systems. We do not discuss infinite dimensional versions of classical methods based on Harris and Doeblin's conditions and on embedded Markov chains mainly because at the moment they do not have too many applications, see however S. Jacquot and G. Royer [93]. Liapunov type techniques are an object of G. Leha and G. Ritter [105], and L. Stettner [147].

Uniqueness of invariant measures is deduced either by a dissipativity argument or by using structural properties of Markov processes like the strong Feller property and irreducibility. Applying Doob's theorem we give sufficient conditions for convergence of transition probabilities to the unique invariant measure. In the final chapter of the second part the regularity of invariant measures is discussed. General conditions are given under which invariant measures are absolutely continuous with respect to a properly chosen Gaussian reference measure. It is shown that under additional requirements the density of the invariant measure belongs to suitably defined Sobolev spaces. We also examine the so–called gradient systems for which explicit formulae for the densities exist. For a different way of investigating uniqueness of invariant measures based on the idea of coupling we refer to C. Mueller [120], where a specific case is treated.

Methods and results developed in the first two parts are applied to specific models in Part III. In many instances the general theory had to be modified to cover interesting cases. Existence and uniqueness of invariant measures are discussed first for various classes of Ornstein–Uhlenbeck processes including wave equations, some equations of financial mathematics and processes in random environments. Next, two chapters are devoted respectively to delay equations and to reaction–diffusion equations in both bounded and unbounded domains. Then invariant measures for classical and spin systems are discussed by the dissipativity method. In particular, exponential convergence of transition probabilities to equilibrium is established.

The final two chapters are devoted to stochastic equations of fluid dynamics: Burgers and Navier–Stokes equations. Asymptotic analysis required here rather sophisticated (technically) considerations.

The majority of the results presented in this book is based on recent results by the authors and their collaborators. Theorems on genuinely dissipative and delay equations and on systems perturbed through the boundary as well as the direct proof of the existence of a solution to the stochastic Navier–Stokes equation appear here for the first time in printed form.

Motivated by applications we have discussed only the so–called mild solutions of evolution equations perturbed by the Wiener process. We have not investigated asymptotic properties of more general martingale solutions and we refer to M. Viot [156], [154], [155], D. Gątarek and B. Goldys [77], and F. Flandoli and B. Maslowski [67], for results in this direction.

A complete description of all invariant measures for linear evolution equations with the noise process being a homogeneous process with independent increments is given in A. Chojnowska–Michalik [25], see also J. Zabczyk [167], for an earlier work on the finite dimensional case, and V. I Bogach ev, M. Röckner and B. Schmuland [12], for connections with the general Dirichlet spaces.

Existence and uniqueness of invariant measures on Polish spaces are discussed in A. G. Bhatt and R. L. Karandikar [9], see also P. E. Echeveria [60] and S. N. Ethier and T. G. Kurz [63], by using the concept of characteristic operators (generators) of the Markov process. This general approach is, however, not applicable to the examples studied in Part III of the book. On the other hand characteristic operators are used in Chapter 8 of the book devoted to the densities of invariant measures.

The authors acknowledge the financial support of the Italian National Project MURST "Problemi non lineari nell'Analisi e nelle applicazioni fisiche, chimiche e biologiche: aspetti analitici, modellistici e computazionali." and the KBN grant No 2 PO3A 082 08 "Ewolucyjne Równania Stochastyczne", during the preparation of the book. They also thank S. Cerrai, D. Gątarek, M. Fuhrman, P. Guiotto, A. Karczewska and L. Stettner, for pointing out some errors and mistakes in earlier versions of the book.

The authors would like to thank their home institutions Scuola Normale Superiore and the Polish Academy of Sciences for good working conditions.

Part I

Markovian Dynamical Systems

Chapter 1

General Dynamical Systems

In this chapter we recall basic concepts from the theory of dynamical systems which will play an important role in the sequel. We also state Birkhoff's ergodic theorem and give a proof of the Koopman–von Neumann theorem on weakly mixing systems.

1.1   Basic concepts

Let $S = (\Omega, \mathcal{G}, \mathbb{P}, \theta_t)$ be a *dynamical system* consisting of a probability space $(\Omega, \mathcal{G}, \mathbb{P})$ and a group of invertible, measurable transformations θ_t, $t \in \mathbb{R}$, from Ω into Ω, preserving measure $\mathbb{P}$:

$$\mathbb{P}(\theta_t A) = \mathbb{P}(A), \text{ for arbitrary } A \in \mathcal{G} \text{ and } t \in \mathbb{R}. \qquad (1.1.1)$$

The group θ_t, $t \in \mathbb{R}$, induces a group of linear transformations U_t, $t \in \mathbb{R}$, either on the real Hilbert space $\mathcal{H} = L^2(\Omega, \mathcal{G}, \mathbb{P})$ or on the complex Hilbert space $\mathcal{H}_{\mathbb{C}} = L^2_{\mathbb{C}}(\Omega, \mathcal{G}, \mathbb{P})$, by the formula

$$U_t \xi(\omega) = \xi(\theta_t \omega), \ \xi \in \mathcal{H} \ (\text{resp. } (\mathcal{H}_{\mathbb{C}})), \ \omega \in \Omega, \ t \in \mathbb{R}. \qquad (1.1.2)$$

We shall denote by $\langle \cdot, \cdot \rangle$ the scalar product in $\mathcal{H}$ (resp. $\mathcal{H}_{\mathbb{C}}$). It is clear that the operators U_t, $t \in \mathbb{R}$, are unitary and that $U_t^* = U_{-t}$, since, by the invariance of $\mathbb{P}$,

$$\langle U_t\xi,\eta\rangle \;=\; \int_\Omega \xi(\theta_t\omega)\eta(\omega)\,\mathbb{P}(d\omega)$$

$$=\; \int_\Omega \xi(\omega)\eta(\theta_{-t}\omega)\,\mathbb{P}(d\omega)$$

$$=\; \langle \xi, U_{-t}\eta\rangle,\; \xi \in \mathcal{H} \;\text{(resp. } \mathcal{H}_{\mathbb{C}}),\; t \in \mathbb{R}.$$

A dynamical system $S = (\Omega,\mathcal{G},\mathbb{P},\theta_t)$ is said to be *continuous* if

$$\lim_{t\to 0} U_t\xi = \xi \text{ for arbitrary } \xi \in \mathcal{H} \;\text{(resp. } \mathcal{H}_{\mathbb{C}}). \qquad (1.1.3)$$

We will restrict our considerations only to continuous systems S.

A dynamical system $S = (\Omega,\mathcal{G},\mathbb{P},\theta_t)$ is called *ergodic* if

$$\lim_{T\to+\infty} \frac{1}{T}\int_0^T \mathbb{P}(\theta_{-t}A\cap B)dt = \mathbb{P}(A)\mathbb{P}(B),\text{ for all } A,B \in \mathcal{G}. \quad (1.1.4)$$

Equivalently, in terms of the group U_t, $t \in \mathbb{R}$, a system S is ergodic if

$$\lim_{T\to+\infty} \frac{1}{T}\int_0^T \langle U_t\xi,\eta\rangle dt = \langle\xi,1\rangle\langle 1,\eta\rangle,\text{ for all } \xi,\eta \in \mathcal{H}_{\mathbb{C}}. \quad (1.1.5)$$

In fact (1.1.5) with $\xi = \chi_A$ and $\eta = \chi_B$ implies (1.1.4). Conversely from (1.1.4) it follows that (1.1.5) holds when ξ and η are simple, and so for all $\xi,\eta \in \mathcal{H}_{\mathbb{C}}$.

A dynamical system $S = (\Omega,\mathcal{G},\mathbb{P},\theta_t)$ is called *weakly mixing* if there exists a set $I \subset [0,+\infty[$ of relative measure 1 such that

$$\lim_{t\to+\infty,t\in I} \mathbb{P}(\theta_{-t}A\cap B) = \mathbb{P}(A)\mathbb{P}(B),\text{ for all } A,B \in \mathcal{G}. \quad (1.1.6)$$

A set $I \subset [0,+\infty[$ is said to have relative measure 1 if

$$\lim_{T\to\infty} \frac{1}{T}\ell_1(I\cap[0,T]) = 1, \qquad (1.1.7)$$

where ℓ_1 denotes the Lebesgue measure on $\mathbb{R}$. Equivalently, a system S is weakly mixing if for a set $I \subset [0,+\infty[$ of relative measure 1

$$\lim_{t\to+\infty,t\in I} \langle U_t\xi,\eta\rangle = \langle\xi,1\rangle\langle 1,\eta\rangle,\text{ for all } \xi,\eta \in \mathcal{H}_{\mathbb{C}}. \quad (1.1.8)$$

Finally, a system S is said to be *strongly mixing* if

$$\lim_{t \to +\infty} \mathbb{P}(\theta_{-t}A \cap B) = \mathbb{P}(A)\mathbb{P}(B), \text{ for all } A, B \in \mathcal{G}, \qquad (1.1.9)$$

or equivalently, if

$$\lim_{t \to +\infty} \langle U_t\xi, \eta \rangle = \langle \xi, 1 \rangle \langle 1, \eta \rangle, \text{ for all } \xi, \eta \in \mathcal{H}_{\mathbb{C}}. \qquad (1.1.10)$$

It is clear that a strongly mixing system is weakly mixing and that a weakly mixing system is ergodic.

Remark 1.1.1 For a thorough discussion and motivation of the concepts introduced we refer to K. Petersen [125]. See in particular Chapter 2.

1.2 Ergodic Systems and the Koopman–von Neumann Theorem

Since for any (continuous) dynamical system S the corresponding group U_t, $t \in \mathbb{R}$, is a C_0–group of unitary transformations on $\mathcal{H}_{\mathbb{C}}$, therefore, by Stone's theorem, see e.g. M. Reed and B. Simon [127, page 274] the infinitesimal generator of U_t, $t \in \mathbb{R}$, is of the form $i\mathcal{A}$ where $\mathcal{A}$ is a self–adjoint operator acting on $\mathcal{H}_{\mathbb{C}}$. $\mathcal{A}$ is called the *infinitesimal generator* of S. We will need the following characterization of ergodic and weakly mixing systems:

Theorem 1.2.1 *Let S be a continuous dynamical system, and let $\mathcal{A}$ be its infinitesimal generator.*

(i) S is ergodic if and only if 0 is a simple eigenvalue of $\mathcal{A}$.

(ii) S is weakly mixing if and only if the operator $\mathcal{A}$ has no eigenvalues $\lambda \neq 0$ and 0 is a simple eigenvalue of $\mathcal{A}$.

The characterization of weakly mixing systems given in (ii) is called the Koopman–von Neumann theorem.

Remark 1.2.2 Since $U_t 1 = 1$ for $t \in \mathbb{R}$, 0 is a simple eigenvalue of $\mathcal{A}$ if and only if the only elements $\xi \in \mathcal{H}_{\mathbb{C}}$ (resp. $\mathcal{H}$) such that

$$U_t\xi = \xi \text{ for all } t \in \mathbb{R}, \qquad (1.2.1)$$

are constant functions.

In a similar manner the operator $\mathcal{A}$ has no eigenvalues $\lambda \neq 0$ and 0 is a simple eigenvalue of $\mathcal{A}$, if and only if from the identity

$$U_t\xi = e^{i\lambda t}\xi \qquad (1.2.2)$$

valid for some $\lambda \in \mathbb{R}, \xi \in \mathcal{H}_\mathbb{C}$ and all $t \in \mathbb{R}$ it follows that $\lambda = 0$ and ξ is a constant function. ∎

To prove Theorem 1.2.1 we will take for granted Birkhoff's ergodic theorem, whose proof can be found in any textbook on ergodic theory, see e.g. K. Petersen [125, Theorem 2.3].

Theorem 1.2.3 *Let $(\Omega, \mathcal{G}, \mathbb{P})$ be a probability space, $\Theta : \Omega \to \Omega$ a measure preserving transformation and $\xi \in \mathcal{H}_\mathbb{C}$. Then for all $\xi \in \mathcal{H}_\mathbb{C}$ there exists $\xi^* \in \mathcal{H}_\mathbb{C}$ such that*

$$\lim_{n\to\infty} \frac{1}{n} \sum_{k=0}^{n-1} \xi(\Theta^k(\omega)) = \xi^*(\omega), \omega \in \Omega, \qquad (1.2.3)$$

$\mathbb{P}$*-a.s. and in* $\mathcal{H}_\mathbb{C}$.
Moreover

$$\xi^*(\omega) = \xi^*(\Theta(\omega)), \ \ for \ \mathbb{P}\text{-}a.s. \ \omega \in \Omega, \qquad (1.2.4)$$

and

$$\mathbb{E}\xi = \mathbb{E}\xi^*, \qquad (1.2.5)$$

where $\mathbb{E}\xi = \int_\Omega \xi(\omega)\mathbb{P}(d\omega)$ denotes the expectation of ξ.

Proof of Theorem 1.2.1 — (i) Assume that 0 is a simple eigenvalue of $\mathcal{A}$. We will show that (1.1.5) holds. Without any loss of generality we can assume that $\xi \geq 0$, $\mathbb{P}$-a.s. For an arbitrary positive number h define

$$\xi_h = \int_0^h U_s\xi ds, \ \ \xi \in \mathcal{H}_\mathbb{C}, \qquad (1.2.6)$$

and consider θ_h, a fixed measure preserving transformation on Ω. Then

$$\frac{1}{n} \sum_{k=0}^{n-1} \xi_h(\theta_h^k(\omega)) = \frac{1}{n} \int_0^{nh} U_s\xi(\omega)ds, \ \omega \in \Omega,$$

and therefore, by Theorem 1.2.3, for arbitrary $h > 0$ there exists $\xi_h^* \in \mathcal{H}_C$ such that

$$\lim_{n \to \infty} \frac{1}{n} \int_0^{nh} U_s \xi \, ds = \xi_h^* \quad \text{in } \mathcal{H}_C \text{ and } \mathbb{P}\text{-}qeda.\text{s.} \tag{1.2.7}$$

For arbitrary $T \geq 0$ let $n_T = [T/h]$ be the maximal nonnegative integer less or equal to T/h. Then $n_T h \leq T < (n_T + 1)h$ and $\mathbb{P}$-a.s.

$$\frac{n_T}{(n_T + 1)h} \frac{1}{n_T} \int_0^{n_T h} U_s \xi \, ds \;\leq\; \frac{1}{T} \int_0^T U_s \xi \, ds$$

$$\leq \; \frac{n_T + 1}{n_T h} \frac{1}{n_T + 1} \int_0^{(n_T+1)h} U_s \xi \, ds.$$

Consequently, for arbitrary $h > 0$

$$\lim_{T \to \infty} \frac{1}{T} \int_0^T U_s \xi \, ds = \frac{1}{h} \xi_h^* \quad \text{in } \mathcal{H}_C. \tag{1.2.8}$$

In particular it follows that $\xi_h^* = h \xi_1^*$; since, from (1.2.4), $U_h \xi_1^* = \xi_h^*$ this implies $U_h \xi_1^* = \xi_1^*$ for all $h \geq 0$. Therefore ξ_1^* is a constant function equal to $\langle \xi, 1 \rangle$. This proves (i) in one direction.

Let now a system S be ergodic and let $U_t \xi = \xi$ for all $t \geq 0$. By (1.1.5)

$$\langle \xi, \eta \rangle = \langle \xi, 1 \rangle \langle 1, \eta \rangle = \langle \langle \xi, 1 \rangle 1, \eta \rangle \,,$$

for all $\eta \in \mathcal{H}_C$, and therefore $\xi = (\langle \xi, 1 \rangle)1$ is a constant <u>function</u>.

(ii) Assume that the system S is ergodic and that A has no eigenvalues $\lambda \neq 0$. We recall that this means that the spectral measure $E(\cdot)$ determined by the operator A has no atoms except at 0. Moreover $E(\{0\}) = E_0$ is a projection operator onto constant functions and $E(\{\lambda\}) = 0$ for all $\lambda \neq 0$. It is enough to prove the result for $\xi, \eta \in \text{Ker } E_0$ and for $\xi = \eta$. Let $\{\xi_n\}$ be a basis of $(\text{Ker } E_0)^\perp$ on $\mathcal{H}_C$. We will prove that there exists a set I of relative measure 1 such that, for all $n \in \mathbb{N}$,

$$\lim_{t \in I, t \to +\infty} \langle U_t \xi_n, \xi_n \rangle = 0. \tag{1.2.9}$$

Now we compute the integral

$$\frac{1}{2T} \int_{-T}^T |\langle U_t \xi, \xi \rangle|^2 \, dt.$$

From the equality

$$\langle U_t \xi, \xi \rangle = \int_{-\infty}^{+\infty} e^{i\lambda t} \| E(d\lambda)\xi \|^2,$$

we find

$$
\begin{aligned}
|\langle U_t\xi,\xi\rangle|^2 &= \int_{-\infty}^{+\infty}\int_{-\infty}^{+\infty} e^{i(\lambda-\mu)t}\|E(d\lambda)\xi\|^2\|E(d\mu)\xi\|^2, \\[2mm]
&= \int_{-\infty}^{+\infty}\int_{-\infty}^{+\infty} \cos[(\lambda-\mu)t]\|E(d\lambda)\xi\|^2\|E(d\mu)\xi\|^2.
\end{aligned}
$$

It follows that

$$\frac{1}{2T}\int_{-T}^{T}|\langle U_t\xi,\xi\rangle|^2\,dt = \int_{-\infty}^{\infty}\int_{-\infty}^{\infty} \frac{\sin[(\lambda-\mu)T]}{(\lambda-\mu)T}\|E(d\lambda)\xi\|^2\|E(d\mu)\xi\|^2,$$

$$(1.2.10)$$

and, since the measure $\|E(\cdot)\xi\|^2$ has no atoms,

$$\lim_{T\to+\infty} \frac{1}{2T}\int_{-T}^{T}|\langle U_t\xi,\xi\rangle|^2\,dt = 0$$

Define

$$\gamma(t) = \sum_{n=1}^{\infty} \frac{1}{2^n\|\xi_n\|^4}\,|\langle U_t\xi_n,\xi_n\rangle|^2\,;$$

then $0 \le \gamma(t) \le 1$ for all $t \in\,]-\infty,+\infty[$ and

$$\lim_{T\to+\infty}\frac{1}{2T}\int_{-T}^{T}\gamma(t)dt = 0.$$

This implies that there exists a set I, of relative measure 1, such that

$$\lim_{\substack{|t|\to+\infty \\ t\in I}} \gamma(t) = 0;$$

this yields (1.2.9).

Conversely, assume that a system S is weakly mixing; then it is ergodic and therefore, by (i), 0 is a simple eigenvalue of $\mathcal{A}$. If for some real λ such that $\lambda \ne 0$ there exists $\xi \in \mathcal{H}_C$, $\xi \ne 0$, such that $U_t\xi = e^{i\lambda t}\xi$ for $t \ge 0$, then

$$\langle U_t\xi,\xi\rangle = e^{i\lambda t}|\xi|^2, \qquad\qquad (1.2.11)$$

and therefore (1.1.8) cannot be true for a set I of relative measure 1. This finishes the proof of Theorem 1.2.1.   ∎

Let S be a continuous dynamical system.

(i) A set $A \in \mathcal{G}$ is said to be *invariant* with respect to S if, for arbitrary $t \in \mathbb{R}$,

$$U_t \chi_A = \chi_A, \ \mathbb{P}\text{–a.s.}, \qquad (1.2.12)$$

or, equivalently, if for every $t \in \mathbb{R}$

$$\mathbb{P}(\theta_t A \cap A) = \mathbb{P}(A) = \mathbb{P}(\theta_t A). \qquad (1.2.13)$$

(ii) A measurable function $\alpha : \Omega \to [0, 2\pi[$ is said to be an *angle variable* for a system S if there exists $\lambda \in \mathbb{R}$ such that for every $t \in \mathbb{R}$,

$$U_t \alpha \equiv \lambda t + \alpha \ (\mathrm{mod}\ 2\pi), \mathbb{P}\text{–a.s.} \qquad (1.2.14)$$

As a consequence of Theorem 1.2.1 we derive the following important result.

Theorem 1.2.4 *Let S be a continuous dynamical system.*
(i) S is ergodic if and only if for any invariant set A, either $\mathbb{P}(A) = 0$ or $\mathbb{P}(A) = 1$.
(ii) S is weakly mixing if and only if any angle variable is constant and corresponds to $\lambda = 0$.

Proof — (i) Assume that S is an ergodic system and A is an invariant set. By the very definition,

$$\mathbb{P}^2(A) = \lim_{T \to +\infty} \frac{1}{T} \int_0^T \mathbb{P}(\theta_t A \cap A) dt = \mathbb{P}(A),$$

and therefore $\mathbb{P}(A) = 0$ or $\mathbb{P}(A) = 1$.

To prove the converse implication assume that for a $\xi \in \mathcal{H}_\mathbb{C}$ and for all $t \in \mathbb{R}$, $U_t \xi = \xi$. Without any loss of generality we can assume that ξ is real valued. If ξ is not a constant function then for a number $\alpha \in \mathbb{R}$ the sets $A = \{\omega : \xi(\omega) > \alpha\}$ and $A^c = \{\omega : \xi(\omega) \le \alpha\}$ are of positive $\mathbb{P}$–measure. This is a contradiction since the sets A and A^c are invariant. We have in fact, for any $\omega \in \Omega$,

$$U_t \chi_A(\omega) = \chi_A(\theta_t \omega) = \begin{cases} 1 & \text{if } \xi(\theta_t \omega) > \alpha, \\ 0 & \text{if } \xi(\theta_t \omega) \leq \alpha. \end{cases}$$

Since $\xi(\theta_t \omega) = \xi(\omega)$ by hypothesis, we have $U_t \chi_A = \chi_A$ so that A is invariant. This shows that ξ is a constant function.

(ii) Assume that S is weakly mixing and that α is its angle variable corresponding to λ, then, for $\xi = e^{i\alpha}$, we have

$$U_t \xi = e^{i\lambda t} \xi, \quad \text{for all } t \in \mathbb{R}. \tag{1.2.15}$$

By Theorem 1.2.1, $\lambda = 0$ and ξ (and thus α as well) is constant. To show the converse implication one can assume that the system S is ergodic. Suppose that (1.2.15) holds for some $\lambda \in \mathbb{R}$ and $\xi \in \mathcal{H}_\mathbb{C}$. Then $U_t |\xi| = |\xi|$, $t \in \mathbb{R}$. Therefore $|\xi|$ is a constant function and we can assume that $|\xi| = 1$. Consequently $\xi = e^{i\alpha}$ where α is a real function with values on $[0, 2\pi[$. From (1.2.15)

$$U_t \alpha \equiv \lambda t + \alpha \ (\mathrm{mod}\ 2\pi), \quad \mathbb{P}\text{–a.s.},$$

and the result follows.   ∎

Remark 1.2.5 The content of Theorem 1.2.4 can be phrased shortly as follows:

A system S is ergodic if and only if it has only trivial invariant sets.

A system S is weakly mixing if and only if it has only trivial angle variables.

Chapter 2

Canonical Markovian Systems

We introduce here dynamical systems determined by Markovian transition semigroups on Polish spaces and give conditions for their continuity.

2.1 Markovian semigroups

Let us first give some notation. In all this chapter E represents a Polish space with metric ρ and, for any $x_0 \in E, \delta > 0$, $B(x_0, \delta)$ is the ball

$$B(x_0, \delta) = \{x \in E : \rho(x, x_0) < \delta\}.$$

We denote by $\mathcal{E} = \mathcal{B}(E)$ the σ–field of all Borel subsets of E, and for any $\Gamma \in \mathcal{E}$, by χ_Γ the characteristic function

$$\chi_\Gamma(x) = \begin{cases} 1 & \text{if } x \in \Gamma, \\ 0 & \text{if } x \in \Gamma^c, \end{cases}$$

where $\Gamma^c = E \backslash \Gamma$.

Moreover $B_b(E)$ (resp. $C_b(E)$, $UC_b(E)$, $\mathrm{Lip}(E)$) is the set of all real (or complex) bounded Borel functions (resp. continuous and bounded functions, uniformly continuous and bounded functions, Lipschitz continuous functions) on E, and $\mathcal{M}_1(E)$ is the set of all probability measures defined on $(E, \mathcal{E})$.

11

We say that $P_t(x, \Gamma)$, $t \geq 0$, $x \in E, \Gamma \in \mathcal{E}$, is a *Markovian transition function* (short a *transition function*) on E, if:

(i) $P_t(x, \cdot)$ is a probability measure on $(E, \mathcal{E})$ for each $t \geq 0$, $x \in E$.

(ii) $P_t(\cdot, \Gamma)$ is an $\mathcal{E}$–measurable function for each $t \geq 0$, $\Gamma \in \mathcal{E}$.

(iii) $P_{t+s}(x, \Gamma) = \int_E P_t(x, dy) P_s(y, \Gamma)$,

   for each $t, s \geq 0$, $x \in E, \Gamma \in \mathcal{E}$.

(iv) $P_0(x, \Gamma) = \chi_\Gamma(x)$, for each $x \in E, \Gamma \in \mathcal{E}$.

Any transition function $P_t(x, \Gamma)$, $t \geq 0$, $x \in E, \Gamma \in \mathcal{E}$, on E defines a semigroup of linear operators P_t, $t \geq 0$, on the space $B_b(E)$, by the formula

$$P_t\varphi(x) = \int_E P_t(x, dy)\varphi(y), \; t \geq 0, \; x \in E, \; \varphi \in B_b(E).$$

P_t, $t \geq 0$, is called the *Markovian transition semigroup* associated to the transition function $P_t(x, \Gamma)$, $t \geq 0$, $x \in E, \Gamma \in \mathcal{E}$. For brevity we also call P_t, $t \geq 0$, a *Markovian semigroup* or a *transition semigroup*.

For any $t \geq 0$ and $\mu \in \mathcal{M}_1(E)$ we set

$$P_t^*\mu(\Gamma) = \int_E P_t(x, \Gamma)\mu(dx), \; t \geq 0, \; \Gamma \in \mathcal{E}.$$

A probability measure $\mu \in \mathcal{M}_1(E)$ is said to be *invariant* or *stationary* with respect to P_t, $t \geq 0$, if and only if

$$P_t^*\mu = \mu \text{ for each } t \geq 0.$$

A Markovian semigroup P_t, $t \geq 0$ is said to be *stochastically continuous* if

$$\lim_{t \to 0} P_t(x, B(x, \delta)) = 1, \text{ for all } x \in E, \; \delta > 0. \tag{2.1.1}$$

The following proposition, see e. g. E. B. Dynkin [59, Chapter 2, §1], gives equivalent characterizations of stochastically continuous semigroups.

Proposition 2.1.1 *A Markovian semigroup P_t, $t > 0$, is stochastically continuous if and only if one of the following equivalent conditions holds.*

(i) $\lim\limits_{t \to 0} P_t f(x) = f(x)$, *for all* $f \in C_b(E), x \in E$.

(ii) $\lim\limits_{t \to 0} P_t f(x) = f(x)$, *for all* $f \in UC_b(E), x \in E$.

(iii) $\lim\limits_{t \to 0} P_t f(x) = f(x)$, *for all* $f \in \text{Lip}(E), x \in E$.

Proof — It is enough to show that (2.1.1) implies (i) and (iii) implies (2.1.1).

Let $f \in C_b(E), x \in E$ and $\delta > 0$. Then

$$|P_t f(x) - f(x)|$$

$$= \left| \int_{B(x,\delta)} (f(y) - f(x)) P_t(x, dy) + \int_{B^c(x,\delta)} (f(y) - f(x)) P_t(x, dy) \right|$$

$$\leq \sup\nolimits_{y \in B(x,\delta)} |f(y) - f(x)| + 2 \sup\nolimits_E |f| \, (1 - P_t(x, B(x, \delta))).$$

Then $\lim\limits_{t \to 0} P_t f(x) = f(x)$ and (2.1.1) implies (i).

To show that (iii) implies (2.1.1) note first that if $f, g \in \text{Lip}(E)$ then also cf, $f \vee g \in \text{Lip}(E)$. For arbitrary $x_0 \in E$ and $\delta > 0$ define

$$f(x) = \frac{1}{\delta}[(\delta - \rho(x, x_0)) \vee 0], \quad x \in E,$$

or, equivalently,

$$f(x) = \begin{cases} 1 - \frac{\rho(x,x_0)}{\delta} & \text{if } x \in B(x_0, \delta), \\ 0 & \text{if } x \in B^c(x_0, \delta). \end{cases}$$

Then $f \in \text{Lip}(E)$ and

$$f(x_0) - P_t f(x_0) = 1 - \int_E P_t(x_0, dy) f(y)$$

$$= 1 - \int_{B(x_0,\delta)} P_t(x_0, dy) f(y)$$

$$\geq 1 - P_t(x_0, B(x_0, \delta)).$$

consequently (iii) implies (2.1.1). ∎

2.2   Canonical systems and their continuity

With a given Markovian semigroup P_t, $t \geq 0$, and with an invariant
measure $\mu \in \mathcal{M}_1(E)$ we will associate now, in a unique way, a dynam-
ical system $(\Omega, \mathcal{F}, \theta_t, \mathbb{P}_\mu)$ on the space $\Omega = E^{\mathbb{R}}$ of all the E–valued
functions. The canonical (coordinate) process $X(t)$, $t \in \mathbb{R}$, will be
Markovian, with transition probabilities $P_t(x, \cdot)$, $t \geq 0$, $x \in E$; it
will also be stationary and such that $\mathcal{L}(X(t)) = \mu$, $t \in \mathbb{R}$. This way
the general theory of Chapter 1 will become applicable to Markov
processes studied in these lecture notes.

Write $\Omega = E^{\mathbb{R}}$, $\Omega_+ = E^{\mathbb{R}_+}$, $\mathcal{F} = \mathcal{E}^{\mathbb{R}}$ and $\mathcal{F}_+ = \mathcal{E}^{\mathbb{R}_+}$ and define for
an arbitrary finite set $I \subset \mathbb{R}_+$, $I = \{t_1, \cdots, t_n\}$, $0 \leq t_1 < t_2 < \cdots < t_n$,
and $x \in E$ a probability measure $\mathbb{P}_I^x$ on $(E^I, \mathcal{E}^I)$ as follows:

$$\mathbb{P}_I^x(\Gamma) = \int_E P_{t_1}(x, dx_1) \cdots \int_E P_{t_{n-1}-t_{n-2}}(x_{n-2}, dx_{n-1})$$

$$\times \int_E P_{t_n-t_{n-1}}(x_{n-1}, dx_n) \chi_\Gamma(x_1, \cdots, x_n), \; \Gamma \in \mathcal{E}^I.$$

Then the system $\{\mathbb{P}_I^x : I \subset \mathbb{R}_+, I \text{ finite }\}$ is projective over $(E^{\mathbb{R}_+}, \mathcal{E}^{\mathbb{R}_+})$,
and then, by the Kolmogorov extension theorem, there exists a unique
probability measure $\mathbb{P}^x$ on $(E^{\mathbb{R}_+}, \mathcal{E}^{\mathbb{R}_+})$, such that

$$\mathbb{P}^x\left(\{\omega : (\omega_{t_1}, \cdots, \omega_{t_n}) \in \Gamma\}\right) = \mathbb{P}_{\{t_1, \dots, t_n\}}^x(\Gamma), \; \Gamma \in \mathcal{E}^{\{t_1, \dots, t_n\}}.$$

By the usual π–systems technique one can easily show that for arbi-
trary $\Gamma \in \mathcal{E}^{\mathbb{R}_+}$ the mapping $x \to \mathbb{P}^x(\Gamma)$ is $\mathcal{E}$–measurable.

In a similar manner for an arbitrary invariant measure $\mu \in \mathcal{M}_1(E)$
and an arbitrary finite set $I = \{t_1, \cdots, t_n\} \subset \mathbb{R}$, $t_1 < t_2 < \cdots < t_n$,
we can define a probability measure $\mathbb{P}_I^\mu$ on $(E^I, \mathcal{E}^I)$ by the formula

$$\mathbb{P}_I^x(\Gamma) = \int_E P_{t_1}(x, dx_1) \int_E P_{t_2-t_1}(x_1, dx_2) \cdots \int_E P_{t_{n-1}-t_{n-2}}(x_{n-2}, dx_{n-1})$$

$$\times \int_E P_{t_n-t_{n-1}}(x_{n-1}, dx_n) \chi_\Gamma(x_1, \cdots, x_n), \; \Gamma \in \mathcal{E}^I.$$

Again, by the Kolmogorov extension theorem, there exists a unique
probability measure $\mathbb{P}^\mu$ on $(\Omega, \mathcal{F})$ such that

$$\mathbb{P}^\mu\left(\{\omega : (\omega_{t_1}, \cdots, \omega_{t_n}) \in \Gamma\}\right) = \mathbb{P}_I^\mu(\Gamma), \; \Gamma \in \mathcal{E}^{\{t_1, \dots, t_n\}}.$$

Any subset Γ of Ω (resp. Ω_+) such that there exist a set $\{t_1, t_2, \cdots, t_n\} \subset \mathbb{R}$, $t_1 < t_2 < \cdots < t_n$, and a Borel subset $C \subset E^n$ such that

$$\Gamma = \{\omega \in \Omega : \{\omega(t_1), \cdots, \omega(t_n)\} \in C\}$$

will be called *cylindrical*. It is easy to see that cylindrical sets form a π–system. We also have the following important proposition.

Proposition 2.2.1 *For arbitrary $\Gamma \in \mathcal{F}$ and arbitrary $\varepsilon > 0$ there exists a cylindrical set C such that*

$$\mathbb{P}^{\mu}(\Gamma \backslash C) + \mathbb{P}^{\mu}(C \backslash \Gamma) < \varepsilon. \tag{2.2.1}$$

Proof — Let $\mathcal{G}$ be the family of all sets Γ from $\mathcal{F}$ such that for each $\varepsilon > 0$ there exists a cylindrical set C such that (2.2.1) holds. It is clear that $\mathcal{G}$ is a σ–field containing all cylindrical sets and therefore $\mathcal{G} = \mathcal{F}$. ∎

Write
$$X_t(\omega) = \omega(t), \ \ \omega \in \Omega, \ t \in \mathbb{R},$$
$$\mathcal{F}_t = \sigma(X_s : s \leq t), \ t \in \mathbb{R}.$$

The process X_t, $t \in \mathbb{R}$, is Markovian in the sense that

$$\mathbb{P}^{\mu}(X_{t+h} \in \Gamma | \mathcal{F}_t) = \mathbb{P}^{\mu}(X_{t+h} \in \Gamma | \sigma(X_t))$$

$$= P_h(X_t, \Gamma), \ \mathbb{P}^{\mu}\text{–a.s.} \ \Gamma \in \mathcal{E}.$$

More generally

$$\mathbb{P}^{\mu}(X_{t+\cdot} \in \Gamma | \mathcal{F}_t) = \mathbb{P}^{\mu}(X_{t+\cdot} \in \Gamma | \sigma(X_t))$$

$$= \mathbb{P}^{X_t}(\Gamma), \ \mathbb{P}^{\mu}\text{–a.s.} \ \Gamma \in \mathcal{F}_+.$$

We now introduce a group of invertible, measurable tranformations θ_t, $t \in \mathbb{R}$, from Ω to Ω :

$$(\theta_t \omega)(s) = \omega(t + s), \ t, s \in \mathbb{R}.$$

If μ is invariant, then the process X_t, $t \in \mathbb{R}$, is stationary:

$$\mathbb{P}^{\mu}(X \in \theta_t \Gamma) = \mathbb{P}^{\mu}(X \in \Gamma) \ \text{for all } t \in \mathbb{R} \ \text{and } \Gamma \in \mathcal{E}^{\mathbb{R}},$$

where
$$\theta_t \Gamma = \{\omega : \theta_t^{-1}\omega \in \Gamma\},$$

and the transformations θ_t, $t \in \mathbb{R}$, preserve the measure $\mathbb{P}^\mu$. Consequently, if μ is invariant, the quadruplet $S^\mu = (\Omega, \mathcal{F}, \theta_t, \mathbb{P}^\mu)$ defines a dynamical system, called the *canonical dynamical system* associated with P_t, $t \geq 0$, and μ. We recall that the group θ_t, $t \in \mathbb{R}$, induces a group of linear transformations U_t, $t \in \mathbb{R}$, on $\mathcal{H}_\mathbb{C}^\mu = L_\mathbb{C}^2(\Omega, \mathcal{F}, \mathbb{P}^\mu)$ (resp. $\mathcal{H}^\mu = L^2(\Omega, \mathcal{F}, \mathbb{P}^\mu)$) by the formula

$$U_t\xi(\omega) = \xi(\theta_t\omega), \ \xi \in \mathcal{H}_\mathbb{C}^\mu \ (\text{resp. } (\mathcal{H}^\mu)), \ \omega \in \Omega, \ t \in \mathbb{R}. \quad (2.2.2)$$

The following theorem links the concepts of stochastic continuity introduced before.

Theorem 2.2.2 *Let P_t, $t \geq 0$, be a stochastically continuous Markovian semigroup and let μ be an invariant measure. Then the corresponding canonical process $X(t)$, $t \in \mathbb{R}$, on $(\Omega, \mathcal{F}, \mathbb{P}^\mu)$ is stochastically continuous.*

Proof — Assume that P_t, $t \geq 0$, is stochastically continuous: for $t > s$ and $\delta > 0$ we have

$$\mathbb{P}^\mu(\rho(X(t), X(s)) \geq \delta) = \mathbb{E}^\mu(\mathbb{P}^\mu(\rho(X(t), X(s)) \geq \delta | \mathcal{F}_s))$$

$$= \mathbb{E}^\mu(P_{t-s}(X(s), B^c(X(s), \delta))),$$
$$(2.2.3)$$

by the Markov property. Since P_t, $t \geq 0$, is stochastically continuous,

$$\lim_{t \downarrow s} \mathbb{P}^\mu(\rho(X(t), X(s)) \geq \delta) = 0$$

as required.   ∎

We now prove the following

Proposition 2.2.3 *If the semigroup P_t, $t \geq 0$, is stochastically continuous then the system S^μ is continuous:*

$$\lim_{s \to t} U_s\xi = U_t\xi, \ \xi \in \mathcal{H}^\mu. \quad (2.2.4)$$

Proof — It is enough to check (2.2.4) for all ξ forming a linearly dense subset of $\mathcal{H}^{\mu}$. Let $\mathcal{H}_0$ be the set of all ξ of the form

$$\xi(\omega) = f(\omega(t_1), \cdots, \omega(t_n)), \ \omega \in \Omega, \tag{2.2.5}$$

where f is a bounded and uniformly continuous function on E^n and $t_1 < t_2 < \cdots < t_n$ is an increasing finite sequence in $\mathbb{R}$. Let C be a closed subset of E^n. For arbitrary $m \in \mathbb{N}$ define

$$f_m(x) = \frac{\rho(x, C_{1/m}^c)}{\rho(x, C) + \rho(x, C_{1/m}^c)}, \ x \in E^n,$$

where $\rho(x, C)$ denotes the distance from x to C, $C_{1/m}$ denotes the $(1/m)$-neighbourhood of C and $C_{1/m}^c$ the complement of $C_{1/m}$. Functions f_m are nonnegative, nonincreasing, bounded by 1 and

$$\lim_{m \to \infty} f_m(x) = \chi_C(x), \ x \in E^n.$$

Consequently, if

$$\begin{cases} \widehat{\xi}(\omega) = \chi_C(\omega(t_1), \cdots, \omega(t_n)), \\ \xi_m(\omega) = f_m(\omega(t_1), \cdots, \omega(t_n)), \ \omega \in \Omega, \ m \in \mathbb{N}, \end{cases} \tag{2.2.6}$$

then

$$\lim_{n \to \infty} \mathbb{E}^{\mu} |\xi_m - \widehat{\xi}|^2 = 0.$$

Since functions ξ of the form (2.2.5) are linearly dense in $\mathcal{H}^{\mu}$, the set $\mathcal{H}_0$ is linearly dense in $\mathcal{H}^{\mu}$ as well. If ξ, given by (2.2.6), is in $\mathcal{H}_0$, then

$$U_t\xi(\omega) = f(\omega(t_1 + t), \cdots, \omega(t_n + t)), \ \omega \in \Omega, t \in \mathbb{R},$$

and

$$U_t\xi(\omega) - U_s\xi(\omega)$$

$$= f(\omega(t_1 + t), \cdots, \omega(t_n + t)) - f(\omega(t_1 + s), \cdots, \omega(t_n + s))$$

$$= f(X(t_1 + t), \cdots, X(t_n + t)) - f(X(t_1 + s), \cdots, X(t_n + s)),$$

for $\omega \in \Omega, t, s \in \mathbb{R}$.

Let $\varepsilon > 0, \delta > 0$ be such that

$$|f(x_1,\cdots,x_n)-f(y_1,\cdots,y_n)| < \varepsilon \text{ provided } \rho(x_i-y_i) < \delta,\ i=1,...,n.$$

Then

$$\mathbb{E}^\mu |U_t\xi - U_s\xi|^2$$

$$\leq \mathbb{E}^\mu |f(X(t_1+t),\cdots,X(t_n+t)) - f(X(t_1+s),\cdots,X(t_n+s))|^2$$

$$\leq \varepsilon^2 \mathbb{P}^\mu \left(\rho(X(t_i+t),X(t_i+s))\right) < \delta \text{ for arbitrary } i=1,...,n)$$

$$+2\sup|f|\ \mathbb{P}^\mu \left(\rho(X(t_i+t),X(t_i+s))\right) \geq \delta \text{ for some } i=1,...,n)$$

$$\leq \varepsilon^2 + 2\sup|f| \sum_{i=1}^n \mathbb{P}^\mu \left(\rho(X(t_i+t),X(t_i+s)) \geq \delta\right).$$

Since X_t, $t \in \mathbb{R}$, is stochastically continuous, we have

$$\lim_{t-s\to 0} \mathbb{E}^\mu |U_t\xi - U_s\xi|^2 = 0. \blacksquare$$

Remark 2.2.4 It follows from (2.2.3) that for $t > 0$

$$\mathbb{P}^\mu \left(\rho(X(t),X(s)) < \delta\right) = \int_E P_{t-s}(x, B^c(x,\delta))\mu(dx).$$

So stochastic continuity of X implies that for $\delta > 0$

$$\lim_{t\downarrow 0} P_t(\cdot, B(\cdot,\delta)) = 1$$

in $L^1(E,\mu)$. So, under rather weak additional conditions, stochastic continuity of X is equivalent to stochastic continuity of P_t, $t \geq 0$.

We finish this section by introducing an important concept of *symmetric* Markovian semigroups, see M. Fukushima [74] and Z–M. Ma and M. Röckner [109]. Let P_t, $t \geq 0$, be an invariant measure for P_t, $t \geq 0$. The operators P_t, $t \geq 0$ have unique extensions to contraction operators on $L^2(\mu)$, see K. Yosida [162, Theorem 1, page 381]. The extended operators also form a C_0–semigroup on $L^2(\mu)$.

If they are symmetric then the semigroup is called μ-*symmetric* or for short *symmetric*. Its generator $\mathcal{A}$ is then a self–adjoint negative definite operator and the bilinear form

$$\mathcal{E}(\varphi,\psi) = \left\langle (-\mathcal{A})^{1/2}\varphi, (-\mathcal{A})^{1/2}\psi \right\rangle_{L^2(E,\mu)}, \quad \varphi,\psi \in D\left((-\mathcal{A})^{1/2}\right),$$

is the associated *Dirichlet form*. Markov processes X corresponding to symmetric semigroups are important because they are *reversible*. This means that if the initial distribution of $X(0)$ is μ and T is an arbitrary positive number then the processes $X(t)$, $t \in [0,T]$, and $X(T-t)$, $t \in [0,T]$, have the same finite–dimensional distributions. see M. Fukushima [74, page 96].

Chapter 3

Ergodic and mixing measures

This chapter is devoted to general properties of invariant measures for Markovian semigroups. We first prove the Krylov–Bogoliubov existence result and then give several characterizations of ergodic and mixing measures. Limit properties of the transition semigroups are discussed as well. The strong law of large numbers and recurrence properties of the corresponding Markov processes are established as well.

3.1 The Krylov–Bogoliubov existence theorem

Before we investigate various properties of invariant measures in the following sections, we prove here a general result on existence of invariant measures which will often be used.

A stochastically continuous Markovian semigroup P_t, $t \geq 0$, is called a *Markovian Feller semigroup*, if for any $\varphi \in C_b(E)$ and $t \geq 0$ one has $P_t \varphi \in C_b(E)$. Such semigroups will be called Feller semigroups for short. If P_t, $t \geq 0$ is a stochastically continuous Markovian semigroup, then for every $x \in E$ and $T > 0$ the formula

$$\frac{1}{T} \int_0^T P_t(x, \Gamma) dt = R_T(x, \Gamma), \quad \Gamma \in \mathcal{E},$$

defines a probability measure. For any $\nu \in \mathcal{M}_1(E)$, $R_T^* \nu$ is defined in the obvious way:

$$R_T^* \nu(\Gamma) = \int_E R_T(x, \Gamma)\nu(dx), \ \Gamma \in \mathcal{E}.$$

It is clear that for any $\varphi \in B_b(E)$

$$\langle R_T^* \nu, \varphi \rangle = \frac{1}{T} \int_0^T \langle P_t^* \nu, \varphi \rangle \, dt.$$

In this sense we write that

$$R_T^* \nu = \frac{1}{T} \int_0^T P_t^* \nu dt.$$

The method of constructing an invariant measure described in the following theorem is due to Krylov–Bogoliubov.

Theorem 3.1.1 *Assume that P_t, $t \geq 0$ is a Feller semigroup. If for some $\nu \in \mathcal{M}_1(E)$ and some sequence $T_n \uparrow +\infty$, $R_{T_n}^* \nu \to \mu$ weakly as $n \to \infty$, then μ is an invariant measure for P_t, $t \geq 0$.*

Proof— Fix $r > 0$ and $\varphi \in C_b(E)$. Then $P_r \varphi \in C_b(E)$ and

$$\langle \varphi, P_r^* \mu \rangle = \langle P_r \varphi, \mu \rangle = \langle P_r \varphi, \lim_{n \to \infty} R_{T_n}^* \nu \rangle$$

$$= \lim_{n \to \infty} \frac{1}{T_n} \langle P_r \varphi, \int_0^{T_n} P_s^* \nu ds \rangle$$

$$= \lim_{n \to \infty} \frac{1}{T_n} \langle \varphi, \int_r^{T_n + r} P_s^* \nu ds \rangle$$

$$= \lim_{n \to \infty} \Big[\frac{1}{T_n} \langle \varphi, \int_0^{T_n} P_s^* \nu ds \rangle + \frac{1}{T_n} \langle \varphi, \int_{T_n}^{T_n + r} P_s^* \nu ds \rangle$$

$$- \frac{1}{T_n} \langle \varphi, \int_0^r P_s^* \nu ds \rangle \Big] = \langle \varphi, \mu \rangle.$$

Consequently $P_r^* \mu = \mu$ for arbitrary $r > 0$ and the result follows. ∎

Corollary 3.1.2 *If for some $\nu \in \mathcal{M}_1(E)$ and some sequence $T_n \uparrow$ $+\infty$ the sequence $\{R_{T_n}^* \nu\}$ is tight, then there exists an invariant measure for P_t, $t \geq 0$.*

Proof — For the proof it is enough to remark that any tight sequence of measures contains a weakly convergent sub–sequence and apply Theorem 3.1.1. ∎

Remark 3.1.3 A stronger result has been recently obtained by A. Lasota [100, Theorem 6.1]. If for some $\nu \in \mathcal{M}_1(E)$ there exists a compact set $K \subset E$ and a sequence $T_n \uparrow +\infty$ such that

$$\liminf_{n \to +\infty} R_{T_n}^* \nu(K) > 0,$$

then there exist an invariant measure for P_t, $t \geq 0$.

3.2 Characterizations of ergodic measures

Let μ be an invariant measure with respect to the semigroup P_t, $t \geq 0$. Then μ is called *ergodic* if the corresponding dynamical system S^μ is ergodic.

Before we establish basic properties of ergodic measures, we will prove the following result, useful also on other occasions.

Proposition 3.2.1 *Let $\lambda \in \mathbb{C}$, $|\lambda| = 1$ and $t > 0$.*
(i) If $\varphi \in L_{\mathbb{C}}^2(E, \mu)$ and $P_t \varphi = \lambda \varphi$, μ–a.s., then $\xi \in \mathcal{H}_{\mathbb{C}}^\mu$ given by

$$\xi(\omega) = \varphi(\omega(0)), \ \omega \in \Omega,$$

satisfies $U_t \xi = \lambda \xi$, $\mathbb{P}^\mu$–a.s.
(ii) If $\xi \in \mathcal{H}_{\mathbb{C}}^\mu$ and $U_t \xi = \lambda \xi$ then there exists $\varphi \in L_{\mathbb{C}}^2(E, \mu)$ such that $P_t \varphi = \lambda \varphi$ and $\xi(\omega) = \varphi(\omega(0))$, $\mathbb{P}^\mu$–a.s.

Proof — To prove (i) we first remark that if $\xi(\omega) = \varphi(\omega(0))$, $\omega \in \Omega$, then

$$U_t \xi(\omega) = \xi(\theta_t \omega) = \varphi(\theta_t \omega(0)) = \varphi(\omega(t)), t > 0,$$

so the condition,

$$U_t \xi = \lambda \xi, \ t > 0$$

is equivalent to
$$\varphi(\omega(t)) = \lambda\varphi(\omega(0)), \ t \geq 0,$$
and therefore to
$$\varphi(X(t)) = \lambda\varphi(X(0)), \ \mathbb{P}^\mu\text{–a.s}, \ t \geq 0, \tag{3.2.1}$$
where $X(t)$, $t \in \mathbb{R}$, is the canonical process.
To prove (3.2.1) note that

$$\mathbb{E}^\mu[|\varphi(X(t)) - \lambda\varphi(X(0))|^2] = \mathbb{E}^\mu[|\varphi(X(t))|^2] + \mathbb{E}^\mu[|\varphi(X(0))|^2]$$

$$-\bar{\lambda}\mathbb{E}^\mu[\varphi(X(t)) \overline{\varphi(X(0))}] - \lambda \, \mathbb{E}^\mu[\overline{\varphi(X(t))} \, \varphi(X(0))].$$
$$\tag{3.2.2}$$

Since the process $X(\cdot)$ is stationary, we have

$$\mathbb{E}^\mu[|\varphi(X(t))|^2] = \mathbb{E}^\mu[|\varphi(X(0))|^2],$$

and since it is a Markov semigroup

$$\mathbb{E}^\mu[\varphi(X(t)) \overline{\varphi(X(0))}] = \mathbb{E}^\mu \left\{ \mathbb{E}^\mu[\varphi(X(t)) \overline{\varphi(X(0))}|\mathcal{F}_0] \right\}$$

$$= \mathbb{E}^\mu \left\{ \overline{\varphi(X(0))} \, \mathbb{E}^\mu[\varphi(X(t))|\mathcal{F}_0] \right\} = \mathbb{E}^\mu \left\{ \overline{\varphi(X(0))} \, P_t\varphi(X(0)) \right\}$$

$$= \lambda\mathbb{E}^\mu \left\{ |\varphi(X(0))|^2 \right\}.$$

By substituting this in (3.2.2), (3.2.1) follows and so (i) is proved.

We now prove (ii). Let us recall that $\mathcal{F}_t = \sigma\{X(s) : s \leq t\}, t \in \mathbb{R}$. We will also use the following notation:

$$\mathcal{F}_{[s,t]} = \sigma\{X(u) : s \leq u \leq t\}.$$

We need the following lemma.

Lemma 3.2.2 *Assume that $U_t\xi = \lambda\xi$, $\mathbb{P}^\mu$–a.s., where $|\lambda| = 1$. Then, for an arbitrary random variable $\tilde{\xi}$ which is $\mathcal{F}_{[-t,t]}$–measurable, we have*

$$\mathbb{E}^\mu \left(\left| \mathbb{E}^\mu \left(\lambda^{-1} U_t \tilde{\xi} | \mathcal{F}_{[0,0]} \right) - \xi \right|^2 \right) \leq 10 \, \mathbb{E}^\mu(|\xi - \tilde{\xi}|^2).$$

Proof — We have in fact

$$\mathbb{E}^\mu \left(\left| \mathbb{E}^\mu \left(\lambda^{-1} U_t \widetilde{\xi} | \mathcal{F}_{[0,0]} \right) - \xi \right|^2 \right)$$

$$\leq 2\mathbb{E}^\mu \left(\left| \mathbb{E}^\mu (\lambda^{-1} U_t \widetilde{\xi} | \mathcal{F}_{[0,0]}) - \lambda U_{-t} \widetilde{\xi} \right|^2 \right) + 2\mathbb{E}^\mu \left(|\lambda U_{-t} \widetilde{\xi} - \xi|^2 \right)$$

$$\leq 2\mathbb{E}^\mu \left(\left| \mathbb{E}^\mu (\lambda^{-1} U_t \widetilde{\xi} | \mathcal{F}_0) - \mathbb{E}^\mu (\lambda U_{-t} \widetilde{\xi} | \mathcal{F}_0) \right|^2 \right) + 2\mathbb{E}^\mu |\xi - \widetilde{\xi}|^2,$$

where in the final estimate we have used that X is a Markov process, that $U_t \widetilde{\xi}$ and $U_{-t} \widetilde{\xi}$ are respectively $\mathcal{F}_{[0,2t]}^-$ and $\mathcal{F}_0$–measurable random variables and that U_t is an isometric transformation on $\mathcal{H}_{\mathbb{C}}$.

By the Jensen inequality

$$\left| \mathbb{E}^\mu (\lambda^{-1} U_t \widetilde{\xi} - \lambda U_{-t} \widetilde{\xi}) | \mathcal{F}_0) \right|^2$$

$$\leq \mathbb{E}^\mu \left(|\lambda^{-1} U_t \widetilde{\xi} - \lambda U_{-t} \widetilde{\xi}|^2 | \mathcal{F}_0) | \right).$$

Therefore

$$\mathbb{E}^\mu \left(|\mathbb{E}^\mu (\lambda^{-1} U_t \widetilde{\xi} | \mathcal{F}_{[0,0]}) - \xi|^2 \right)$$

$$\leq 2\mathbb{E}^\mu |\lambda^{-1} U_t \widetilde{\xi} - \lambda U_{-t} \widetilde{\xi}|^2 + 2\mathbb{E}^\mu |\xi - \widetilde{\xi}|^2$$

$$\leq 2\mathbb{E}^\mu |\lambda^{-1} U_{2t} \widetilde{\xi} - \lambda \widetilde{\xi}|^2 + 2\mathbb{E}^\mu |\xi - \widetilde{\xi}|^2$$

$$\leq 4\mathbb{E}^\mu |\lambda^{-1} U_{2t} \widetilde{\xi} - \lambda^{-1} U_{2t} \xi|^2 + 4\mathbb{E}^\mu |\lambda^{-1} U_{2t} \xi - \lambda \widetilde{\xi}|^2$$

$$+ 2\mathbb{E}^\mu |\xi - \widetilde{\xi}|^2.$$

Since $U_{2t} \xi = \lambda^2 \xi$ it follows that

$$\mathbb{E}^\mu \left(|\mathbb{E}^\mu (\lambda^{-1} U_t \widetilde{\xi} | \mathcal{F}_{[0,0]}) - \xi|^2 \right)$$

$$\leq 4\mathbb{E}^\mu |U_{2t} (\widetilde{\xi} - \xi)|^2 + 4\mathbb{E}^\mu |\lambda(\xi - \widetilde{\xi})|^2 + 2\mathbb{E}^\mu |\xi - \widetilde{\xi}|^2$$

$$\leq 10\mathbb{E}^\mu |\xi - \widetilde{\xi}|^2,$$

and the lemma is proved.

We go back now to the proof of Proposition 3.2.1–(ii). Let us remark that by Proposition 2.2.1 and Lemma 3.2.2 there exists a sequence $\{\tilde{\xi}_n\}$ of $\mathcal{F}_{[-nt,nt]}$–measurable elements of $\mathcal{H}_\mathbb{C}^\mu$ such that

$$\lim_{n\to\infty} \lambda^{-n}\mathbb{E}^\mu(U_{nt}\tilde{\xi}_n|\mathcal{F}_{[0,0]}) = \xi \text{ in } \mathcal{H}_\mathbb{C}^\mu.$$

Moreover, there exists $\varphi_n \in L_\mathbb{C}^2(E,\mu)$ such that

$$\lambda^{-n}\mathbb{E}^\mu(U_{nt}\tilde{\xi}_n|\mathcal{F}_{[0,0]}) = \varphi_n(X(0)), \ \mathbb{P}^\mu\text{–a.s.}$$

Without any loss of generality we can assume that

$$\lim_{n\to\infty} \varphi_n(X(0)) = \xi, \ \mathbb{P}^\mu\text{–a.s. and in } \mathcal{H}_\mathbb{C}^\mu.$$

It is therefore enough to define

$$\varphi(x) = \begin{cases} \lim_{n\to\infty} \varphi_n(x), & \text{if the limit exists} \\ 0, & \text{otherwise.} \end{cases}$$

We have in fact $\xi = \varphi(X(0))$. ∎

Remark 3.2.3 If μ is an invariant measure with respect to P_t, $t \geq 0$, then all operators P_t, $t \geq 0$, have unique extensions to contraction operators on $L^p(E,\mu)$, and on $L_\mathbb{C}^p(E,\mu)$, for arbitrary $p \geq 1$, see also the end of §2.2. The extended operators will also be denoted by P_t, $t \geq 0$. They form a semigroup of operators, which is strongly continuous under stochastic continuity assumptions.

We now prove the main result concerning ergodicity of an invariant measure $\mu \in \mathcal{M}_1(E)$.

Theorem 3.2.4 *Let* P_t, $t \geq 0$, *be a stochastically continuous Markovian semigroup and* μ *an invariant measure with respect to* P_t, $t \geq 0$. *Then the following conditions are equivalent:*

(i) μ *is ergodic.*

(ii) *If* $\varphi \in L_\mathbb{C}^2(E,\mu)$, *or* $\varphi \in L^2(E,\mu)$ *and*

$$P_t\varphi = \varphi, \ \mu \text{ a.s. for all } t > 0,$$

then φ is constant μ a.s.

   (iii) If for a set $\Gamma \in \mathcal{E}$ and all $t > 0$

$$P_t \chi_\Gamma = \chi_\Gamma \quad \mu \ a.s.$$

then either $\mu(\Gamma) = 0$ or $\mu(\Gamma) = 1$.

   (iv) For arbitrary $\varphi \in L^2_{\mathbb{C}}(E, \mu)$ (resp. $\varphi \in L^2(E, \mu)$)

$$\lim_{T \to +\infty} \frac{1}{T} \int_0^T P_s \varphi \, ds = \langle \varphi, 1 \rangle \quad in \ L^2_{\mathbb{C}}(E, \mu)(resp. \ L^2(E, \mu)).$$

Proof —   Equivalences (i)$\Leftrightarrow$ (ii)$\Leftrightarrow$(iii) follow directly from Theorem 1.2.1, Theorem 1.2.4–(i) and Proposition 3.2.1.

   We will show now that (i)$\Leftrightarrow$(iv). Assume first that μ is ergodic. Let $\varphi, \psi \in L^2_{\mathbb{C}}(E, \mu)$ and define

$$\xi(\omega) = \varphi(\omega(0)), \quad \eta(\omega) = \psi(\omega(0)), \quad \omega \in \Omega.$$

Then, by the ergodicity of the dynamical system S^μ,

$$\lim_{T \to \infty} \frac{1}{T} \int_0^T \langle U_s \xi, \eta \rangle_{\mathcal{H}^\mu_{\mathbb{C}}} ds = \langle \xi, 1 \rangle_{\mathcal{H}^\mu_{\mathbb{C}}} \langle 1, \eta \rangle_{\mathcal{H}^\mu_{\mathbb{C}}}.$$

However,

$$\langle U_s \xi, \eta \rangle_{\mathcal{H}^\mu_{\mathbb{C}}} = \mathbb{E}^\mu \left[\varphi(X(s)) \overline{\psi(X(0))} \right]$$

$$= \langle P_s \varphi, \psi \rangle_{L^2_{\mathbb{C}}(E, \mu)}.$$

It follows that

$$\lim_{T \to \infty} \frac{1}{T} \int_0^T \langle P_s \varphi, \psi \rangle_{L^2_{\mathbb{C}}(E, \mu)} ds = \langle \varphi, 1 \rangle_{L^2_{\mathbb{C}}(E, \mu)} \langle 1, \psi \rangle_{L^2_{\mathbb{C}}(E, \mu)}.$$

Consequently

$$\lim_{T \to \infty} \frac{1}{T} \int_0^T P_s \varphi \, ds = \langle \varphi, 1 \rangle \quad \text{weakly in } L^2_{\mathbb{C}}(E, \mu).$$

Now we prove that the convergence is strong. Let $\varphi = \chi_\Gamma$, $\Gamma \in \mathcal{E}$. Then we have $P_s \varphi(x) = P_s(x, \Gamma)$ and, by the invariance of μ,

$$\left\| \frac{1}{T} \int_0^T P_s \varphi \, ds \right\|_{L^2_{\mathbb{C}}(E, \mu)} \le \frac{1}{T} \int_0^T \| P_s \varphi \|_{L^2_{\mathbb{C}}(E, \mu)} ds$$

$$= \mu(\Gamma) = \| \varphi \|_{L^2_{\mathbb{C}}(E, \mu)}.$$

Therefore for $\varphi = \chi_\Gamma$, $\Gamma \in \mathcal{E}$,

$$\lim_{T \to \infty} \frac{1}{T} \int_0^T P_s \varphi \, ds = \langle \varphi, 1 \rangle \text{ strongly in } L_{\mathbb{C}}^2(E, \mu).$$

Since indicator functions form a linearly dense set in $L_{\mathbb{C}}^2(E, \mu)$, statement (iv) holds for all $\varphi \in L_{\mathbb{C}}^2(E, \mu)$.

Conversely, assume that (iv) holds and that

$$P_t \varphi = \varphi \text{ for all } t \geq 0, \ \varphi \in L_{\mathbb{C}}^2(E, \mu).$$

Then

$$\varphi = \frac{1}{T} \int_0^T P_s \varphi \, ds \to \langle \varphi, 1 \rangle_{L_{\mathbb{C}}^2(E,\mu)} \text{ as } T \to \infty.$$

So φ is a constant function. ∎

As an application of Theorem 3.2.4 we deduce the following important corollary:

Proposition 3.2.5 *Let* P_t, $t \geq 0$, *be a stochastically continuous Markovian semigroup. If* μ *and* ν *are ergodic measures with respect to* P_t, $t \geq 0$, *and if* $\mu \neq \nu$, *then* μ *and* ν *are singular.*

Proof — Let $\Gamma \in \mathcal{E}$ be such that

$$\nu(\Gamma) \neq \mu(\Gamma).$$

By Theorem 3.2.4 there exists a sequence $\{T_N \uparrow +\infty\}$ such that

$$\lim_{N \to \infty} \frac{1}{T_N} \int_0^{T_N} P_s \chi_\Gamma \, ds = \mu(\Gamma), \ \mu - \text{a.s.},$$

$$\lim_{N \to \infty} \frac{1}{T_N} \int_0^{T_N} P_s \chi_\Gamma \, ds = \nu(\Gamma), \ \nu - \text{a.s.}.$$

Define

$$A = \left\{ x \in E : \lim_{N \to \infty} \frac{1}{T_N} \int_0^{T_N} P_s(x, \Gamma) \, ds = \mu(\Gamma) \right\},$$

$$B = \left\{ x \in E : \lim_{N \to \infty} \frac{1}{T_N} \int_0^{T_N} P_s(x, \Gamma) \, ds = \nu(\Gamma) \right\}.$$

It is clear that $A \cap B = \emptyset$ and that $\mu(A) = \nu(B) = 1$. Therefore μ and ν are singular. ∎

Theorem 3.2.6 *If $\mu \in \mathcal{M}_1(E)$ is the unique invariant measure for the semigroup P_t, $t > 0$, then it is ergodic.*

Proof — Assume in contradiction that μ is the unique invariant measure for the semigroup P_t, $t > 0$, and that there exists a set $\Gamma \in \mathcal{E}$ with $\mu(\Gamma) \in]0,1[$, such that, for all $t > 0$

$$P_t \chi_\Gamma = \chi_\Gamma, \quad \mu\text{–a.s.} \tag{3.2.3}$$

We will check that the measure $\tilde{\mu}$

$$\tilde{\mu}(A) = \frac{1}{\mu(\Gamma)} \mu(A \cap \Gamma), \ A \in \mathcal{E},$$

is also invariant for the semigroup P_t, $t > 0$. Note that for arbitrary $A \in \mathcal{E}$ and $t \geq 0$

$$P_t^* \tilde{\mu}(A) = \int_E P_t(x, A) \tilde{\mu}(dx) = \frac{1}{\mu(\Gamma)} \int_\Gamma P_t(x, A) \mu(dx)$$

$$= \frac{1}{\mu(\Gamma)} \int_\Gamma P_t(x, A \cap \Gamma) \mu(dx) + \frac{1}{\mu(\Gamma)} \int_\Gamma P_t(x, A \cap \Gamma^c) \mu(dx).$$

It follows from (3.2.3) that

$$P_t(x, A \cap \Gamma^c) \leq P_t(x, \Gamma^c) = 0, \ \mu\text{–a.s. on } \Gamma,$$

and

$$P_t(x, A \cap \Gamma) \leq P_t(x, \Gamma) = 0, \ \mu\text{–a.s. on } \Gamma^c.$$

Therefore, by the invariance of μ,

$$P_t^* \tilde{\mu}(A) = \frac{1}{\mu(\Gamma)} \int_E P_t(x, A \cap \Gamma) \mu(dx)$$

$$= \frac{1}{\mu(\Gamma)} \mu(A \cap \Gamma) = \tilde{\mu}(A),$$

and so $\tilde{\mu}$ is invariant for P_t, $t > 0$. ∎

We finish this section by giving another characterization of ergodic measures. Note that Theorem 3.2.6 is an immediate consequence of that characterization.

Proposition 3.2.7 *An invariant probability measure for the semigroup P_t, $t \geq 0$, is ergodic if and only if it is an extremal point of the set of all the invariant probability measures for the semigroup P_t, $t \geq 0$.*

Proof — We show first that if μ is an ergodic measure and if ν is an invariant probability measure for the semigroup P_t, $t \geq 0$ which is absolutely continuous with respect to μ, then we have $\nu = \mu$.

Let $A \in \mathcal{E}$, then by Theorem 3.2.4 it follows that

$$\lim_{T \to +\infty} \frac{1}{T} \int_0^T P_s \chi_A \, ds = \mu(A), \quad \text{in } L^2(E, \mu).$$

Consequently, for a sequence $T_n \uparrow +\infty$

$$\lim_{n \to +\infty} \frac{1}{T_n} \int_0^{T_n} P_s \chi_A \, ds = \mu(A), \quad \mu\text{–a.s.} ,$$

and therefore

$$\lim_{n \to +\infty} \frac{1}{T_n} \int_0^{T_n} P_s \chi_A \, ds = \mu(A), \quad \nu\text{–a.s.}$$

In particular

$$\lim_{n \to +\infty} \int_E \left(\frac{1}{T_n} \int_0^{T_n} P_s \chi_A(x) ds \right) \nu(dx) = \mu(A).$$

However,

$$\frac{1}{T_n} \int_E \left(\int_0^{T_n} P_s \chi_A(x) ds \right) \nu(dx) = \frac{1}{T_n} \int_0^{T_n} \langle P_s^* \nu, \chi_A \rangle \, ds$$

$$= \frac{1}{T_n} \int_0^{T_n} \langle \nu, \chi_A \rangle \, ds = \nu(A).$$

So $\nu(A) = \mu(A)$.

Assume now that μ is ergodic and that we have

$$\mu = \alpha \mu_1 + (1 - \alpha) \mu_2,$$

for some $\alpha \in \,]0, 1[$ and some invariant measures μ_1, μ_2. Then μ_1 is absolutely continuous with respect to μ and therefore $\mu_1 = \mu$ and $\mu_2 = \mu$.

Conversely, let μ be an extremal point and assume, in contradiction, that it is not ergodic. Then there exists a set $\Gamma \in \mathcal{E}$ such that $0 < \mu(\Gamma) < 1$ and

$$P_t \chi_\Gamma = \chi_\Gamma, \text{ for every } t \geq 0, \ \mu\text{–a.s..}$$

Define measures μ_1 and μ_2 by the formula

$$\mu_1(A) = \frac{1}{\mu(\Gamma)}\mu(A \cap \Gamma), \quad \mu_2(A) = \frac{1}{\mu(\Gamma^c)}\mu(A \cap \Gamma^c).$$

It is easy to check that both the measures μ_1 and μ_2 are invariant. Since

$$\mu = \mu(\Gamma)\mu_1 + (1 - \mu(\Gamma))\mu_2$$

and $\mu_1 \neq \mu_2$, μ cannot be extremal, a contradiction.   ∎

3.3   The strong law of large numbers

We deduce now, as a probabilistic motivation of our study, important consequences of the existence of an invariant ergodic measure and of its uniqueness.

Assume that P_t, $t \geq 0$, is a stochastically continuous Markovian semigroup on a Polish space (E, ρ) with invariant measure μ.

Let $Z(t)$, $t \geq 0$, be any stationary Markov process, with the transition semigroup P_t, $t > 0$, defined on a probability space

$$(\Omega_1, \mathcal{F}_1, \mathcal{F}_{1,t}, \mathbb{P}_1),$$

such that

$$\mathcal{L}(Z(t)) = \mu, \ t \geq 0.$$

Since $Z(t)$, $t \geq 0$, is stochastically continuous, it possesses a version which is progressively measurable. Therefore, without any loss of generality, we can assume that $Z(\cdot)$ is progressively measurable.

Theorem 3.3.1 *Assume that $\varphi \in L^2(E, \mu)$. Then there exists $\eta^* \in L^2(\Omega_1, \mathcal{F}_1, \mathbb{P}_1)$ such that*

$$\lim_{T \to +\infty} \frac{1}{T} \int_0^T \varphi(Z(s))ds = \eta^*, \tag{3.3.1}$$

$\mathbb{P}_1$-*a.s. and in* $L^2(\Omega_1, \mathcal{F}_1, \mathbb{P}_1)$.

If in addition μ *is ergodic, then*

$$\lim_{T \to +\infty} \frac{1}{T} \int_0^T \varphi(Z(s)) ds = \int_E \varphi(x) \mu(dx) = \langle \mu, \varphi \rangle, \quad \mathbb{P}_1\text{--}a.s. \quad (3.3.2)$$

Proof — For the proof we denote by $(\Omega, \mathcal{F}, \theta_t, \mathbb{P}^\mu)$ the canonical system introduced in §1.1, where

$$\theta_t \omega(s) = \omega(t + s), \ t, s \in \mathbb{R}, \ \omega \in \Omega.$$

Let moreover $X(t)$, $t \geq 0$, be the canonical Markov process

$$X(t)(\omega) = \omega(t), \ t \in \mathbb{R}, \ \omega \in \Omega,$$

and U_t, $t \in \mathbb{R}$, the group of transformations on $\mathcal{H}^\mu = L^2(\Omega, \mathcal{F}, \mathbb{P}^\mu)$ defined by

$$U_t \xi(\omega) = \xi(\theta_t \omega), \ t \in \mathbb{R}, \ \omega \in \Omega, \ \xi \in \mathcal{H}^\mu.$$

Without any loss of generality we can assume that $\varphi \geq 0$. Define

$$\xi(\omega) = \varphi(\omega(0)), \ \omega \in \Omega.$$

Let us fix $h > 0$ and put

$$\xi_h = \int_0^h U_s \xi ds.$$

Since $U_s \xi$, $s \geq 0$, is a continuous function of $s \geq 0$, ξ_h is well defined as a Riemann integral in $L^2(\Omega_1, \mathcal{F}_1, \mathbb{P}_1)$. Let $\tilde{X}(s)$, $s \geq 0$, be a progessively measurable modification of $X(s)$, $s \geq 0$. Then

$$\xi_h(\omega) = \int_0^h \varphi(\tilde{X}(s, \omega)) ds, \text{ for } \mathbb{P}^\mu\text{--}a.s. \ \omega \in \Omega,$$

where the integral on the right hand side is for each $\omega \in \Omega$ the Lebesgue integral.

Now consider θ_h as a fixed measure preserving transformation in Ω. Then we have, for all $n \in \mathbb{N}$,

$$\frac{1}{n} \sum_{k=0}^{n-1} \xi_h \left(\theta_h^k(\omega) \right) = \frac{1}{n} \int_0^{nh} \varphi(\tilde{X}(s, \omega)) ds.$$

By Theorem 1.2.3 there exists $\xi_h^* \in L^2(\Omega, \mathcal{F}, \mathbb{P}^\mu)$ such that the limit

$$\lim_{n\to\infty} \int_0^{nh} \varphi((\tilde{X}(s,\omega))ds = \xi_h^*(\omega), \quad \omega \in \Omega,$$

does exist $\mathbb{P}^\mu$–a.s. and in $L^2(\Omega, \mathcal{F}, \mathbb{P}^\mu)$. Moreover

$$\xi_h^*(\tilde{X}(h)) = \xi_h^*(\tilde{X}(0)), \quad \mathbb{P}^\mu\text{–a.s.}$$

Note that for arbitrary $0 \leq s_1 \leq \ldots \leq s_k$, $k \in \mathbb{N}$, the laws of the random vectors

$$(\tilde{X}(s_1), ..., \tilde{X}(s_k)), \ (Z(s_1), ..., Z(s_k)),$$

are identical and therefore the laws of the vectors

$$\left(\int_0^h \varphi(\tilde{X}(s))ds, \ \int_h^{2h} \varphi(\tilde{X}(s))ds, \ \int_{(k-1)h}^{kh} \varphi(\tilde{X}(s))ds \right)$$

and

$$\left(\int_0^h \varphi(Z(s))ds, \ \int_h^{2h} \varphi(Z(s))ds, \ \int_{(k-1)h}^{kh} \varphi(Z(s))ds \right),$$

whose coordinates can be treated as Riemann integrals or as Lebesgue integrals (for each $\omega \in \Omega$), are identical for arbitrary $k \in \mathbb{N}$. Consequently $\mathbb{P}_1$–a.s. and in $L^2(\Omega_1, \mathcal{F}_1, \mathbb{P}_1)$ there exists the limit

$$\lim_{n\to\infty} \frac{1}{n} \int_0^{nh} \varphi(Z(s))ds = \eta_h^*,$$

and

$$\mathcal{L}(\xi_h^*) = \mathcal{L}(\eta_h^*). \tag{3.3.3}$$

Taking into account that $\varphi \geq 0$ and denoting by n_T the maximal integer less than or equal to T/h we have (compare Theorem 1.2.1), for all $\omega_1 \in \Omega_1$ and $T > 0$,

$$\frac{n_T}{(n_T + 1)h} \frac{1}{n_T} \int_0^{n_T h} \varphi(Z(s))ds \ \leq \ \frac{1}{T} \int_0^T \varphi(Z(s))ds$$

$$\leq \ \frac{n_T+1}{n_T h} \frac{1}{n_T + 1} \int_0^{(n_T+1)h} \varphi(Z(s))ds.$$

Therefore the limit

$$\lim_{T \to +\infty} \frac{1}{T} \int_0^T \varphi(Z(s)) ds = \eta_h^*$$

exists $\mathbb{P}$–a.s. and in $L^2(\Omega, \mathcal{F}, \mathbb{P})$ and is equal to $\frac{1}{h} \eta_h^*$. So, in order to prove (3.3.1), it is enough to take $\eta^* = \frac{1}{h} \eta_h^*$.

To prove the last statement in the theorem it is enough to show that ξ_1^* is a constant random variable (then by (3.3.3) η_1^* is constant as well). It is obvious that for arbitrary $h > 0$, $\frac{1}{h} \xi_h^* = \xi_1^*$, $\mathbb{P}_1$–a.s. and

$$U_h(\xi_1^*) = \frac{1}{h} U_h \xi_h^* = \xi_1^*, \ \mathbb{P}_1\text{–a.s.}.$$

Consequently, by Theorem 1.2.1, ξ_1^* is a constant random variable. This completes the proof. ∎

Remark 3.3.2 With a similar proof, defining $\xi_h = \int_0^h U_{-s} \xi ds$, and replacing Θ_h by Θ_{-h}, one can show that if μ is the unique invariant measure for P_t, $t \geq 0$, and $Z(t)$, $t \in \mathbb{R}$, is a stationary Markov process on $(\Omega, \mathcal{F}, \mathbb{P}_1)$ we also have

$$\lim_{T \to +\infty} \frac{1}{T} \int_0^T \varphi(Z(-s)) ds = \langle \mu, \varphi \rangle, \ \mathbb{P}_1\text{–a.s.} \qquad (3.3.4)$$

3.4 Mixing and recurrence

We are given a Markovian semigroup P_t, $t \geq 0$, and an invariant probability measure $\mu \in \mathcal{M}_1(E)$. The measure μ is said to be *weakly mixing* or *strongly mixing* if the corresponding canonical system S^μ is weakly mixing or strongly mixing respectively.

The proof of the following theorem is similar to that of Theorem 3.2.4.

Theorem 3.4.1 *Let P_t, $t \geq 0$ be a stochastically continuous Markovian semigroup and μ an invariant measure with respect to P_t, $t \geq 0$. Then the following conditions are equivalent:*

(i) μ is weakly mixing.

(ii) If $\varphi \in L_\mathbb{C}^2(E, \mu)$ and $\lambda \in \mathbb{R}$ are such that

$$P_t \varphi = e^{i\lambda t} \varphi, \ \mu\text{–a.s. for all } t > 0,$$

then $\lambda = 0$ and φ is constant μ-a.s.

(iii) There exists a set $I \subset [0, +\infty[$ of relative measure 1 such that

$$\lim_{\substack{|t| \to +\infty \\ t \in I}} P_t \varphi = \langle \varphi, 1 \rangle \ \ in \ L^2_{\mathbb{C}}(E, \mu).$$

Proof— Equivalence (i)⇔(ii) follows directly from Theorem 1.2.4–(ii) . We will prove that $(i) \Leftrightarrow (iii)$.

Assume that (i) holds; let $\varphi, \psi \in L^2_{\mathbb{C}}(E, \mu)$ and define

$$\xi(\omega) = \varphi(\omega(0)), \ \eta(\omega) = \psi(\omega(0)), \ \omega \in \Omega,$$

and let I be the set from the definition of weak mixing. Then

$$\lim_{\substack{|t| \to +\infty \\ t \in I}} \langle U_t \xi, \eta \rangle_{\mathcal{H}^{\mu}_{\mathbb{C}}} = \lim_{t \to +\infty, t \in I} \langle P_t \varphi, \psi \rangle_{L^2_{\mathbb{C}}(E, \mu)}$$

$$= \langle \varphi, 1 \rangle_{L^2_{\mathbb{C}}(E, \mu)} \langle 1, \psi \rangle_{L^2_{\mathbb{C}}(E, \mu)}.$$

Therefore

$$P_t \varphi \to \langle \varphi, 1 \rangle_{L^2_{\mathbb{C}}(E, \mu)} \ \text{weakly as } t \to +\infty, t \in I \ \text{ in, } L^2_{\mathbb{C}}(E, \mu),$$

and (iv) is proved.

Conversely assume that (iv) holds. If $\varphi = \chi_\Gamma, \Gamma \in \mathcal{E}$, we have

$$\| P_t \varphi \|_{L^2_{\mathbb{C}}(E, \mu)} \leq |\langle \varphi, 1 \rangle_{L^2_{\mathbb{C}}(E, \mu)}|, \ t > 0,$$

and in this case

$$\lim_{\substack{|t| \to +\infty \\ t \in I}} P_t \varphi = \langle \varphi, 1 \rangle_{L^2_{\mathbb{C}}(E, \mu)} \ \text{strongly in } \ L^2_{\mathbb{C}}(E, \mu).$$

As before this implies

$$\lim_{\substack{|t| \to +\infty \\ t \in I}} P_t \varphi = \langle \varphi, 1 \rangle_{L^2_{\mathbb{C}}(E, \mu)} \ \text{strongly in } L^2_{\mathbb{C}}(E, \mu),$$

for all $\varphi \in L^2_{\mathbb{C}}(E, \mu)$, so that μ is weakly mixing. ∎

Theorem 3.4.2 *Let P_t, $t \geq 0$, be a stochastically continuous Markovian semigroup and μ an invariant measure with respect to P_t, $t \geq 0$. Then the following conditions are equivalent:*

(i) μ is strongly mixing.

(ii) For arbitrary $\varphi \in L^2_{\mathbb{C}}(E, \mu)$ (resp. $\varphi \in L^2(E, \mu)$) we have

$$\lim_{t \to +\infty} P_t \varphi = \langle \varphi, 1 \rangle, \quad in \ L^2_{\mathbb{C}}(E, \mu) \ (\ resp. \ L^2(E, \mu)).$$

Proof — (i)$\Rightarrow$(ii). Assume that μ is strongly mixing and let $\varphi, \psi \in L^2_{\mathbb{C}}(E, \mu)$. Define

$$\xi(\omega) = \varphi(\omega(0)), \quad \eta(\omega) = \psi(\omega(0)), \quad \omega \in \Omega.$$

Then

$$\begin{aligned}
\mathbb{E}(U_t \xi \, \overline{\eta}) &= \mathbb{E}(\varphi(X(t)) \, \overline{\psi}(X(0))) \\
&= \mathbb{E}(P_t \varphi(X(0)) \, \overline{\psi}(X(0))) \\
&= \langle P_t \varphi, \psi \rangle_{L^2_{\mathbb{C}}(E, \mu)}.
\end{aligned}$$

Strong mixing of μ implies that

$$\lim_{t \to +\infty} \langle P_t \varphi, \psi \rangle = \langle \varphi, 1 \rangle_{L^2_{\mathbb{C}}(E, \mu)} \langle 1, \psi \rangle_{L^2_{\mathbb{C}}(E, \mu)}, \quad in \ L^2_{\mathbb{C}}(E, \mu),$$

and therefore $P_t \varphi \to \langle \varphi, 1 \rangle_{L^2_{\mathbb{C}}(E, \mu)}$ weakly as $t \to +\infty$. In the same way as in the proof of Theorem 3.2.4 this implies $P_t \varphi \to \langle \varphi, 1 \rangle$ strongly as $t \to +\infty$.

(ii)$\Rightarrow$(i) Assume now that $P_t \varphi \to \langle \varphi, 1 \rangle_{L^2_{\mathbb{C}}(E, \mu)}$ strongly as $t \to +\infty$, for arbitrary $\varphi \in L^2_{\mathbb{C}}(E, \mu)$. Let $\xi = f(X(t_1), ..., X(t_k))$ for some fixed $t_1, ..., t_k$ and a measurable and bounded $f : \mathbb{R}^n \to \mathbb{R}$, and let $\eta \in \mathcal{H}^\mu$ be $\mathcal{F}_h$-measurable for some fixed h. To show the remaining implication it is enough to prove that

$$\lim_{t \to +\infty} \mathbb{E}(U_t \xi \eta) = \mathbb{E}\xi \, \mathbb{E}\eta.$$

But

$$\mathbb{E}(U_t \xi \eta) = \mathbb{E}(f(X(t + t_1), ..., X(t + t_k))\eta)$$

and without any loss of generality we can assume that $0 < t_1 < ... < t_k$. By an easy induction argument

$$\mathbb{E}(f(X(t + t_1), ..., X(t + t_k))|\mathcal{F}_t) = \hat{f}(X(t)), \quad \mathbb{P}\text{-a.s.}$$

for a bounded measurable function $\widehat{f}$ such that

$$\mathbb{E}(\widehat{f}(X(t))) = \mathbb{E}(f(X(t+t_1), ..., X(t+t_k))).$$

Consequently, if $t > h$,

$$\mathbb{E}(U_t\xi\eta) = \mathbb{E}(\widehat{f}(X(t))\eta) = \mathbb{E}(P_{t-h}\widehat{f}(X(h))\eta).$$

Moreover, if $t_n \uparrow +\infty$ is a sequence such that

$$P_{t_n-h}\widehat{f} \to \langle \overline{f}, 1 \rangle, \quad \mu\text{--a.s.}, \tag{3.4.1}$$

then

$$\mathbb{E}(U_{t_n}\xi\eta) \to \mathbb{E}(\xi)\mathbb{E}(\eta), \tag{3.4.2}$$

by the Lebesgue dominated convergence theorem ($\overline{f}$ is bounded).

Since for every sequence $s_n \uparrow +\infty$ one can construct a subsequence $t_n \to +\infty$ such that (3.4.1) (and thus (3.4.2)) holds, the result follows. ∎

Corollary 3.4.3 *Let μ be an invariant measure with respect to P_t, $t \geq 0$. Assume that*

$$\lim_{t\to+\infty} P_t(x,\cdot) = \mu \ \ weakly, \ \ x \in E. \tag{3.4.3}$$

Then μ is strongly mixing.

Proof —   To prove the result take first $\varphi \in C_b(E)$, then

$$\lim_{t\to+\infty} P_t\varphi = \langle \mu, \varphi \rangle, \ \mu\text{--a.s. and in } L^2(E,\mu).$$

Since $C_b(E)$ is dense in $L^2(E,\mu)$, the result follows. ∎

Remark 3.4.4 For a given Feller semigroup P_t, $t \geq 0$, there might be many strongly mixing measures. To see this consider an ordinary differential equation

$$z'(t) = F(z(t)), \ \ z(0) = x \in E, \tag{3.4.4}$$

where F is a Lipschitz continuous mapping from E into E. Let $z(\cdot, x)$ be the unique solution of (3.4.4) and set

$$P_t\varphi(x) = \varphi(z(t, x)), \ t \geq 0, \ \varphi \in C_b(H).$$

If, for $x_0 \in H$, $F(x_0) = 0$, then trivially the Dirac mass δ_{x_0} is strongly mixing for P_t, $t \geq 0$. So the required example is provided by a function F possessing several stationary points.

Results discussed in this section are related to the concept of *recurrence*. Let $X(t)$, $t \geq 0$, be a (progressively measurable) Markov process on E. We say that X is *recurrent* with respect to a set $\Gamma \in \mathcal{E}$ if

$$\mathbb{P}(X(t) \in \Gamma, \text{ for an unbounded set of } t > 0) = 1. \tag{3.4.5}$$

Note that if μ is the unique invariant measure for P_t, $t \geq 0$, and X is a stationary Markov process with the semigroup P_t, $t \geq 0$, then by Theorem 3.3.1, the process X is recurrent with respect to all sets $\Gamma \in \mathcal{E}$ such that $\mu(\Gamma) > 0$. However, other (non-stationary) Markov processes with the semigroup P_t, $t \geq 0$, may not be recurrent with respect to Γ, $\mu(\Gamma) > 0$.

We have the following proposition.

Proposition 3.4.5 *Assume that for a set $\Gamma \in \mathcal{E}$ there exists a positive number γ such that*

$$\liminf_{t \to +\infty} P_t(x, \Gamma) \geq \gamma \text{ for all } x \in E. \tag{3.4.6}$$

Then for an arbitrary Markov process X with respect to P_t, $t \geq 0$, and an arbitrary sequence $\{t_n\} \to +\infty$

$$\mathbb{P}(X(t_n) \in \Gamma, \text{ for infinitely many } n \in \mathbb{N}) = 1. \tag{3.4.7}$$

Proof — We follow J. Zabczyk [164]. It is enough to show that

$$\mathbb{P}(X(t_n) \in \Gamma, \text{ for some } n \in \mathbb{N}) = 1 \tag{3.4.8}$$

and then apply the Markov property. We will prove first that there exists a subsequence $\{s_k\}$ of $\{t_n\}$ such that

$$\mathbb{P}(X(s_1) \in \Gamma^c, ..., X(s_k) \in \Gamma^c, X(s_{k+1}) \in \Gamma)$$
$$\geq \frac{\gamma}{2}\mathbb{P}(X(s_1) \in \Gamma^c, ..., X(s_k) \in \Gamma^c). \tag{3.4.9}$$

Define $s_1 = t_1$. Let ν be a measure defined by the formula

$$\nu(\Delta) = \mathbb{P}(X(s_1) \in \Delta \cap \Gamma^c), \quad \Delta \in \mathcal{E}.$$

It follows from (3.4.6) that there exist a set $\Delta \subset \Gamma^c$ and a number $t_n > s_1$ such that

$$\nu(\Delta) \geq \frac{2}{3}\nu(\Gamma^c) \qquad (3.4.10)$$

and

$$P_{t_n - s_1}(x, \Gamma) \geq \frac{3}{4}\gamma \text{ for all } x \in \Delta. \qquad (3.4.11)$$

We set $s_2 = t_n$. Then by the Markov property, (3.4.10) and (3.4.11)

$$\mathbb{P}(X(s_1) \in \Gamma^c, X(s_2) \in \Gamma) \;=\; \mathbb{E}(\chi_{\{X(s_1)\in\Gamma^c\}} P_{s_2-s_1}(X(s_1),\Gamma))$$

$$\geq \; \mathbb{E}(\chi_{\{X(s_1)\in\Delta\}} P_{s_2-s_1}(X(s_1),\Gamma))$$

$$\geq \; \frac{3}{4}\mathbb{P}(X(s_1) \in \Delta) \geq \frac{\gamma}{2}\mathbb{P}(X(s_1) \in \Gamma^c).$$

Generalizing this argument one easily obtains by induction that (3.4.9) holds for all $k \in \mathbb{N}$. It follows from (3.4.9) that

$$\mathbb{P}(X(s_1) \in \Gamma^c, ..., X(s_{k+1}) \in \Gamma^c)$$

$$\geq \left(1 - \frac{\gamma}{2}\right) \mathbb{P}(X(s_1) \in \Gamma^c, ..., X(s_k) \in \Gamma^c)$$

$$\geq \left(1 - \frac{\gamma}{2}\right)^k \mathbb{P}(X(s_1) \in \Gamma^c).$$

Consequently

$$\mathbb{P}(X(s_k) \in \Gamma, \text{ for some } n \in \mathbb{N}) = 1. \quad \blacksquare$$

Corollary 3.4.6 *If for a set $\Gamma \subset \mathcal{E}$,*

$$\lim_{t \to +\infty} P_t(x, \Gamma) = \mu(\Gamma) > 0,$$

where μ is an invariant measure for P_t, $t \geq 0$, then any Markov process X with the transition semigroup P_t, $t \geq 0$, is recurrent with respect to Γ.

3.5 Limit behaviour of P_t, $t \geq 0$

Ergodicity, weak mixing and strong mixing of an invariant measure μ imply some kind of convergence of the operators P_t, $t \geq 0$, as $t \to +\infty$, to the projection operator

$$\pi\varphi = \langle \varphi, 1 \rangle 1, \quad \varphi \in L^2(E, \mu).$$

In many instances, however, the convergence is stronger than Theorem 3.2.4, or Theorem 3.4.1 implies.

For our future discussion it is useful to distinguish the following cases.

(i) $P_t(x, \cdot) \to \mu$ as $t \to +\infty$, weakly, for any $x \in E$.

(ii) $P_t(x, \Gamma) \to \mu(\Gamma)$ as $t \to +\infty$, for any $x \in E$, $\Gamma \in \mathcal{B}(E)$.

(iii) $\mathrm{Var}(P_t(x, \cdot) - \mu) \to 0$ as $t \to +\infty$, for any $x \in E$.

(iv) There exist $\omega > 0$ and a positive function $c(\cdot)$ such that, for any bounded Lipschitz continuous function φ, all $t > 0$ and all $x \in E$,

$$|P_t\varphi(x) - \langle \varphi, \mu \rangle| \leq c(x)e^{-\omega t}\|\varphi\|_{\mathrm{Lip}}.$$

It is clear, from Corollary 3.4.3 that any of the conditions (i)–(iv) implies strong mixing. Moreover (ii)$\Rightarrow$(i), (iii)$\Rightarrow$(ii) and (iv)$\Rightarrow$(ii). For the so–called *regular* case discussed in the next chapter we will always have convergence of (ii) type, and for the so–called dissipative systems convergence of (iv) type. Convergence (iii) takes place in many interesting situations, see B. Maslowski [117] and G. Leha and G. Ritter [105].

Let us recall, see R. Fortet and B. Mourier [68], that the Fortet–Mourier norm of a signed measure ν is defined by the formula

$$\|\nu\|_{F-M} = \sup\{|\langle \nu, \varphi \rangle| : \|\varphi\|_0 \leq 1, \|\varphi\|_{\mathrm{Lip}} \leq 1\}.$$

It is well known, see e.g. A. Lasota and J. A. Yorke [101], that weak convergence of a sequence $\{\mu_n\}$ of probability measures to μ is equivalent to the convergence in the Fortet–Mourier norm:

$$\lim_{n \to \infty} \|\mu_n - \mu\|_{F-M} = 0.$$

Note that, using this concept, property (iv) can be rephrased equivalently as existence of $\omega > 0$ and of a function $c(\cdot)$ such that for all

$x \in E$ and $t > 0$

$$\|P_t(x, \cdot) - \mu\|_{F-M} \leq c(x)e^{-\omega t}, \ t > 0. \tag{3.5.1}$$

If for an invariant measure μ, estimate (3.5.1) holds (equivalently (iv) holds), the measure μ will be called *exponentially mixing*.

Chapter 4

Regular Markovian systems

This chapter is concerned with an important class of Markovian semigroups generating strongly mixing invariant measures. It starts with an introduction on regular, strong Feller and irreducible transition semigroups. Then Doob's theorem on existence and uniqueness of invariant measures as well as on limit behaviour of transition semigroups is proved. An essential role is played here by the Koopman–von Neumann theorem.

4.1 Regular, strong Feller and irreducible semigroups

A Markovian semigroup P_t, $t \geq 0$, is said to be t_0-*regular* if all transition probabilities $P_{t_0}(x, \cdot), x \in E$, are mutually equivalent. By the semigroup property, if a semigroup P_t, $t \geq 0$, is regular at time $t_0 > 0$ it is s-regular for arbitrary $s > t_0$ and all transition probability measures $P_s(x, \cdot)$, $s \geq t_0$ and $x \in E$, are equivalent. If μ is an invariant measure for a t_0-regular semigroup P_t, $t \geq 0$, then all transition probability measures $P_t(x, \cdot)$, $t \geq t_0, x \in E$, are equivalent to μ.

Defining $E = [0, t_0]$ and

$$P_t(x, \cdot) = \delta_{(x-t) \vee 0}(\cdot), \ t \geq 0, \ x \in E,$$

41

one can easily see that regularity at time $t_0 > 0$ does not imply regularity at times $t \in [0, t_0[$.

A sufficient condition for t_0–regularity can be phrased in terms of the so–called strong Feller and irreducible semigroups.

A Markovian semigroup P_t, $t \geq 0$, is said to be a *strongly Feller* semigroup at time $t_0 > 0$ if for arbitrary $\varphi \in B_b(E)$, $P_{t_0}\varphi \in C_b(E)$.

A Markovian semigroup P_t, $t \geq 0$, is said to be *irreducible* at time $t_0 > 0$ if, for arbitrary non empty open set Γ and all $x \in \Gamma$,

$$P_{t_0}(x, \Gamma) = P_{t_0}\chi_\Gamma(x) > 0.$$

Again by the semigroup property if a Markovian semigroup P_t, $t \geq 0$, is a strongly Feller (resp. irreducible) semigroup at time $t_0 > 0$ then it is a strongly Feller (resp. irreducible) semigroup at arbitrary time $t > t_0$.

The following result is due to R.Z. Khas'minskii.

Proposition 4.1.1 *If a Markovian semigroup P_t, $t \geq 0$, is a strongly Feller semigroup at time $t_0 > 0$ and irreducible at time $s_0 > 0$, then it is regular at time $t_0 + s_0$.*

Proof — Assume that for some $x_0 \in E$ and $\Gamma \in \mathcal{E}$, $P_{t_0+s_0}(x_0, \Gamma) > 0$. Since

$$P_{t_0+s_0}(x_0, \Gamma) = \int_E P_{s_0}(x_0, dy) P_{t_0}(y, \Gamma),$$

for some $y_0 \in E$ we have $P_{t_0}(y_0, \Gamma) > 0$. Since $P_{t_0}(y, \Gamma) = P_{t_0}(\chi_\Gamma)(y_0)$ and $P_{t_0}(\chi_\Gamma) \in C_b(E)$, because the semigroup P_t, $t \geq 0$ is a strongly Feller semigroup at time $t_0 > 0$, there exists $r_0 > 0$ such that $P_{t_0}(y, \Gamma) > 0$ for all $y \in B(y_0, r_0)$.

Consequently, for arbitrary $x \in E$,

$$P_{t_0+s_0}(x, \Gamma) = \int_E P_{s_0}(x, dy) P_{t_0}(y, \Gamma),$$

$$\geq \int_{B(y_0, r_0)} P_{s_0}(x, dy) P_{t_0}(y, \Gamma) > 0.$$

We used here the fact that $P_{s_0}(x, B(y_0, r_0)) > 0$ and $P_{t_0}(y, \Gamma) > 0$ for all $y \in B(y_0, r_0)$. Thus if $P_{t_0+s_0}(x_0, \Gamma) > 0$ for some $x_0 \in E$ then $P_{t_0+s_0}(x, \Gamma) > 0$ for all $x \in E$ and the $(t_0 + s_0)$–regularity follows. ∎

4.2 Doob's theorem

The main result concerned with regular semigroups is due to J. L. Doob [56] and can be formulated as follows.

Theorem 4.2.1 *Let P_t, $t \geq 0$, be a stochastically continuous Markovian semigroup and μ an invariant measure with respect to P_t, $t \geq 0$. If P_t, $t \geq 0$, is t_0-regular for some $t_0 > 0$, then*
 (i) μ is strongly mixing and for arbitrary $x \in E$ and $\Gamma \in \mathcal{E}$

$$\lim_{t \to +\infty} P_t(x, \Gamma) = \mu(\Gamma).$$

 (ii) μ is the unique invariant probability measure for the semigroup P_t, $t \geq 0$.
 (iii) μ is equivalent to all measures $P_t(x, \cdot)$, for all $x \in E$ and all $t > t_0$.

Proof —
Step 1 — μ is ergodic
 Let $\Gamma \in \mathcal{E}$ be an invariant set

$$P_t \chi_\Gamma = \chi_\Gamma, \quad \text{for all } t > 0, \tag{4.2.1}$$

such that $\mu(\Gamma) > 0$; we have to show that $\mu(\Gamma) = 1$. From (4.2.1)

$$P_{t_0}(x, \Gamma) = 1, \ x \in \Gamma, \ \mu\text{-a.s.},$$

and for $\mu(\Gamma) > 0$ there exists $x_0 \in \Gamma$ such that $P_{t_0}(x_0, \Gamma) = 1$. Since all probabilities $P_{t_0}(x, \cdot)$ are equivalent, we have $P_{t_0}(x, \Gamma) = 1, \forall\, x \in E$. It follows that

$$\mu(\Gamma) = \int_E P_{t_0}(x, \Gamma)\mu(dx) = 1.$$

Step 2— μ is the unique invariant measure
 If ν is another invariant probability measure for P_t, $t \geq 0$, then ν is ergodic by Step 1. Since the two measures μ and ν are equivalent with respect to any measure $P(t_0, x, \cdot)$, $x \in E$, they are equivalent. By Proposition 3.2.5 we have $\mu = \nu$.

Step 3— P_t, $t \geq 0$ has no angle variable

Assume that for some $\lambda \in \mathbb{R}$, $\lambda \neq 0$, we have

$$P_t f = e^{i\lambda t} f, \quad t > 0. \tag{4.2.2}$$

We will show first that $|f(\cdot)|$ is constant.

Since μ is ergodic, it is enough to prove that

$$P_t |f| = |f|, \quad \mu\text{--a.s.} \tag{4.2.3}$$

By (4.2.2) it follows that

$$|f(x)| = |P_t f(x)| \leq P_t |f|(x).$$

Set now

$$K = \{x \in E : |f(x)| < P_t |f|(x)\},$$

and assume in contradiction that $\mu(K) > 0$. Then, taking into account that μ is invariant,

$$\int_E |f(x)| \mu(dx) = \int_K |f(x)| \mu(dx) + \int_{K^c} |f(x)| \mu(dx)$$

$$< \int_K P_t |f|(x) \mu(dx) + \int_{K^c} P_t |f|(x) \mu(dx)$$

$$= \int_E P_t |f|(x) \mu(dx) = \int_E |f(x)| \mu(dx),$$

a contradiction. So $|f(\cdot)|$ is constant.

Now we show that

$$f(X(t)) = e^{i\lambda t} f(X(0)), \quad t > 0, \ \mathbb{P}^\mu\text{--a.s.}, \tag{4.2.4}$$

where $X(t), t \in \mathbb{R}$, is the canonical process. We have in fact, using (4.2.3),

$$\mathbb{E}^\mu[|f(X(t)) - e^{i\lambda t} f(X(0))|^2] = \mathbb{E}^\mu[|f(X(t))|^2] + \mathbb{E}^\mu[|f(X(0))|^2]$$

$$-e^{-i\lambda t} \mathbb{E}^\mu[f(X(t)) \overline{f(X(0))}] - e^{i\lambda t} \mathbb{E}^\mu[\overline{f(X(t))} \, f(X(0))]. \tag{4.2.5}$$

Since the process $X(t), t \in \mathbb{R}$, is stationary, we have

$$\mathbb{E}^\mu[|f(X(t))|^2] = \mathbb{E}^\mu[|f(X(0))|^2],$$

and since it is a Markov process

$$\mathbb{E}^\mu[f(X(t))\,\overline{f(X(0))}] = \mathbb{E}^\mu\left\{\mathbb{E}^\mu[f(X(t))\,\overline{f(X(0))}|\mathcal{F}_0]\right\}$$

$$= \mathbb{E}^\mu\left\{\overline{f(X(0))}\,\mathbb{E}^\mu[f(X(t))|\mathcal{F}_0]\right\} = \mathbb{E}^\mu\left\{\overline{f(X(0))}\,P_t f(X(0))\right\}$$

$$= e^{i\lambda t}\mathbb{E}^\mu\left\{|f(X(0))|^2\right\}.$$

Now the conclusion (4.2.4) follows on substituting this in (4.2.5).

Let Γ be a Borel set on the circumference $\{\lambda : |\lambda| = 1\}$. We show now that if $\mathbb{P}^\mu(f(X(t)) \in \Gamma) > 0$, then we have

$$P_t(X(0), f^{-1}(\Gamma)) = 1, \quad P^\mu\text{–a.s.}$$

In fact

$$\mathbb{P}^\mu(f(X(t)) \in \Gamma) = \mathbb{P}^\mu(X(t) \in f^{-1}(\Gamma))$$

$$= \mathbb{E}^\mu\left[P_t(X(0), f^{-1}(\Gamma))\right],$$

and so, by (4.2.4),

$$\mathbb{P}^\mu(f(X(t)) \in \Gamma) = \mathbb{P}^\mu(e^{i\lambda t} f(X(0)) \in \Gamma)$$

$$= \mathbb{P}^\mu(f(X(t)) \in \Gamma, e^{i\lambda t} f(X(0)) \in \Gamma)$$

$$= \int_{\{e^{i\lambda t} f(X(0)) \in \Gamma\}} P_t(X(0), f^{-1}(\Gamma)) d\mu.$$

So if $\mathbb{P}^\mu(f(X(t)) \in \Gamma) > 0$ we have $P_t(X(0), f^{-1}(\Gamma)) = 1$, P^μ–a.s., as required.

To complete the proof of Step 3 set $\nu = \mathcal{L}(f(X(t)))$. Then ν is concentrated on the circle $\{\lambda \in \mathbb{C} : |\lambda| = 1\}$. Let e^{ic} belong to the support of ν. Then for all $n \in \mathbb{N}$

$$\mathbb{P}^\mu\left(f(X(t)) \in S(c - \frac{1}{n}, c + \frac{1}{n})\right) > 0,$$

where

$$S\left(c - \frac{1}{n}, c + \frac{1}{n}\right) = \left\{\lambda \in \mathbb{C} : c - \frac{1}{n} \leq \arg\lambda \leq c + \frac{1}{n}\right\}.$$

By (4.2.3) we have

$$P_t\left(X(0), f^{-1}\left(S\left(c - \frac{1}{n}, c + \frac{1}{n}\right)\right)\right) = 1, \ \mathbb{P}^\mu\text{--a.s.}.$$

Consequently we have

$$P_t\left(x, f^{-1}\left(S\left(c - \frac{1}{n}, c + \frac{1}{n}\right)\right)\right) = 1, \ \mathbb{P}^\mu\text{--a.s.},$$

for one (and then for all) $x \in E$, by the hypothesis of equivalence of all transition probabilities. It follows that

$$P_t(x, f^{-1}(c)) = 1, \ t > 0, \ x \in H,$$

and so

$$\mu(f^{-1}(c)) = \int_E P(t, x, f^{-1}(c))\mu(dx) = 1$$

and f is a.s. constant.

Step 4 — For arbitrary $x \in E$ and $\Gamma \in \mathcal{E}$, we have

$$\lim_{t \to \infty} P_t(x, \Gamma) = \mu(\Gamma). \tag{4.2.6}$$

From the absolute continuity of $P_u(x, dy)$, $u > t_0, x \in E$, with respect to μ there exists a density $p(u, x, z)$ such that

$$P_u(x, \Gamma) = \int_\Gamma p(u, x, z)\mu(dz), \ \forall \ \Gamma \in \mathcal{E}, \ u > t_0.$$

On the other hand, for $t > u > t_0$ and $\Gamma \in \mathcal{E}$

$$\begin{aligned}
P_t(x, \Gamma) &= \int_E P_u(x, dy) P_{t-u}(y, \Gamma) \\
&= \int_E P_{t-u}(y, \Gamma) p(u, x, y)\mu(dy) \\
&= \mathbb{E}^\mu\left(P_{t-u}(X(u), \Gamma)p(u, x, X(u))\right) = \mathbb{E}^\mu(\beta \ U_t \alpha),
\end{aligned} \tag{4.2.7}$$

where

$$\alpha = \chi_\Gamma(X(0)), \ \ \beta = p(u, x, X(u)).$$

We have in fact

$$\mathbb{E}^{\mu}(\beta U_t(\alpha)) = \mathbb{E}^{\mu}(\beta \chi_\Gamma(X(t))) = \mathbb{E}^{\mu}[\beta \mathbb{E}^{\mu}(\chi_\Gamma(X(t))|\mathcal{F}_u)]$$

$$= \mathbb{E}^{\mu}[\beta P_{t-u}\chi_\Gamma(X(u))] = \mathbb{E}^{\mu}[\beta P_{t-u}(X(u),\Gamma)].$$

By Proposition 3.2.1 and Step 3, the canonical dynamical system is weakly mixing. Thus there exists a set $I \subset [0,+\infty[$ of relative measure 1 such that

$$\lim_{\substack{|t|\to+\infty \\ t\in I}} P_t(x,\Gamma) = \mu(\Gamma),$$

for arbitrary $x \in E, \Gamma \in \mathcal{B}(E)$.

We show finally that the limit in (4.2.6) does exist without any restriction on t. Let $t \to +\infty$, then there exists $s = s(t)$ such that $\frac{t}{3} < s(t) < \frac{2t}{3}$ and $s(t) \in I, t - s(t) \in I$. This is true for sufficiently large t. Since

$$P_t(x,\Gamma) = \int_E P_{s(t)}(x,dy)P_{t-s(t)}(y,A),$$

the result follows from the previous considerations and by Lemma 4.2.2 below.

Lemma 4.2.2 *Let Γ be a Borel set in $[0,1]$ with $\lambda(\Gamma) > \frac{1}{2}$ where λ is the Lebesgue measure. Then there exists $s \in \Gamma$ such that $1 - s \in \Gamma$.*

Proof — Let
$$\tilde{\Gamma} = \{1 - s : s \in \Gamma\}.$$
Since $\lambda(\tilde{\Gamma}) = \lambda(\Gamma) > \frac{1}{2}$ we have $\lambda\left(\Gamma \cap \tilde{\Gamma}\right) > 0$. ∎

Remark 4.2.3 Using different methods J. Seidler [136] and L. Stettner [147] were able to show that, under the conditions of Theorem 4.2.1, the convergence in (i) can be replaced by the convergence in the variation norm.

Part II

Invariant measures for stochastic evolution equations

Chapter 5

Stochastic Differential Equations

Part II of these lecture notes deals with general methods of studying invariant measures with respect to Markovian semigroups P_t, $t \geq 0$, determined by solutions of stochastic evolution equations on a Hilbert or a Banach space.

In this chapter we summarize basic results on stochastic integration and on stochastic evolution equations on infinite dimensional spaces. We recall important inequalities for stochastic integrals and stochastic convolutions and prove the existence of a regular solution to a class of evolution equations with Lipschitz or locally Lipschitz drift and diffusion coefficients. Then dissipative systems are discussed in some detail. Regular dependence of solutions on initial data are established as well.

5.1 Introduction

We deal here with equations of the form:

$$\left. \begin{array}{l} dX = (AX + F(X))dt + B(X)dW(t), \\[2mm] X(0) = \xi, \end{array} \right\} \qquad (5.1.1)$$

where ξ is a random variable on a given probability space $(\Omega, \mathcal{F}, \mathbb{P})$.

As a rule the process $W(t)$, $t \geq 0$, will be a cylindrical Wiener process on a Hilbert space U. Moreover F and B are nonlinear transformations and A the infinitesimal generator of a strongly continuous semigroup $S(t)$, $t \geq 0$. Precise definitions of all the data related to problem (5.1.1), and to the solution X to (5.1.1), will be given in §5.2 and in §5.3.

If $X(t,x)$, $t \geq 0$, $x \in E$, denotes the unique solution to (5.1.1) with $\xi = x$ constant, then the transition semigroup P_t, $t \geq 0$, is given by

$$P_t\varphi(x) = \mathbb{E}[\varphi(X(t,x))], \ t \geq 0, \ x \in E, \ \varphi \in B_b(E). \qquad (5.1.2)$$

If there exists an invariant measure μ for P_t, $t \geq 0$, given by (5.1.2) then the solution to (5.1.1) corresponding to a random variable ξ such that $\mathcal{L}(\xi) = \mu$ is a stationary solution to (5.1.1), and this solution has important limit properties, see §3.5.

In the following chapters we shall concentrate on basic principles leading to the existence of invariant measures for P_t, $t \geq 0$, given by (5.1.2) and to other related properties. These include uniqueness, ergodicity, mixing and regularity of the density of the invariant measure with respect to a natural reference measure. The present chapter, however, contains a selection of basic results on stochastic evolution equations to which we will often refer.

5.2 Wiener and Ornstein–Uhlenbeck processes

Let U be a separable Hilbert space and μ a Gaussian measure on U. The measure μ is completely determined by its mean value $m \in U$ and its covariance operator Q which is a trace–class nonnegative linear operator. Such a measure will be denoted by $\mathcal{N}(m, Q)$.

A U–valued stochastic process $W(t), t \geq 0$, defined on a probability space $(\Omega, \mathcal{F}, \mathbb{P})$ is called a Q–Wiener process if it has the following properties.

(i) $W(\cdot)$ has continuous paths and $W(0) = 0$.

(ii) $W(\cdot)$ has independent increments.

(iii) For arbitrary $t \geq 0$ and $h > 0$

$$\mathcal{L}(h^{-1/2}(W(t+h) - W(t))) = \mathcal{N}(m, Q).$$

We will always assume that the probability space is equipped with a right–continuous filtration $\{\mathcal{F}_t\}_{t \geq 0}$ such that $\mathcal{F}_0$ contains all sets of $\mathbb{P}$–measure zero. The Wiener process is assumed to be adapted to $\{\mathcal{F}_t\}_{t \geq 0}$ and for every $t > s$ the increments $W(t) - W(s)$ are independent of $\mathcal{F}_t$.

Let $U_0 = $ Image $Q^{1/2}$ be a Hilbert space equipped with the scalar product $\langle \cdot, \cdot \rangle_0$,

$$\langle u, v \rangle_0 = \langle Q^{-1/2}u, Q^{-1/2}v \rangle, \ u, v \in U_0,$$

where $Q^{-1/2}$ is the pseudo–inverse of $Q^{1/2}$, and let $\{e_k\}$ be an arbitrary, complete, orthonormal basis in U. Then the formula

$$W(t) = \sum_{k=1}^{\infty} \beta_k(t)e_k, \quad t \geq 0, \tag{5.2.1}$$

where β_k, $k = 1, ...$, are independent real processes defines a Q–Wiener process on U. The series (5.2.1) converges in U, $\mathbb{P}$–a.s. uniformly on an arbitrary finite time interval $[0, T]$. Note that for arbitrary $g, h \in U_0$ and $t, s \geq 0$

$$\mathbb{E}\langle g, W(t) \rangle_0 \langle h, W(s) \rangle_0 = \langle h, g \rangle_0 \, t \wedge s.$$

This is why the process W is also called a *cylindrical* Wiener process on U_0. Its covariance operator with respect to U_0 is the identity operator. If W is a Wiener process on U with the covariance operator different from the identity then it is called a *coloured* Wiener process.

Example 5.2.1 An important example of a coloured Wiener process is a process on $U = L^2(\mathbb{R}^d)$ used to describe "Random environment", see D. A. Dawson and H. Salehi [53]. The covariance operator is defined as the convolution operator

$$Qu(\xi) = \int_{\mathbb{R}^d} q(\xi - \eta)u(\eta)d\eta, \ u \in U, \tag{5.2.2}$$

where q is a continuous, positive definite function:

$$\sum_{i,j=1}^{N} q(\xi_i - \xi_j)\lambda_i\overline{\lambda_j} \geq 0, \ \xi_1, ..., \xi_N \in \mathbb{R}^d, \ \lambda_1, ..., \lambda_N \in \mathbb{C}, N \in \mathbb{N}.$$

Moreover by Bochner's theorem

$$q(\lambda) = \int_{\mathbb{R}^d} e^{-\langle\lambda,x\rangle}\pi(dx), \lambda \in \mathbb{R}^d,$$

where π is a nonnegative, finite measure, the so-called *spectral measure*. The Wiener process $W(t,\xi), t \geq 0, \xi \in \mathbb{R}^d$, corresponding to Q has the property that for fixed t, $W(t,\cdot)$ is a stationary Gaussian field with the correlation function $\sqrt{t}q(\cdot)$,

$$\mathbb{E}W(t,\xi)W(t,\eta) = \sqrt{t}q(\xi - \eta), \ \xi,\eta \in \mathbb{R}^d.$$

Typical examples: $q(\xi) = e^{-|\xi|^\alpha}$, $\xi \in \mathbb{R}^d$, for arbitrary $\alpha \in]0,2]$. The corresponding spectral measures are then isotropic α–stable distributions, and in particular for $\alpha = 2$ the spectral measure is Gaussian. Note that the operator Q is never compact on U and thus never of trace class.

A natural generalization of this example consists in replacing $\mathbb{R}^d$ by a group G with a translation invariant measure ν and in considering covariance operators Q on $U = L^2(G,\nu)$ as generalized convolutions

$$Qu(\xi) = \int_G q(\xi - \eta)u(\eta)\nu(d\eta), \ \xi \in G, u \in L^2(G,\nu).$$

It is sometimes convenient to consider Wiener processes represented in a form more general than (5.2.1):

$$W(t) = \sum_{k=1}^{\infty} \alpha_k e_k \beta_k(t), \ t \geq 0, \qquad (5.2.3)$$

where $\{\alpha_k\} \subset \mathbb{R}$ and the basis $\{e_k\}$ consists of eigenvectors of a self–adjoint operator A.

Example 5.2.2 The sequence

$$\frac{1}{\sqrt{2\pi}}, \ \frac{1}{\sqrt{\pi}}\cos n\xi, \ \frac{1}{\sqrt{\pi}}\sin n\xi, \ n \in \mathbb{N}, \ \xi \in [0,2\pi],$$

forms a complete orthonormal basis in $U = L^2(0, 2\pi)$. Its elements are eigenfunctions of the operator

$$A = \frac{d^2}{d\xi^2}, \ D(A) = \{x \in H^2(0, 2\pi) : \ x(0) = x(2\pi), \ x'(0) = x'(2\pi)\}.$$

In this case the series (5.2.3) can be equivalently written as follows:

$$W(t, \xi) = \sum_{n=0}^{+\infty}(a_n \beta_n^1(t) \cos n\xi + b_n \beta_n^2(t) \ \sin n\xi), \ t \geq 0, \xi \in [0, 1],$$

$$(5.2.4)$$

where $\{a_n\}$ and $\{b_n\}$ are sequences of real numbers and $\{\beta_n^1\}$, $\{\beta_n^2\}$ are mutually independent Wiener processes.

Denote by G the interval $[0, 2\pi]$ treated as a group with addition calculated mod 2π. Covariance operators Q on $L^2(G)$ are of convolution type with a kernel q if and only if, by Bochner's Theorem, the positive definite function q is of the form

$$q(\xi) = \gamma_0 + \sum_{n=1}^{\infty} \gamma_n \cos n\xi, \ \xi \in [0, 2\pi], \qquad (5.2.5)$$

with $\gamma_n \geq 0$, $n \in \mathbb{N}$ and $\sum_{n=1}^{\infty} \gamma_n < +\infty$.

It is easy to see that if

$$a_n = b_n = \sqrt{\gamma_n}, \ n \in \mathbb{N},$$

then the covariance operator Q of (5.2.4) is of convolution type with the kernel q of the form (5.2.5). Thus, in some cases, Wiener processes defined by (5.2.3) can have a physical meaning.

Similar considerations can be made if G is a d dimensional torus and A is either a Laplace–Beltrami operator on G or a Stokes operator.

Assume that U and H are separable Hilbert spaces, A is the infinitesimal generator of a C_0–semigroup $S(t), t \geq 0$, on H and B is a bounded linear operator from U into H. Let $W(t), t \geq 0$, be a cylindrical Wiener process on U and ξ an H–valued $\mathcal{F}_0$–measurable random variable.

An H–valued $\mathcal{F}_t$–adapted stochastic process $Z(t), t \geq 0$, is said to be a *weak solution* to the equation

$$dZ = AZ dt + B dW, \; Z(0) = \xi, \tag{5.2.6}$$

if for arbitrary $h \in D(A^*)$ and all $t \geq 0$, $\mathbb{P}$–a.s.

$$\langle h, Z(t) \rangle = \langle h, \xi \rangle + \int_0^t \langle A^* h, Z(s) \rangle ds + \langle B^* h, W(t) \rangle. \tag{5.2.7}$$

Note that all terms in (5.2.7) have well defined meanings. One can show that there exists a solution to (5.2.6) if and only if the operators

$$Q_t = \int_0^t S(r) B B^* S^*(r) dr, \; t \geq 0, \tag{5.2.8}$$

are of trace class. In this case the solution is given by the formula

$$Z(t) = S(t)\xi + \int_0^t S(t - s) B dW(s), \; t \geq 0. \tag{5.2.9}$$

For the definition of stochastic integral see next section. The process Z is called an *Ornstein–Uhlenbeck* process and the process

$$W_A(t) = \int_0^t S(t - s) B dW(s), \; t \geq 0, \tag{5.2.10}$$

a *stochastic convolution*. The process Z is Gaussian and Markovian with transition semigroup R_t, $t \geq 0$, of the form

$$R_t \varphi(x) = \int_H \varphi(y) \mathcal{N}(S(t)x, Q_t)(dy), \; \varphi \in B_b(H). \tag{5.2.11}$$

The operator BB^* will often be denoted by Q:

$$BB^* = Q.$$

5.2.1   Stochastic integrals and convolutions

We assume, see earlier, that $W(t)$, $t \geq 0$, is a cylindrical Wiener process on a separable Hilbert space U, given by a formal expansion

$$W(t) = \sum_{n=1}^{\infty} \beta_n(t) e_n, \; t \geq 0,$$

where e_n, $n \in \mathbb{N}$, is an orthonormal basis on U. Very often the space U will be identical with H. The random variable ξ is assumed to be $\mathcal{F}_0$–measurable.

By $\|R\|_{HS}$ or $\|R\|_2$ we denote the Hilbert–Schmidt norm of the operator $R \in L(U, H)$. The space of all Hilbert–Schmidt operators from U into H (endowed with the Hilbert–Schmidt norm) will be denoted by $L_2(U, H)$. This is again a separable Hilbert space.

The stochastic integral

$$\int_0^t \Phi(s)dW(s), \ \ t \geq 0,$$

is well defined, see G. Da Prato and J. Zabczyk [44], for any $L_2(U, H)$–valued adapted process Φ such that

$$\mathbb{P}\left(\int_0^t \|\Phi(s)\|_2^2 \, ds < +\infty, t \geq 0\right) = 1.$$

Moreover

$$\mathbb{E}\left|\int_0^t \Phi(s)dW(s)\right|^2 \leq \mathbb{E}\int_0^t \|\Phi(s)\|_{HS}^2 \, ds, \ t \geq 0, \qquad (5.2.12)$$

with the equality applying in (5.2.12) if the right hand side is finite. The following generalization of (5.2.12) will be used, see G. Da Prato and J. Zabczyk [44, Lemma 7.7, page 194]:

$$\sup_{s\in[0,t]} \mathbb{E}\left|\int_0^s \Phi(\sigma)dW(\sigma)\right|^p \leq c_p \left(\int_0^t (\mathbb{E}\|\Phi(\sigma)\|_{HS}^p)^{2/p}d\sigma\right)^{p/2},$$

$$(5.2.13)$$

where $t \in [0, T]$ and $c_p = (p(p-1)/2)^{p/2}$, valid for arbitrary $p \geq 2$, $t > 0$ and an arbitrary predictable process $\Phi(s)$, $s \in [0, T]$, with values in $L_2(U, H)$.

We will also need the so–called maximal inequalities formulated in the following two theorems.

Theorem 5.2.3 (Doob's inequalities) *Assume that*

$$\mathbb{E}\int_0^T \|\Phi(s)\|_{HS}^2 ds < +\infty.$$

(i) For arbitrary $p \geq 1$ and $\lambda > 0$,

$$\mathbb{P}\left(\sup_{t \leq T}\left|\int_0^t \Phi(s)dW(s)\right| \geq \lambda\right) \leq \frac{1}{\lambda^p}\,\mathbb{E}\left|\int_0^T \Phi(s)dW(s)\right|^p.$$

(ii) For arbitrary $p > 1$,

$$\mathbb{E}\left(\sup_{t \leq T}\left|\int_0^t \Phi(s)dW(s)\right|^p\right) \leq \frac{p}{p-1}\,\mathbb{E}\left|\int_0^T \Phi(s)dW(s)\right|^p.$$

Theorem 5.2.4 (Burkholder–Davis–Gundy) *For arbitrary $p > 0$ there exists a constant $c_p > 0$ such that*

$$\mathbb{E}\left(\sup_{t \leq T}\left(\int_0^t \Phi(s)dW(s)\right)^p\right) \leq c_p\mathbb{E}\left|\int_0^T \|\Phi(s)\|_{HS}^2 ds\right|^{p/2}.$$

Let A be the infinitesimal generator of a C_0–semigroup $S(t)$, $t \geq 0$, on H. If $\Phi(t)$, $t \in [0,T]$, is an $L_2(U,H)$–valued adapted process such that the stochastic integrals

$$\int_0^t S(t-s)\Phi(s)dW(s) = W_A^\Phi(t), \ t \in [0,T],$$

are well defined, then the process W_A^Φ is called a *stochastic convolution*. If $\Phi(t) = B$, $t \in [0,T]$, we write for short W_A instead W_A^Φ.

Let $\alpha \in {]0,1]}$ and $Y_\alpha^\Phi(t)$, $t \in [0,T]$, be the following stochastic process,

$$Y_\alpha^\Phi(t) = \int_0^t (t-s)^{-\alpha}S(t-s)\Phi(s)dW(s), \ t \in [0,T].$$

The following result, which will often be referred to as the *factorization formula*, is an easy corollary of the stochastic Fubini theorem, see e.g. G. Da Prato and J. Zabczyk [Theorem 4.18][44].

Theorem 5.2.5 *Assume that for all $t \in [0,T]$*

$$\int_0^t (t-s)^{\alpha-1}\left[\int_0^s (s-\sigma)^{-2\alpha}\mathbb{E}\left(\|S(t-\sigma)\Phi(\sigma)\|_2^2\right)d\sigma\right]^{1/2} ds < +\infty. \tag{5.2.14}$$

Then

$$\int_0^t S(t-s)\Phi(s)dW(s) = \frac{\sin\alpha\pi}{\pi}\int_0^t (t-s)^{\alpha-1}S(t-s)Y_\alpha^\Phi(s)ds, \ t \in [0,T]. \tag{5.2.15}$$

We note that condition (5.2.14) is only sufficient for the factorization (5.2.15) to hold. Different sufficient conditions can be more useful in specific cases.

We will need the following result on stochastic convolutions whose proof is a typical application of the factorization formula.

Theorem 5.2.6 *Assume that*
(i) The operator A generates an analytic semigroup $S(t)$, $t \geq 0$, on H such that

$$\|S(t)\| \leq Me^{-\omega t}, \ t \geq 0,$$

with ω and M positive numbers ([1]).
(ii) There exists $\alpha \in \,]0, 1/2[$ such that for arbitrary $T > 0$

$$\int_0^T t^{-2\alpha} \|S(t)B\|_{HS}^2 dt < +\infty. \tag{5.2.16}$$

Then for arbitrary $\gamma \in [0, \alpha[$ there exists a version of W_A which is Hölder continuous with values in $D((-A)^\gamma)$ with arbitrary exponent smaller than $\alpha - \gamma$. ([2])

Proof — For any $\alpha \in \,]0, 1[$, $\gamma \in [0, \alpha[$ and $p > 1$, define the integral operator

$$R_{\alpha,\gamma}\varphi(t) = \int_0^t (t - \sigma)^{\alpha-1}(-A)^\gamma S(t - \sigma)\varphi(\sigma)d\sigma, \ t \in [0, T],$$

on the space $L^p(0, T; H)$.

Let us choose $p > \frac{1}{\alpha}$. By Proposition A.1.1 and Theorem 5.2.5 on the factorization, it is enough to show that the process

$$Y_\alpha(s) = \int_0^s (s - \sigma)^{-\alpha} S(s - \sigma)BdW(\sigma), \ s \in [0, T],$$

has p–integrable trajectories. By Theorem 5.2.4

$$\mathbb{E}|Y_\alpha(s)|^p \leq c_p \left(\int_0^s \|(s - \sigma)^{-\alpha} S(s - \sigma)B\|_2^2 d\sigma \right)^{p/2}, \ s \in [0, T],$$

---

[1]The assumption $\omega > 0$ is not essential in Theorems 5.2.6 and 5.2.7
[2]$(-A)^\gamma$ means the fractional power of $-A$, see e.g. A. Lunardi [108, §2.2.2]

and therefore

$$\mathbb{E}\int_0^T |Y_\alpha(s)|^p ds \le c_p \int_0^T \left(\int_0^s (\sigma)^{-2\alpha}\|S(\sigma)B\|_2^2 d\sigma\right)^{p/2} ds < +\infty.$$

The proof is complete.   ∎

With exactly the same proof the following result can be established.

Theorem 5.2.7 *Assume that*
(i) The operator A generates an analytic semigroup $S(t)$, $t \ge 0$, on H such that

$$\|S(t)\| \le Me^{-\omega t}, \ t \ge 0,$$

with ω and M positive numbers,
(ii) there exists $\alpha \in \,]0, 1/2[$ such that for arbitrary $T > 0$

$$\int_0^T t^{-2\alpha}\,\|S(t)\|_{HS}^2 dt < +\infty, \qquad (5.2.17)$$

(iii) there exists a constant $C > 0$ such that $\mathbb{P}$-a.s.

$$\|\Phi(t)\| \le C \ \text{for all } t \ge 0.$$

Then for arbitrary $\gamma \in [0, \alpha[$ there exists a version of W_A^Φ which is Hölder continuous with values in $D((-A)^\gamma)$ with arbitrary exponent smaller than $\alpha - \gamma$.

Example 5.2.8 Let $H = L^2(0, 1) = U$, $B = I$,

$$A = \frac{d^2}{d\xi^2}, \quad D(A) = H^2(0, 1) \cap H_0^1(0, 1).$$

The functions

$$e_n(\xi) = \sqrt{2}\sin \pi n\xi, \ n \in \mathbb{N}, \ \xi \in \,]0, 1[,$$

form an orthonormal sequence of eigenfunctions of the operator A corresponding to the eigenvalues

$$-\pi^2 n^2, \ n \in \mathbb{N}.$$

The stochastic convolution $Z(t) = W_A(t)$, $t \geq 0$, is in this case a weak solution of the equation

$$dZ(t) = AZ(t)dt + dW(t), \ Z(0) = 0.$$

Note that

$$\|S(t)\|_{HS}^2 = \sum_{n=1}^{\infty} e^{-2\pi^2 n^2 t}.$$

The condition (5.2.16) of Theorem 5.2.6 is

$$\int_0^T t^{-2\alpha} \sum_{n=1}^{\infty} e^{-2\pi^2 n^2 t} dt < +\infty.$$

It is fulfilled if and only if $\alpha < \frac{1}{4}$. Consequently the process Z is Hölder continuous in H with any exponent smaller than $\frac{1}{4}$.

We will now prove a more general result concerned with the stochastic convolution on $L^2(\mathcal{O})$ where $\mathcal{O}$ is a bounded domain in $\mathbb{R}^d$ and the semigroup $S(t)$, $t \geq 0$, is given by

$$S(t)x = \sum_{n=1}^{\infty} e^{-\alpha_n t} \langle e_n, x \rangle e_n, \ x \in H, \qquad (5.2.18)$$

and $\{e_n\}$ is a complete orthonormal set on $L^2(\mathcal{O})$. Moreover we will assume that

$$W(t) = \sum_{n=1}^{\infty} \sqrt{\lambda_n} \beta_n(t) e_n, \ t \geq 0, \qquad (5.2.19)$$

where $\beta_n(\cdot)$, $n \in \mathbb{N}$, are independent Wiener processes. As far as the domain $\mathcal{O}$ is concerned we will assume that

$$\kappa(\mathcal{O}) > 0 \qquad (5.2.20)$$

where the constant $\kappa(\mathcal{O})$ is defined in Appendix B, by (B.1.1). Concerning the functions $\{e_n\}$ we shall assume that

$$e_k \in C_0(\overline{\mathcal{O}}), \qquad (5.2.21)$$

$$|e_k(\xi)| \leq C, \ |D_k e_k(\xi)| \leq C\sqrt{\alpha_k}, \ \xi \in \mathcal{O}, \ k \in \mathbb{N}. \qquad (5.2.22)$$

In the formulation of the theorem below $\|\cdot\|_a$ stands for the Hölder semi-norm on $\mathcal{O}$:

$$\|x\|_a = \sup_{\xi \neq \eta \in \mathcal{O}} \frac{|x(\xi) - x(\eta)|}{|\xi - \eta|^a}.$$

Theorem 5.2.9 *Assume that* (5.2.18) *and* (5.2.19)–(5.2.22) *hold and for some* $\gamma \in \,]0,1[$

$$\sum_{k=1}^{\infty} \frac{\lambda_k}{\alpha_k^{1-\gamma}} < +\infty. \tag{5.2.23}$$

Then there exists a continuous version of the process $W_A(\cdot)$ *with values in* $C_0(\overline{\mathcal{O}})$. *Moreover, if* $r > 0$, $\gamma \in \,]0,1[$, $a > 0$ *are numbers such that*

$$\gamma - \frac{d}{2r} > \frac{\beta - 2d}{2r} > 0,$$

then the version has the additional property that

$$\mathbb{E}\|W_A(t)\|_{\frac{a-2d}{r}} \leq C \left(\sum_{k=1}^{\infty} \frac{\lambda_k}{\alpha_k^{1-\gamma}} \right)^{r/2}.$$

Proof — Fix $T > 0$ and define

$$Z_N(t,\xi) = \sum_{n=1}^{N} \sqrt{\lambda_n} \int_0^t S(t-s) e_n(\xi) d\beta_n(s)$$

$$= \sum_{n=1}^{N} \sqrt{\lambda_n} e_n(\xi) \int_0^t e^{-\alpha_n(t-s)} d\beta_n(s), \ t \in [0,T].$$

Set also

$$Z_{N,M} = Z_M - Z_N, \ M, N \in \mathbb{N}.$$

It is clear that we can assume that Z_N is a continuous function on $[0,T] \times \overline{\mathcal{O}}$ and that $Z_N(t,\xi) = 0$ for $t \in [0,T]$ and ξ belonging to the boundary $\partial\mathcal{O}$ of $\mathcal{O}$.

By a straightforward interpolation argument it follows that

$$|e_k(\xi) - e_k(\eta)| \leq C\alpha_k^{1/2}|\xi - \eta|^{\gamma}, k \in \mathbb{N},$$

for all $\gamma \in [0,1]$ and $\xi, \eta \in \overline{\mathcal{O}}$.

Let $M > N$ then for all $t \in [0,T]$, $\xi, \eta \in \mathcal{O}$

$$\mathbb{E}\,|Z_{N,M}(t,\xi) - Z_{N,M}(t,\eta)|^2$$

$$= \sum_{k=N+1}^{M} \lambda_k \int_0^t e^{-2\alpha_k(t-s)}|e_k(\xi) - e_k(\eta)|^2 ds$$

$$\leq \frac{C}{2} \sum_{k=N+1}^{M} \frac{\lambda_k}{\alpha_k} \alpha_k^\gamma |\xi - \eta|^{2\gamma}$$

and therefore

$$\mathbb{E}\,|Z_{N,M}(t,\xi) - Z_{N,M}(t,\eta)|^2 \leq \frac{C}{2} \sum_{k=N+1}^{M} \frac{\lambda_k}{\alpha_k^{1-\gamma}}|\xi - \eta|^{2\gamma}. \quad (5.2.24)$$

Moreover for $0 \leq s < t \leq T$, $\xi \in \mathcal{O}$

$$\mathbb{E}\,|Z_{N,M}(t,\xi) - Z_{N,M}(t,\eta)|^2$$

$$= \sum_{k=N+1}^{M} \lambda_k \int_s^t e^{-2(t-\sigma)\alpha_k}|e_k(\xi)|^2 d\sigma$$

$$+ \sum_{k=N+1}^{M} \lambda_k \int_0^s \left| e^{-(t-\sigma)\alpha_k} - e^{-(s-\sigma)\alpha_k} \right| |e_k(\xi)|^2 d\sigma$$

$$= I_1(t,s,\xi) + I_2(t,s,\xi).$$

Let $c_\gamma > 0$ be a number such that for arbitrary $u, v \geq 0$, $|e^{-u} - e^{-v}| \leq c_\gamma |u - v|^\gamma$. Then

$$I_1(t,s,\xi) \leq \frac{1}{2}C^2 \sum_{k=N+1}^{M} \frac{\lambda_k}{\alpha_k} \left(1 - e^{-2(t-s)\alpha_k}\right)$$

$$\leq \frac{c_\gamma}{2^{1-\gamma}} \sum_{k=N+1}^{M} \frac{\lambda_k}{\alpha_k^{1-\gamma}} |t - s|^\gamma.$$

Moreover

$$
I_2(t,s,\xi) \;\leq\; C^2 \sum_{k=N+1}^{M} \lambda_k \int_0^s e^{-2(s-\sigma)\alpha_k} \left(1 - e^{-2(t-s)\alpha_k}\right)^2 d\sigma
$$

$$
\leq\; \frac{c_\gamma}{2} C^2 \sum_{k=N+1}^{M} \frac{\lambda_k}{\alpha_k^{1-\gamma}}\, |t-s|^\gamma.
$$

Collecting all the estimates and taking into account that the random variables $Z_N(t,\xi)$ are Gaussian we obtain that for arbitrary $r > 0$ there exists c_r such that

$$
\mathbb{E}\,|Z_{N,M}(t,\xi) - Z_{N,M}(t,\eta)|^r
$$

$$
\leq C_r \left(\sum_{k=N+1}^{M} \frac{\lambda_k}{\alpha_k^{1-\gamma}} \right)^{r/2} \left(|\xi-\eta|^2 + |t-s|^2\right)^{r\gamma/4}. \tag{5.2.25}
$$

From Theorem B.1.1 we have

$$
\frac{|Z_{M,N}(u,\kappa) - Z_{M,N}(v,\zeta)|^r}{\left(|u-v|^2 + |\kappa-\zeta|^2\right)^{\frac{a-2d}{r}}}
$$

$$
\leq C \int_0^T \int_0^T \int_{\mathcal{O}} \int_{\mathcal{O}} \frac{|Z_{M,N}(t,\xi) - Z_{M,N}(s,\eta)|^r}{\left(|t-s|^2 + |\xi-\eta|^2\right)^{\frac{a}{2}}} \, d\xi\,d\eta\,dt\,ds,
$$

or

$$
\|Z_{N,M}\|_{\frac{a-2d}{r}}^r \leq C \int_0^T \int_0^T \int_{\mathcal{O}} \int_{\mathcal{O}} \frac{|Z_{M,N}(t,\xi) - Z_{M,N}(s,\eta)|^r}{\left(|t-s|^2 + |\xi-\eta|^2\right)^{\frac{a}{2}}} \, d\xi\,d\eta\,dt\,ds,
$$

where $\|\cdot\|_a$ denotes the a–Hölder norm on $[0,T] \times \overline{\mathcal{O}}$.

Taking expectation, using (5.2.24) and assuming that

$$
\frac{\gamma}{2} - \frac{d}{r} > \frac{a-2d}{r} > 0, \text{ and } r > 1,
$$

we obtain that there exists a finite constant C such that

$$
\mathbb{E}\|Z_{N,M}\|_{\frac{a-2d}{r}}^r \leq C \left(\sum_{k=N+1}^{M} \frac{\lambda_k}{\alpha_k^{1-\gamma}} \right)^{r/2}.
$$

So the sequence of random variables $\{Z_N\}$ satisfies the Cauchy condition in the space $\tilde{L}$ of random variables Z whose values are $\frac{a-2d}{r}$-Hölder continuous functions on $[0,T] \times \overline{\mathcal{O}}$, vanishing on $[0,T] \times \partial\mathcal{O}$. Therefore $\{Z_N\}$ has a limit Z on this Banach space which is the desired version. In a similar way one proves the second part of the theorem using (5.2.23) instead of (5.2.24). ∎

Remark 5.2.10 It follows from the proof that if

$$\frac{\gamma}{2} - \frac{d}{r} > \frac{a - 2d}{r} > 0, \text{ and } r > 1,$$

then

$$\mathbb{E}\|W_A(\cdot,\cdot)\|^r_{\frac{a-2d}{r}} \leq C \left(\sum_{k=N+1}^{M} \frac{\lambda_k}{\alpha_k^{1-\gamma}} \right)^{r/2}.$$

5.3 Stochastic evolution equations

In this section we will recall basic existence and uniqueness results on stochastic evolution equations of the form

$$\left. \begin{array}{l} dX(t) = (AX + F(X))dt + B(X)dW(t) \\[2mm] X(0) = \xi, \end{array} \right\} \tag{5.3.1}$$

on a separable Hilbert space H. We will also study regular dependence of solutions on initial data and Kolmogorov's equation for transition functions.

We will recall now existence results for problem (5.3.1), under suitable Lipschitz continuous conditions on F and B.

It is convenient to introduce the following assumptions:

Hypothesis 5.1 *(i) A is the infinitesimal generator of a strongly continuous semigroup $S(t)$, $t \geq 0$, on H.*

(ii) F is a mapping from H into H and there exists a constant $c_0 > 0$ such that

$$|F(x)| \leq c_0(1 + |x|), \quad x \in H,$$

$$|F(x) - F(y)| \leq c_0|x - y|, \quad x, y \in H.$$

(iii) B is a strongly continuous mapping from H into L(U; H) (^{3}) such that for any $t > 0$ and $x \in H$, $S(t)B(x)$ belongs to $L_2(U; H)$, and there exists a locally square integrable mapping

$$K : [0, +\infty[\to [0, +\infty[, \ t \to K(t),$$

such that

$$\|S(t)B(x)\|_{HS} \le K(t)(1 + |x|), \ t > 0, \ x \in H,$$

$$\|S(t)B(x) - S(t)B(y)\|_{HS} \le K(t)|x - y|, \ t > 0, \ x, y \in H.$$

An $\mathcal{F}_t$–adapted process $X(t)$, $t \ge 0$, is said to be a *mild solution* of (5.3.1) if it satisfies the following integral equation,

$$
\begin{aligned}
X(t) &= S(t)\xi + \int_0^t S(t-s)F(X(s))ds \\
&+ \int_0^t S(t-s)B(X(s))dW(s), \ t \in [0, T].
\end{aligned}
\tag{5.3.2}
$$

We shall denote by $\mathcal{H}_{p,T}$ the Banach space of all (equivalence classes) of predictable H–valued processes $Y(t)$, $t \ge 0$, such that

$$\|Y\|_{p,T} = \sup_{t \in [0,T]} (\mathbb{E}|Y(t)|^p)^{1/p} < +\infty.$$

We have the following result.

Theorem 5.3.1 *Assume Hypothesis 5.1 and let $p \ge 2$. Then for an arbitrary $\mathcal{F}_0$–measurable initial condition ξ such that $\mathbb{E}|\xi|^p < +\infty$ there exists a unique mild solution X of (5.3.1) in $\mathcal{H}_{p,T}$ and there exists a constant C_T, independent of ξ, such that*

$$\sup_{t \in [0,T]} \mathbb{E}|X(t)|^p \le C_T(1 + \mathbb{E}|\xi|^p). \tag{5.3.3}$$

Finally, if there exists $\alpha \in \ ]0, 1/2[$ such that

$$\int_0^1 s^{-2\alpha} K^2(s)ds < +\infty, \tag{5.3.4}$$

where K is the function from Hypothesis 5.1–(iii), then the solution $X(\cdot)$ is continuous $\mathbb{P}$-a.s.

---

[3]Means that for any $u \in U$ the mapping $x \to B(x)u$ from H into H is continuous.

We shall denote by $X(\cdot, \xi)$ the mild solution of (5.3.1).

The proof of Theorem 5.3.1 is similar to that in G. Da Prato and J. Zabczyk [44, Theorem 7.4, page 186]. We will sketch it here for the sake of completeness. A more general result, needed in some applications, see S. Peszat and J. Zabczyk [124], will be given in Part III.

Proof of Theorem 5.3.1 — For arbitrary $\xi \in L^p(\Omega, H)$ and $X \in \mathcal{H}_{p,T}$ define a process $Y = \mathcal{K}(\xi, X)$ by the formula

$$
\begin{aligned}
Y(t) &= S(t)\xi + \int_0^t S(t-s)F(X(s))ds \\
&+ \int_0^t S(t-s)B(X(s))dW(s), \ t \in [0,T].
\end{aligned}
\tag{5.3.5}
$$

We will note first that, by inequality (5.2.13), we have $K(\xi, X) \in \mathcal{H}_{p,T}$ for arbitrary $X \in \mathcal{H}_{p,T}$. Moreover, setting $M_T = \sup_{t \in [0,T]} \|S(t)\|$, we have

$$
\begin{aligned}
\mathbb{E}|Y(t)|^p &\leq 3^{p-1} \Bigg\{ \|S(t)\|^p \mathbb{E}|\xi|^p + \mathbb{E}\left[\left(\int_0^t |S(t-s)F(X(s))|ds \right)^p \right] \\
&+ \mathbb{E}\left[\left\| \int_0^t S(t-s)B(X(s))dW(s) \right\|^p \right\} \right] \\
&\leq 3^{p-1} \Bigg\{ M_T^p \mathbb{E}|\xi|^p + T^{p-1}M_T^p \int_0^t \mathbb{E}|F(X(s))|^p ds \\
&+ c_p \left[\int_0^t (\mathbb{E}\|S(t-s)B(X(s))\|_{HS}^p)^{2/p} ds \right]^{p/2} \Bigg\}.
\end{aligned}
$$

Moreover

$$
\int_0^t \mathbb{E}|F(X(s))|^p ds \leq 2^{p-1}c_0^p \sup_{s \in [0,t]} (1 + \mathbb{E}|X(s)|^p)t,
$$

and

$$\left[\int_0^t \left(\mathbb{E}\|S(t-s)B(X(s))\|_{HS}^p\right)^{2/p} ds\right]^{p/2}$$

$$\le 2^{p-1} \left(\int_0^t K^2(t-s)(1+\mathbb{E}|X(s)|^p)^{2/p}ds\right)^{p/2}$$

$$\le 2^{p-1} \left(\int_0^t K^2(t-s)ds\right)^{p/2} \sup_{s\in[0,t]} (1+\mathbb{E}|X(s)|^p).$$

It is now clear that, for some constants c_1, c_2, c_3,

$$\sup_{t\in[0,T]} \mathbb{E}|Y(t)|^p \le c_1 + c_2\mathbb{E}|\xi|^p + c_3 \sup_{t\in[0,T]} \mathbb{E}|X(t)|^p. \qquad (5.3.6)$$

Thus $Y \in \mathcal{H}_{p,T}$.

In exactly the same way, if $X_1, X_2 \in \mathcal{H}_{p,T}$ and $Y_1 = \mathcal{K}(\xi, X_1)$, $Y_2 = \mathcal{K}(\xi, X_2)$, then

$$\sup_{t\in[0,T]} \mathbb{E}|Y_1(t) - Y_2(t)|^p \le c_3 \sup_{t\in[0,T]} \mathbb{E}|X_1(t) - X_2(t)|^p.$$

It is easy to see that if T is small enough then $c_3 < 1$ and consequently, by the contraction principle, the equation (5.3.1) has a unique solution in $\mathcal{H}_{p,T}$. The case of general $T > 0$ can be treated by considering the equation in intervals $[0, \widetilde{T}]$, $[\widetilde{T}, 2\widetilde{T}], \ldots$ with $\widetilde{T}$ such that $c_3(\widetilde{T}) < 1$. Moreover with such a $\widetilde{T}$ we get from (5.3.6) for the solution of (5.3.1) that

$$\sup_{t\in[0,T]} \mathbb{E}|X(t)|^p \le \frac{1}{1 - c_3(\widetilde{T})} [c_1 + c_2\mathbb{E}|\xi|^p]$$

which is inequality (5.3.3). The case of general $T > 0$ can be easily obtained by iteration as well.

Finally, the continuity of the solution can be proved by using factorization, see Theorem 5.2.5, in a similar way to Theorem 5.2.6 and Theorem 5.2.7. ∎

Remark 5.3.2 If the operator A generates an analytic semigroup, then trajectories of solutions are even Hölder continuous with values in $D((-A)^\alpha)$. Compare the proofs of Theorems 5.2.6 and 5.2.7.

5.4 Regular dependence on initial conditions and Kolmogorov equations

If one assumes that the coefficients F and B of equation (5.3.1) are smooth then the solutions to the equation are also smooth in a proper sense. Note that in the proof of Theorem 5.3.1 we showed that the transformation $x \to X(\cdot, x)$ from H into $\mathcal{H}_{p,T}$ was Lipschitz continuous. We want now to prove that this mapping is, under suitable regularity assumptions, differentiable. For this we need a generalization of the local inversion theorem proved in Appendix C.

5.4.1 Differentiable dependence on initial datum

We prove here the result.

Theorem 5.4.1 *Assume that the mappings A, F and B satisfy Hypothesis 5.1.*

(i) If F and B have first Fréchet derivatives bounded and continuous then the solution $X(\cdot, x)$ to problem (5.3.1) is continuously differentiable in x as a mapping from H into $\mathcal{H}_{2,T}$. Moreover, for any $h \in H$, the process $\zeta^h(t) = X_x(t, x)h$, $t \in [0, T]$, is a mild solution of the following equation,

$$\left. \begin{array}{ll} d\zeta^h & = (A\zeta^h + F_x(X) \cdot \zeta^h dt + B_x(X) \cdot \zeta^h dW(t), \\ \\ \zeta^h(0) & = h. \end{array} \right\} \tag{5.4.1}$$

In addition there exists a constant $C_{1,T}$, independent of h, such that

$$\sup_{t \in [0,T]} \mathbb{E}|X_x(t, x)h|^2 \leq C_{1,T}|h|^2. \tag{5.4.2}$$

(ii) Assume in addition that F and B have bounded and continuous second Fréchet derivatives and that for any $t > 0$, and $x, y, z \in H$, $S(t)B_{xx}(x)(y, z)$ belongs to $L_2(U; H)$ and there exists a locally square integrable mapping

$$K_1 : [0, +\infty[\to [0, +\infty[, \ t \to K_1(t),$$

such that

$$\|S(t)B_{xx}(x)(y, z)u\|_{HS} \leq K_1(t)|y|\,|z|\,|u|, \ \forall \, x, y, z \in H, \ u \in U.$$

Then the solution $X(\cdot, x)$ to problem (5.3.1) is twice continuously differentiable, and for any $h, g \in H$, the process $\eta^{h,g}(t) = X_{xx}(t, x)(h, g)$, $t \in [0, T]$, is a mild solutions of the following equation,

$$
\left.
\begin{aligned}
d\eta^{h,g} &= (A\eta^{h,g} + F_x(X) \cdot \eta^{h,g})dt + B_x(X) \cdot \eta^{h,g}dW(t), \\
&\quad + F_{xx}(X) \cdot (\zeta^h, \zeta^g))dt + B_{xx}(X) \cdot (\zeta^h, \zeta^g)dW(t) \\
\eta^{h,g}(0) &= 0.
\end{aligned}
\right\}
$$

$$(5.4.3)$$

Proof — Consider the mapping

$$\mathcal{F} : H \times \mathcal{H}_{2,T} \to \mathcal{H}_{2,T},$$

defined by

$$
\begin{aligned}
\mathcal{F}(x, X)(t) &= S(t)x + \int_0^t S(t-s)F(X(s))ds \\
&\quad + \int_0^t S(t-s)B(X(s))dW(s), \ t \in [0, T].
\end{aligned}
$$

It remains to check that, setting

$$\Lambda = H, \ E = \mathcal{H}_{2,T}, \ G = \mathcal{H}_{4,T},$$

$\mathcal{F}$ fulfils the hypotheses of Proposition C.1.3, and the conclusion follows. ∎

5.4.2 Kolmogorov equation

Under the hypotheses of Theorem 5.4.1 one can show, see G. Da Prato and J. Zabczyk [44, Theorem 9.8], that for any $x \in H$, $X(t, x)$, $t \geq 0$, is a Markov process. The corresponding transition semigroup P_t, $t \geq 0$ is defined by

$$P_t\varphi(x) = \mathbb{E}[\varphi(X(t, x))], \ x \in H, \ \varphi \in B_b(H).$$

From Theorem 5.4.1 we can also deduce an important result about the Kolmogorov backward equation associated to (5.3.1) with $\xi = x$:

$$\left.\begin{array}{rcl}
\dfrac{\partial}{\partial t}v(t,x) & = & \dfrac{1}{2}\ \mathrm{Tr}\,[B^*(x)v_{xx}(t,x)B(x)] + \langle Ax + F(x), v_x(t,x)\rangle, \\[2mm]
& & t > 0,\ x \in D(A), \\[3mm]
v(0,x) & = & \varphi(x);\ x \in H.
\end{array}\right\}$$

$$(5.4.4)$$

We need the following hypothesis, stronger than Hypothesis 5.1.

Hypothesis 5.2 *(i) Hypothesis 5.1–(i)–(ii) hold.*
(ii) B is a mapping from H into $L_2(U;H)$, and there exists a constant $c_1 > 0$ such that

$$\|B(x)\|_{HS} \leq c_1(1 + |x|),\ x \in H,$$

$$\|B(x) - B(y)\|_{HS} \leq c_1|x - y|,\ x, y \in H.$$

A *strict solution* of problem (5.4.4) is a continuous function v : $[0, +\infty[\ \times H \to \mathbb{R}$ having continuous first and second partial derivatives with respect to x, such that $u(\cdot, x)$ is continuously differentiable in t for all $x \in D(A)$, and fulfilling equation (5.4.4) for all $x \in D(A)$ and $t \geq 0$.

The following result is proved in G. Da Prato and J. Zabczyk [44, Theorem 9.16].

Theorem 5.4.2 *Assume that the mappings F and B satisfy Hypothesis 5.2. If in addition the first and the second derivatives of F and B are bounded and continuous and $\varphi \in C_b^2(H)$ then equation (5.4.4) has a unique strict solution v and it is given by the formula*

$$v(t,x) = \mathbb{E}(\varphi(X(t,x))) = P_t\varphi(x),\ t \geq 0,\ x \in H. \qquad (5.4.5)$$

5.5 Dissipative stochastic systems

There is an important class of infinite dimensional deterministic dynamical systems for which the asymptotic behaviour is well understood. They are called *dissipative*. After recalling their basic properties we will study their stochastic perturbation described by the evolution equation (5.1.1) with A, F and B having additional properties.

5.5.1   Generalities about dissipative mappings

Let E be a Banach space with the norm $\| \cdot \|$. Let us first recall some properties of the subdifferential of the norm. The subdifferential $\partial \|x\|$ of $\| \cdot \|$ at x is defined as follows,

$$\partial \|x\| = \{x^* \in E^* : \|x + y\| - \|x\| \geq \langle y, x^* \rangle, \ \forall \ y \in E\},$$

where E^* is the dual of E. One can easily prove that the set $\partial \|x\|$ is convex, closed, nonempty and given by

$$\partial \|x\| = \begin{cases} \{x^* \in E^* : \langle x, x^* \rangle = \|x\|, \ \|x^*\| = 1\} & \text{if } x \neq 0, \\[2mm] \{x^* \in E^* : \|x^*\| \leq 1\} & \text{if } x = 0. \end{cases}$$

Moreover:

$$D_+\|x\| \cdot y = \max\{\langle y, x^* \rangle : \ x^* \in \partial \|x\|\},$$

$$D_-\|x\| \cdot y = \min\{\langle y, x^* \rangle : \ x^* \in \partial \|x\|\}.$$

We will use the following chain rule, see for instance G. Da Prato and J. Zabczyk [44, Proposition D.4].

Proposition 5.5.1 *Let $u : [0,T] \to E, t \to u(t)$, differentiable in $t_0 \in [0,T]$. Then the function $\gamma = \|u(\cdot)\|$ is differentiable on the right and on the left at t_0 and we have*

$$\frac{d^+\gamma}{dt}(t_0) \ = \ D_+\|u(t_0)\| \cdot u'(t_0)$$

$$= \ \max\{\langle u'(t_0), x^* \rangle : \ x^* \in \partial \|u(t_0)\|\},$$

$$\frac{d^-\gamma}{dt}(t_0) \ = \ D_-\|u(t_0)\| \cdot u'(t_0)$$

$$= \ \min\{\langle u'(t_0), x^* \rangle; \ x^* \in \partial \|u(t_0)\|\}.$$

A mapping $f : D(f) \subset E \to E$ is said to be *dissipative* if and only if for any $x, y \in D(f)$ there exists $z^* \in \partial\|x - y\|$ such that

$$\langle f(x) - f(y), z^* \rangle \leq 0.$$

We will need the following simple result.

Proposition 5.5.2 *Let $x, y \in E$, then the following assertions are equivalent.*

(i) $\|x\| \leq \|x + \alpha y\|$, $\forall\, \alpha \geq 0$;

(ii) $\exists\, x^* \in \partial\|x\|$ *such that* $\langle y, x^* \rangle \geq 0$.

By Proposition 5.5.2 a mapping $f : D(f) \subset E \to E$ is dissipative if and only if

$$\|x - y\| \leq \|x - y - \alpha(f(x) - f(y))\|, \,\forall\, x, y \in D(f), \,\forall\, \alpha > 0.$$

If, for instance, $E = H$ is a Hilbert space with inner product $\langle \cdot, \cdot \rangle$, a mapping $f : D(f) \subset E \to E$ is dissipative if and only if

$$\langle f(x) - f(y), x - y \rangle \leq 0, \,\forall\, x, y \in D(f).$$

A mapping f is called *strongly dissipative* if there exists $\omega > 0$ such that $f + \omega I$ is dissipative.

A dissipative mapping f is called *m–dissipative* if the range of $\lambda I - f$ is the whole space E for some $\lambda > 0$ (and then for any $\lambda > 0$). Finally a mapping f is called *almost m–dissipative* if $f - \alpha I$ is m–dissipative for some $\alpha \in \mathbb{R}$.

It is well known, see R. Martin [115], that any continuous dissipative mapping is m–dissipative.

The *Yosida approximations* $f_\alpha, \alpha > 0$, of an m–dissipative mapping f are defined by

$$f_\alpha(x) = f(J_\alpha(x)) = \frac{1}{\alpha}(J_\alpha(x) - x), \; x \in E, \tag{5.5.1}$$

where

$$J_\alpha(x) = (I - \alpha f)^{-1}(x), \; x \in E. \tag{5.5.2}$$

We list some useful properties of J_α and f_α.

Proposition 5.5.3 *Let $f : D(f) \to E$ be an m-dissipative mapping in E, and let J_α and f_α be defined by (5.5.1) and (5.5.2) respectively.*
(i) For any $\alpha > 0$ we have

$$\|J_\alpha x - J_\alpha y\| \leq \|x - y\|, \forall \, x, y \in E. \tag{5.5.3}$$

(ii) For any $\alpha > 0$ f_α is dissipative and Lipschitz continuous:

$$\|f_\alpha(x) - f_\alpha(y)\| \leq \frac{2}{\alpha} \|x - y\|, \forall \, x, y \in E, \tag{5.5.4}$$

and

$$\|f_\alpha(x)\| \leq \|f(x)\|, \forall \, x \in D(f). \tag{5.5.5}$$

(iii) We have

$$\lim_{\alpha \to 0} J_\alpha(x) = x, \forall \, x \in \overline{D(f)}. \tag{5.5.6}$$

Proof — (i) follows from the dissipativity of f and then (5.5.4) is clear by (5.5.1). We now prove dissipativity of f_α. Let $x, y \in E$ and $\beta > 0$, we have

$$\|x - y - \beta(f_\alpha(x) - f_\alpha(y))\|$$
$$= \left\|(1 + \tfrac{\beta}{\alpha})(x - y) - \tfrac{\beta}{\alpha}(J_\alpha(x) - J_\alpha(y))\right\| \geq \|x - y\|, \tag{5.5.7}$$

which implies that f_α is dissipative. Finally, since

$$f_\alpha(x) = \frac{1}{\alpha}(J_\alpha(x) - J_\alpha(x - \alpha f(x))),$$

(5.5.5) follows from (5.5.3). Finally we have

$$\|J_\alpha(x) - x\| = \alpha\|f_\alpha(x)\| \leq \alpha\|f(x)\|, \forall \, x \in D(f),$$

which implies (5.5.6).
The following result will be useful in the sequel.

Proposition 5.5.4 *Let E be a Hilbert space, $f : D(f) \to E$ a m-dissipative mapping in E, and let $\alpha, \beta > 0$. Then we have*

$$\langle f_\alpha(x) - f_\beta(y), x - y \rangle \leq (\alpha + \beta)(|f_\alpha(x)| + |f_\beta(y)|)^2, \, \forall \, x, y \in E.$$

Proof — Since

$$x = J_\alpha(x) + \alpha f_\alpha(x), \ y = J_\beta(y) + \beta f_\beta(y),$$

we have

$$
\begin{aligned}
\langle f_\alpha(x) - f_\beta(y), x - y \rangle &= \langle f_\alpha(x) - f_\beta(y), J_\alpha(x) - J_\beta(y) \rangle \\
&\quad + \langle f_\alpha(x) - f_\beta(y), \alpha f_\alpha(x) - \beta f_\beta(y) \rangle.
\end{aligned}
$$

But

$$
\begin{aligned}
\langle f_\alpha(x) - f_\beta(y), J_\alpha(x) - J_\beta(y) \rangle &= \langle f(J_\alpha)(x) - f(J_\beta)(y), J_\alpha(x) - J_\beta(y) \rangle \\
&\leq 0,
\end{aligned}
$$

so that

$$
\begin{aligned}
\langle f_\alpha(x) - f_\beta(y), x - y \rangle &\leq \langle f_\alpha(x) - f_\beta(y), \alpha f_\alpha(x) - \beta f_\beta(y) \rangle \\
&\leq (\alpha + \beta) \left(|f_\alpha(x)| + |f_\beta(x)| \right)^2. \quad \blacksquare
\end{aligned}
$$

5.5.2 Existence of solutions for deterministic equations

Let us consider now the problem

$$
\left.
\begin{aligned}
y'(t) &= Ay(t) + G(t, y(t)), \ t \geq 0, \\
y(0) &= x_0 \in H,
\end{aligned}
\right\}
\tag{5.5.8}
$$

under the following hypothesis.

Hypothesis 5.3 *(i)* $A : D(A) \subset E \to E$ *generates a semigroup* $S(t)$, $t \geq 0$ *on* E *that is strongly continuous in* $]0, +\infty[$.
(ii) There exists $\omega \in \mathbb{R}$ *such that*

$$\|S(t)\| \leq e^{\omega t}, \ t \geq 0.$$

(iii) $G : [0, T] \times E \to E$ *is continuous.*
(iv) There exists $\eta \in \mathbb{R}$ *such that* $A + G(t, \cdot) - \eta$ *is dissipative for any* $t \in [0, T]$.

We say that $u \in C([0,T]; E)$ is a *mild* solution of (5.5.8) if

$$u(t) = S(t)x_0 + \int_0^t S(t-s)G(s, u(s))ds.$$

To solve problem (5.5.8) we need a lemma about the construction of approximate solutions, that is a straightforward extension to the nonautonomous case of a result proved in G. F. Webb [159], see also G. Da Prato [34]. We give the proof here for the reader convenience.

Lemma 5.5.5 *Assume that Hypothesis 5.3 holds and let $x_0 \in E$. Then there exists $T_0 \in \,]0,T]$ such that for arbitrary $\varepsilon > 0$ there exist $u_\varepsilon \in C([0,T_0]; E)$ and θ_ε piecewise continuous on $[0,T_0]$, such that*

$$u_\varepsilon(t) = S(t)x_0 + \int_0^t S(t-s)[G(s, u_\varepsilon(s)) + \theta_\varepsilon(s)]ds, \qquad (5.5.9)$$

and

$$\|\theta_\varepsilon(t)\| \leq \varepsilon, \ \forall \, t \in [0,T_0]. \qquad (5.5.10)$$

Proof — In the proof we set $\omega = \eta = 0$ for simplicity. Let us first define for arbitrary $\varepsilon > 0, t \in [0,T], x \in E$

$$\rho_\varepsilon(t,x) = \sup \Big\{ \delta > 0 : t_1, t_2 \in [0,T], |t_1 - t| < \delta, \ |t_2 - t| < \delta,$$

$$x_1, x_2 \in B(x,\delta) \Longrightarrow \|G(t_1, x_1) - G(t_2, x_2)\| < \varepsilon \Big\}.$$

As is easily checked,

$$|\rho_\varepsilon(t,x) - \rho_\varepsilon(s,y)| \leq |t - s| + \|x - y\|, \ \forall \, t, s \in [0,T], \forall \, x, y \in E.$$

So $\rho_\varepsilon : [0,T] \times E \to E$ is Lipschitz continuous. Moreover $\rho_\varepsilon(t,x) > 0, \ \forall \, t \in [0,T], \ x \in E$, since F is continuous.

Since F is continuous there exist $r > 0$ and $M > 1$ such that

$$\|F(t,x)\| \leq M, \ \forall \, t \in [0,T], \ \forall \, x \in B(x_0, r). \qquad (5.5.11)$$

We now define a function u_ε fulfilling (5.5.9) and (5.5.10). Set $t_0 = 0$ and let $t_1 < t_2 < ...$ be positive numbers. Define by recurrence

$$u_\varepsilon(t) = S(t - t_{n-1})x_{n-1} + \int_{t_{n-1}}^t S(t-s)F(s, x_{n-1})ds, \ t \in [t_{n-1}, t_n],$$

$$(5.5.12)$$

where
$$x_{n-1} = u_\varepsilon(t_{n-1}).$$

Then setting
$$\theta_\varepsilon(t) = G(t, u_\varepsilon(t)) - G(t, x_{n-1}), \ t \in [t_{n-1}, t_n], \tag{5.5.13}$$

we have
$$u_\varepsilon(t) = S(t - t_{n-1})x_{n-1} + \int_{t_{n-1}}^{t} S(t - s)[G(s, u_\varepsilon(s)) + \theta_\varepsilon(s)]ds,$$
$$\tag{5.5.14}$$

for $t \in [t_{n-1}, t_n]$. We want now to show that it is possible to choose $\{t_k\}$ such that
$$\|\theta_\varepsilon(t)\| \leq \varepsilon, \ t \in [t_{n-1}, t_n].$$

Suppose we have chosen t_{n-1}. To choose t_n note that by (5.5.12) we have

$$\|u_\varepsilon(t) - x_{n-1}\| \leq \|S(t - t_{n-1})x_{n-1} - x_{n-1}\| + M(t - t_{n-1}), \ t \in [t_{n-1}, t_n].$$

Now let $t_n = s_n \wedge \lambda_n$ where

$$s_n - t_{n-1} = \frac{1}{2M}\rho_\varepsilon(t_{n-1}, x_{n-1}) \tag{5.5.15}$$

and λ_n is such that

$$\sup_{t \in [t_{n-1}, \lambda_n]} \|S(t - t_{n-1})x_{n-1} - x_{n-1}\| = \frac{1}{2}\rho_\varepsilon(t_{n-1}, x_{n-1}). \tag{5.5.16}$$

Then if $\|u_\varepsilon(t) - x_0\| \leq r$ on $[t_{n-1}, t_n]$ we have

$$\|\theta_\varepsilon(t)\| = \|G(t, u_\varepsilon(t)) - G(t, x_{n-1})\| \leq \varepsilon, \ t \in [t_{n-1}, t_n].$$

It remains to show that the times $\{t_n\}$ can be chosen sufficiently large and such that $\|u_\varepsilon(t) - x_0\| \leq r$ on $[0, t_n]$. Let us distinguish two cases.

First case — There exists $\bar{t} > 0$ such that

$$\|u_\varepsilon(t) - x_0\| \leq r, \ \forall \ t \in [0, \bar{t}], \text{ and } \|u_\varepsilon(\bar{t}) - x_0\| = r.$$

By (5.5.12) we have

$$
\begin{aligned}
u_\varepsilon(t) \;=\; & S(t)x_0 + \sum_{k=1}^{n-1} \int_{t_{k-1}}^{t_k} S(t-s)G(s,x_{k-1})ds \\[2mm]
& + \int_{t_{n-1}}^{t} S(t-s)G(s,x_{n-1})ds.
\end{aligned}
\tag{5.5.17}
$$

It follows that

$$
r = \|u_\varepsilon(\bar t) - x_0\| \le \|S(\bar t)x_0 - x_0\| + M\bar t.
$$

Thus there exists $T_0 > 0$, depending only on x_0, r, M, and not on ε such that $\bar t \ge T_0 > 0$.

Second case — $t_n \uparrow t^*$

If $t^* = +\infty$ the conclusion is obvious. Let assume that t^* is finite. Then by (5.5.17) we have

$$
\begin{aligned}
x_n - x_{n-1} \;=\; & S(t_n)x_0 - S(t_{n-1})x_0 \\[2mm]
& + \sum_{k=1}^{n-2} \int_{t_{k-1}}^{t_k} [S(t_n - s) - S(t_{n-1} - s)]G(s,x_{k-1})ds \\[2mm]
& + \int_{t_{n-1}}^{t_n} S(t_n - s)G(s,x_{n-1})ds.
\end{aligned}
$$

It follows that

$$
\begin{aligned}
\|x_n - x_{n-1}\| \;\le\; & \|S(t_n)x_0 - S(t_{n-1})x_0\| \\[2mm]
& + \sum_{k=1}^{n-2} \int_{t_{k-1}}^{t_k} \|[S(t_n - s) - S(t_{n-1} - s)]F(s,x_{k-1})\|ds \\[2mm]
& + M(t_n - t_{n-1}).
\end{aligned}
$$

Thus there exists $x^* \in E$ such that $x_n \to x^*$, and, recalling that ρ_ε is continuous, we have

$$
\rho_\varepsilon(t_n, x_n) \to \rho_\varepsilon(t^*, x^*).
$$

We notice finally that by (5.5.15)–(5.5.16) it follows that $\rho_\varepsilon(t^*, x^*) = 0$, a contradiction. ∎

Proposition 5.5.6 *Assume that Hypothesis 5.3 holds. Then for any $x_0 \in E$, problem (5.5.8) has a unique mild solution $y(\cdot, x)$ in $[0, +\infty[$* (4).

Proof — For any $\varepsilon > 0$ let u_ε be the continuous function in $[0, T_0]$ defined in Lemma 5.5.5. Then u_ε is the mild solution to the problem

$$\left.\begin{array}{l} u_\varepsilon'(t) = Au_\varepsilon(t) + F(t, u_\varepsilon(t)) + \theta_\varepsilon(t), \ t \in [0, T_0] \\[2mm] u_\varepsilon(0) = x_0. \end{array}\right\}$$

For arbitrary $\varepsilon_1, \varepsilon_2 > 0$ we have, by Hypothesis 5.3

$$\frac{d^+}{dt}\|u_{\varepsilon_1}(t) - u_{\varepsilon_2}(t)\| \le \varepsilon_1 + \varepsilon_2.$$

Thus there exists $u \in C([0, T_0]; E)$ such that $u_\varepsilon \to u$ in $C([0, T_0]; E)$. Passing to the limit for $\varepsilon \to 0$ in (5.5.9) it follows that u is a solution to (5.5.8) in $[0, T_0]$. By a standard extension argument one can show that there is a solution in $[0, T]$. Finally, the uniqueness follows from the dissipativity assumptions. ∎

A stochastic version of Proposition 5.5.6 is the object of the next subsection.

---

[4]In fact one can show that the solution is *strong*, that is for any $T > 0$ there exists a sequence

$$\{y_n\} \subset C^1([0, T]; E) \cap C([0, T]; D(A)),$$

such that

$$y_n \to y(\cdot, x), \frac{d}{dt}y_n - Ay_n - F(y_n) \to 0, \ \text{in} \ C([0, T]; E).$$

5.5.3   Existence of solutions for stochastic equations in Hilbert spaces

In this section we present a method which implies existence and uniqueness of the solutions to the problem

$$\left. \begin{array}{l} dX = (AX + F(X))dt + B\,dW(t), \\[2mm] X(0) = x, \end{array} \right\} \qquad (5.5.18)$$

where A, F satisfy some dissipativity assumptions on appropriate spaces and B is a bounded operator. As a byproduct we will obtain a proof of Proposition 5.5.6 with a slightly weaker concept of solution. Let H be a Hilbert space and let K be a reflexive Banach space included in H. We assume that K is a dense Borel subset of H and such that the embedding of K in H is continuous.

On the mappings A and F we impose the following condition.

Hypothesis 5.4 *(i) There exists $\eta \in \mathbb{R}$ such that the operators $A - \eta$ and $F - \eta$ are m-dissipative on H.*

(ii) The parts on K of $A - \eta$ and $F - \eta$ are m-dissipative on K.

(iii) $D(F) \supset K$ and F maps bounded sets in K into bounded sets of H.

We shall denote by A_K and F_K the parts of A and F respectively, that is

$$D(A_K) = \{x \in D(A) \cap K : A_K x \in K\}, \quad A_K x = Ax, \ x \in D(A_K),$$

$$D(F_K) = \{x \in D(F) \cap K : F_K x \in K\}, \quad F_K(x) = F(x), \ x \in D(F_K).$$

Let $Z(t) = W_A(t)$, $t \geq 0$, be the solution to the linear equation

$$\left. \begin{array}{l} dZ = AZ\,dt + B\,dW(t), \\[2mm] X(0) = 0, \end{array} \right\} \qquad (5.5.19)$$

given by

$$W_A(t) = \int_0^t S(t - s)B\,dW(s), \ t \geq 0,$$

where $S(t)$, $t \geq 0$, is the semigroup generated by A in H.
We will assume the following.

Hypothesis 5.5 *The process* $W_A(t)$, $t \geq 0$, *is continuous in* H, *takes values in the domain* $D(F_K)$ *of the part of* F *in* K, *and for any* $T > 0$ *we have*

$$\sup_{t \in [0,T]} (\|W_A(t)\|_K + \|F(W_A(t))\|_K) < +\infty, \ \mathbb{P}\text{-}a.s.$$

Remark 5.5.7 Notice that Hypotheses 5.4 and 5.5 are satisfied if

(i) $K = H$.

(ii) A is a linear operator almost m–dissipative on H.

(iii) F is either Lipschitz continuous, or continuous and monotone with linear growth.

(iv) $W_A(\cdot)$ is an H–continuous process.

An H–continuous, adapted process $X(t)$, $t \geq 0$, is said to be a *mild solution* to (5.5.18) if it satisfies $\mathbb{P}$-a.s. the integral equation

$$X(t) = S(t)x + \int_0^t S(t-s)F(X(s))ds + W_A(t), \ t \geq 0. \quad (5.5.20)$$

If, for an H–valued process X, there exists a sequence $\{X_n\}$ of mild solutions of (5.5.18) such that $\mathbb{P}$-a.s., $X_n(\cdot) \to X(\cdot)$ uniformly on any interval $[0, T]$, then X is said to be a *generalized* solution to (5.5.18). Note that each mild solution is also a generalized solution.

Theorem 5.5.8 *Assume that Hypotheses 5.4 and 5.5 are fulfilled. Then for arbitrary* $x \in K$ *there exists a unique mild solution of (5.5.18) and for arbitrary* $x \in H$ *there exists a unique generalized solution of (5.5.18).*

Remark 5.5.9 It will follow from the proof that generalized solutions $X(t, x)$, $t \in [0, T], x \in H$, of (5.5.18) are all Markov processes in H with a Feller transition semigroup P_t, $t \geq 0$, given by

$$P_t\varphi(x) = \mathbb{E}[\varphi(X(t, x))], \ x \in H, \varphi \in B_b(H).$$

Proof of Theorem — We will show that the initial value problem

$$y'(t) = Ay(t) + F(y(t) + W_A(t)), \ y(0) = x, \quad (5.5.21)$$

has a mild solution. That is there exists $y(\cdot)$ satisfying the following equation:

$$y(t) = S(t)x + \int_0^t S(t-s)F(y(s) + W_A(s))ds, \quad t \in [0,T]. \quad (5.5.22)$$

Then the solution X to (5.5.18) is given by

$$X(t) = y(t) + W_A(t), \quad t \in [0,T].$$

For arbitrary $\alpha > 0$ we consider the approximating problem

$$y'_\alpha(t) = Ay_\alpha(t) + F_\alpha(y_\alpha(t) + W_A(t)), \quad y_\alpha(0) = x, \quad (5.5.23)$$

where F_α are Yosida approximations of F. Since F_α are Lipschitz continuous, equation (5.5.23) has a unique H–continuous solution. We will show that $\lim_{\alpha \to 0} y_\alpha(t) = y(t)$ exists uniformly for $t \in [0,T]$ and is the required solution of (5.5.22). The proof will be done in several steps. For simplicity we assume $\eta = 0$.

Step 1— A priori estimate in H.

We have, by dissipativity of A and F_α (if necessary replacing A by Yosida approximations A_n of A), that

$$\frac{1}{2}\frac{d}{dt}|y_\alpha(t)|_H^2 = \langle Ay_\alpha(t), y_\alpha(t)\rangle_H + \langle F_\alpha(y_\alpha(t) + W_A(t)), y_\alpha(t)\rangle_H$$

$$= \langle Ay_\alpha(t), y_\alpha(t)\rangle_H + \langle F_\alpha(W_A(t)), y_\alpha(t)\rangle_H$$

$$+ \langle F_\alpha(y_\alpha(t) + W_A(t)) - F_\alpha(W_A(t)), y_\alpha(t)\rangle_H$$

$$\leq \langle F_\alpha(W_A(t)), y_\alpha(t)\rangle_H$$

$$\leq |F_\alpha(W_A(t))|_H |y_\alpha(t)|_H$$

$$\leq |F(W_A(t))|_H |y_\alpha(t)|_H.$$

Now by the differential inequality just obtained, Hypothesis 5.4–(iii) and Hypothesis 5.5 there exists a constant C_1 such that

$$|y_\alpha(t)|_H \leq C_1, \quad t \in [0,T], \quad \alpha > 0. \quad (5.5.24)$$

Step 2 — A priori estimate in K
 In a similar way, but using the subdifferential of the norm in K instead of the scalar product of H, we obtain that there exists a constant $C_2 > 0$, depending on φ, such that

$$\|y_\alpha(t)\|_K \leq C_2 \|x\|_K, \ t \in [0,T], \ \alpha > 0. \qquad (5.5.25)$$

It follows from (5.5.25) that for a constant $C_3 > 0$

$$|F(W_A(t) + y_\alpha(t))|_H \leq C_3, \ t \in [0,T], \ \alpha > 0. \qquad (5.5.26)$$

Step 3 — Convergence in H
 Let $\alpha > 0, \beta > 0$, then we have

$$\frac{1}{2}\frac{d}{dt}|y_\alpha(t) - y_\beta(t)|^2_H = \langle Ay_\alpha(t) - Ay_\beta(t), y_\alpha(t) - y_\beta(t)\rangle_H$$

$$+\langle F_\alpha(y_\alpha(t) + W_A(t)) - F_\beta(y_\beta(t) + W_A(t)), y_\alpha(t) - y_\beta(t)\rangle_H$$

$$\leq \langle F_\alpha(y_\alpha(t) + W_A(t)) - F_\beta(y_\beta(t) + W_A(t)), y_\alpha(t) - y_\beta(t)\rangle_H.$$

By Proposition 5.5.4 it follows that

$$\frac{1}{2}\frac{d}{dt}|y_\alpha(t) - y_\beta(t)|^2_H \ \leq \ (\alpha + \beta)\Big(|F_\alpha(y_\alpha(t) + W_A(t))|$$

$$+ \ F_\beta(y_\beta(t) + W_A(t))\Big)^2 \leq 2(\alpha + \beta)C_3^2,$$

in virtue of (5.5.26)
 It is therefore clear that $y_\alpha(t) \to y(t)$ in H as $\alpha \to 0$ uniformly on $[0,T]$.

**Step 4 — ** We finally prove that we can pass to the limit as $\alpha \to 0$ in the mild version of (5.5.23) :

$$y_\alpha(t) = S(t)x + \int_0^t S(t-s)F_\alpha(y_\alpha(s) + W_A(s))ds, \ t \in [0,T]. \ (5.5.27)$$

Note that by (5.5.25) and reflexivity of K, for arbitrary $t \in [0,T]$, there exists a sub–sequence $\{y_{\alpha,n}(t)\}$ converging weakly in K to an element in K. Since $\{y_{\alpha,n}(t)\}$ is strongly convergent in H, $y(t) \in K$ for all $t \in [0,T]$ and

$$\|y(t)\|_K \leq C_2 \|x\|_K, \ t \in [0,T].$$

Let $h \in H$, then

$$\langle y_\alpha(t), h \rangle_H = \langle S(t)x, h \rangle_H + \int_0^t \langle F(J_\alpha(y_\alpha(s)+W_A(s))), S^*(t-s)h \rangle_H ds.$$
$$(5.5.28)$$

Moreover

$$J_\alpha(y_\alpha(s) + W_A(s)) \to y(s) + W_A(s) \ \text{strongly in } H, \ \text{as } \alpha \to 0,$$

$$F(J_\alpha(y_\alpha(s) + W_A(s))) \to F(y(s) + W_A(s)) \ \text{weakly in } H, \ \text{as } \alpha \to 0.$$

So, letting α tend to 0 in (5.5.28), we arrive at

$$\langle y(t), h \rangle = \langle S(t)x, h \rangle + \int_0^t \langle S(t-s)F(y(s) + W_A(s)), h \rangle ds.$$

The conclusion follows from the arbitrariness of h.

By arguments as above one can show that there exists a constant $c > 0$ such that for arbitrary $x, y \in K$

$$\sup_{t \in [0,T]} |X(t,x) - X(t,y)|_H \leq c|x - y|_H.$$

This easily implies uniqueness of the mild solution of (5.5.18) for arbitrary $x \in K$ and existence and uniqueness of the generalized solution of (5.5.18) for arbitrary $x \in H$.   ∎

Remark 5.5.10 If $F : K \to H$ is continuous, the hypothesis that K is reflexive can be dropped.

Some useful variations of Theorem 5.5.8 are possible. To formulate one of them it is convenient to introduce the following hypothesis.

Hypothesis 5.6 *(i) A separable Banach space K is continuously and densely embedded into a separable Hilbert space H.*

(ii) For some $\omega \in \mathbb{R}$ the operators $A - \omega$ and $A_K - \omega$ are m-dissipative on H and on K respectively.

(iii) The mapping $F : K \to K$ satisfies a Lipschitz condition on bounded sets and there is a continuous function $a : \mathbb{R}^+ \to \mathbb{R}^+$ such that for arbitrary $x, y \in K, x^* \in \partial \|x\|$

$$\langle F(x + y), x^* \rangle \leq a(\|y\|).$$

(iv) The process $W_A(t)$, $t \geq 0$, has a K-continuous version.

Theorem 5.5.11 *Under Hypothesis 5.6 for arbitrary $x \in K$ there exists a unique continuous solution of (5.5.18) in K which determines a Feller transition semigroup on K. If in addition for some $\omega \in \mathbb{R}$ the mapping $F - \omega$ is dissipative in H then for arbitrary $x \in H$ there exists a unique H-continuous generalized solution $X(t, x)$, $t \geq 0$, of (5.5.18) with Feller transition semigroup on H.*

Proof — Write $y(t) = X(t) - W_A(t)$, $t \in [0, T]$ and let $S_K(\cdot)$ be the semigroup generated by A_K. Then

$$y(t) = S_K(t)x + \int_0^t S_K(t - s)F(y(s) + W_A(s))ds, \ t \in [0, T], \ (5.5.29)$$

and for $x \in K$ equation (5.5.18) is equivalent to (5.5.29). It is therefore enough to show that for an arbitrary K-continuous function $z(\cdot)$ the equation

$$y(t) = S_K(t)x + \int_0^t S_K(t - s)F(y(s) + z(s))ds, \ t \in [0, T], \ (5.5.30)$$

has a unique global solution y. Local existence follows by a contraction mapping principle. To obtain global existence it is sufficient to deduce an a priori estimate for $\|y(\cdot)\|$. If a solution $z(\cdot)$ exists on a time interval $[0, T_0]$ then there exists a sequence $\{y_n\} \subset C^1([0, T_0]; D(A_K))$ such that

$$y_n(t) \to y(t), \ \frac{dy_n(t)}{dt} - A_K y_n(t) - F(y_n(t) + z(t)) = \delta_n(t) \to 0$$

as $n \to \infty$, uniformly on $[0, T_0]$. Now for some $x_{t,n}^* \in \partial \|y_n(t)\|$, and $t \in [0, T]$,

$$\frac{d^-}{dt}\|y_n(t)\| \leq \langle A_K y_n(t) + F(y_n(t) + z(t)), x_{t,n}^* \rangle$$

$$+ \langle \delta_n(t), x_{t,n}^* \rangle$$

$$\leq \omega\|y_n(t)\| + a\left(\|z(t)\|\right) + \|\delta_n(t)\|.$$

Consequently

$$\|y_n(t)\| \leq e^{\omega t}\|y_n(0)\| + \int_0^t e^{\omega(t-s)}\left[a\left(\|z(s)\|\right) + \|\delta_n(s)\|\right] ds.$$

Letting n tend to infinity we have

$$\|y(t)\| \leq e^{\omega t}\|x\| + \int_0^t e^{\omega(t-s)} a\left(\|z(s)\|\right) ds,$$

the required a priori estimate.

To prove the final part of the theorem it is enough to notice that if $X(t,a)$ and $X(t,b)$ are two solutions of (5.5.18) with $a, b \in K$ then, going if necessary to smooth approximations of the solution,

$$\frac{1}{2}\frac{d}{dt}|X(t,a) - X(t,b)|^2$$

$$\leq \langle A(X(t,a) - X(t,b)) + F(X(t,a)) - F(X(t,b)), X(t,a) - X(t,b)\rangle$$

$$\leq 2\omega|X(t,a) - X(t,b)|^2.$$

So

$$|X(t,a) - X(t,b)|^2 \leq e^{2\omega t}|a - b|, \ t \geq 0,$$

and the result easily follows. ∎

Remark 5.5.12 For another version of Theorem 5.5.8 we refer to G. Da Prato and J. Zabczyk [43] and [44, Theorem 7.13].

5.5.4 Existence of solutions for stochastic equations in Banach spaces

In this section we consider the problem

$$dX = (AX + F(X))dt + dW(t), \\ X(0) = x \in E, \quad \Bigg\} \qquad (5.5.31)$$

on a Banach space (norm $\| \cdot \|$) $E \subset H$. We assume

Hypothesis 5.7 *(i)* $A : D(A) \subset E \to E$ *generates a semigroup* $S(t)$, $t \geq 0$, *on* E *that is strongly continuous in* $]0, +\infty[$.
(ii) There exists $\omega \in \mathbb{R}$ *such that*

$$\|S(t)\| \leq e^{\omega t}, \ t \geq 0.$$

(iii) $F : E \to E$ *is continuous.*
(iv) There exists $\eta \in \mathbb{R}$ *such that* $A + F - \eta$ *is dissipative.*
(v) $W(\cdot)$ *is a cylindrical Wiener process on* H *such that the stochastic convolution* $W_A(t)$, $t \geq 0$, *belongs to* $C([0, T]; E)$ *for arbitrary* $T > 0$.

We say that $X \in C([0, T]; E)$ is a *mild* solution of (5.5.31) if

$$X(t) = S(t)x_0 + \int_0^t S(t - s)F(X(s))ds + W_A(t). \qquad (5.5.32)$$

Theorem 5.5.13 *Assume that Hypothesis 5.7 holds. Then for any* $x \in E$ *problem (5.5.31) has a unique mild solution.*

Proof — Setting

$$Y(t) = X(t) - W_A(t),$$

equation (5.5.32) reduces to the problem

$$Y'(t) = AX + F(Y(t) + W_A(t)), \\ Y(0) = x \in E. \quad \Bigg\}$$

Now it suffices to set $G(t, z) = F(Y(t) + W_A(t))$ and to apply Proposition 5.5.6. ∎

Chapter 6

Existence of invariant measures

In this chapter we establish existence of invariant measures for stochastic evolution equations by exploiting either compactness or dissipativity properties of the drift parts of the equations. A complete characterization of those linear equations for which an invariant measure exists is given as well. For dissipative systems convergence of the transition semigroups to equilibrium is obtained.

6.1   Existence from boundedness

We are here concerned with the problem

$$\left.\begin{array}{l} dX(t) = (AX + F(X))dt + B(X)dW(t) \\[2mm] X(0) = x \in H, \end{array}\right\} \qquad (6.1.1)$$

under Hypothesis 5.1. We denote by $X(\cdot, x)$ the mild solution of (6.1.1) and by $P_t(x, \cdot)$ the corresponding transition probability:

$$P_t(x, \Gamma) = \mathcal{L}(X(t, x))(\Gamma), \ \Gamma \in \mathcal{B}(H), \ t > 0, \ x \in H.$$

It follows from Corollary 3.1.2 that if for some $x \in E$ the family of measures

$$\frac{1}{T} \int_0^T P_t(x, \cdot) dt, \ T \geq 1, \qquad (6.1.2)$$

89

is tight then there exists an invariant measure for the transition semi-group P_t, $t \geq 0$. That property implies in particular that

$$\forall_{x \in H} \; \forall_{\varepsilon > 0} \; \exists_{R > 0} \; \forall_{T \geq 1} \; \frac{1}{T} \int_0^T \mathbb{P}(|X(t, x)| \geq R) \, dt < \varepsilon. \qquad (6.1.3)$$

We will show in this section that under some additional assumptions the easier to check condition (6.1.3) implies tightness of (6.1.2).

A stochastic process $X(t)$, $t \geq 0$, is said to be *bounded in probability* if

$$\forall_{\varepsilon > 0} \; \exists_{R > 0} \; \forall_{t \geq 0} \; \mathbb{P}(|X(t)| \geq R) \leq \varepsilon.$$

Note that if the process $X(t, x)$, $t \geq 0$, is bounded in probability then (6.1.3) holds.

Remark 6.1.1 It is clear that (6.1.3) implies the existence of an invariant measure in the case dim $H < +\infty$. If dim $H = +\infty$ then the implication is not true even for deterministic equations. In fact I. Vrkoc [157] constructed a bounded and Lipschitz mapping $F : H \to H$ on a separable Hilbert space H such that all solutions $X(\cdot, x)$ of the problem

$$X'(t) = F(X(t)), \quad X(0) = x, \; t \geq 0, \qquad (6.1.4)$$

are bounded. Nevertheless there is no invariant probability measure for problem (6.1.4).

However we have the following theorem.

Theorem 6.1.2 *Assume that*
(i) Hypothesis 5.1 holds.
(ii) There exists $\alpha \in \,]0, 1/2[$ such that

$$\int_0^1 t^{-2\alpha} K^2(s) ds < +\infty,$$

where K is the function from Hypothesis (5.1)–(ii).
(iii) The operators $S(t)$, $t > 0$, are compact.
(iv) Condition (6.1.3) is fulfilled.
Then there exists an invariant measure for (6.1.1).

Remark 6.1.3 Existence of an invariant measure for (6.1.1) with condition (6.1.3) replaced by a stronger one, that for some $x \in H$, $X(t, x)$, $t \geq 0$, is bounded in probability, was proved in G. Da Prato, D. Gątarek and J. Zabczyk [39]. A similar proof, however, goes through under (6.1.3).

Proof — For the proof we use the factorization formula, see Theorem 5.2.5. For any $\alpha \in]0, 1]$ we define operators $G_\alpha : L^p(0, 1; H) \to H$ by the formula

$$G_\alpha f = \int_0^1 (1 - s)^{\alpha - 1} S(1 - s) f(s) ds, f \in L^p(0, 1; H),$$

and we set

$$Y(t, x) = \int_0^t (t - s)^{-\alpha} S(t - s) B(X(s, x)) dW(s). \tag{6.1.5}$$

Then one easily checks the identity

$$X(1, x) = S(1)x + G_1 F(X(\cdot, x)) + \frac{\sin \alpha \pi}{\pi} G_\alpha Y(\cdot, x). \tag{6.1.6}$$

We have the following technical result.

Lemma 6.1.4 *Assume that $S(t)$, $t > 0$, are compact operators and let $p \geq 2$ and $\alpha > \frac{1}{p}$. Then G_α is compact.*

Proof — For $\varepsilon \in]0, 1[$, $f \in L^p(0, 1; H)$ define

$$G_\alpha^\varepsilon f = \int_0^{1-\varepsilon} (1 - s)^{\alpha - 1} S(1 - s) f(s) ds.$$

Then

$$G_\alpha^\varepsilon f = S(\varepsilon) \int_0^{1-\varepsilon} (1 - s)^{\alpha - 1} S(1 - \varepsilon - s) f(s) ds,$$

and from the compactness of $S(\varepsilon)$, it follows that G_α^ε is a compact operator. Set $q = \frac{p}{p-1}$ and note that

$$(\alpha - 1)q + 1 = \frac{\alpha p - 1}{p - 1} > 0.$$

By the Hölder inequality, it follows that

$$|G_\alpha f - G_\alpha^\varepsilon f| = \left| \int_{1-\varepsilon}^1 (1-s)^{\alpha-1} S(1-s) f(s) ds \right|$$

$$\leq \left(\int_{1-\varepsilon}^1 (1-s)^{(\alpha-1)q} \|S(1-s)\|^q ds \right)^{1/q} \left(\int_{1-\varepsilon}^1 |f(s)|^p ds \right)^{1/p}$$

$$\leq M \left(\frac{\varepsilon^{(\alpha-1)q+1}}{(\alpha-1)q+1} \right)^{1/q} |f|_p,$$

where $|f|_p$ is the norm of f in $L^p(0,1;H)$ and $M = \sup_{s \in [0,1]} \|S(s)\|$. Consequently $G_\alpha^\varepsilon \to G_\alpha$ as $\varepsilon \to 0$ in the operator norm so that G_α is compact. ∎

It follows from the lemma that the mapping

$$\gamma : H \times L^p(0,1;H) \times L^p(0,1;H) \to X, \ (y,g,h) \to S(1)y + G_1 g + G_\alpha h,$$
$$(6.1.7)$$

is compact. Consequently, for arbitrary $r > 0$ the set

$$K(r) = \Big\{ x \in H : x = S(1)y + G_1 g + G_\alpha h,$$

$$|y| \leq r, \ |g|_p \leq r, \ |h|_p \leq r \Big\}$$

is relatively compact in H.

The key to the proof of the theorem is the following lemma.

Lemma 6.1.5 *Assume that $p > 2, \alpha \in]\frac{1}{p}, \frac{1}{2}[$, and that the Hypotheses of Theorem 6.1.2 hold. Then there exists a constant $c > 0$ such that for arbitrary $r > 0$ and all $x \in H$ such that $|x| \leq r$,*

$$\mathbb{P}(X(1,x) \in K(r)) \geq 1 - cr^{-p}(1 + |x|^p), r > 0.$$

Proof —   Let $Y(\cdot, x)$ be defined by (6.1.5). Then, by Theorem 5.2.4, there exists a constant $k > 0$ such that

$$\mathbb{E} \int_0^1 |Y(s,x)|^p ds = \mathbb{E} \int_0^1 \left| \int_0^s (s-u)^{-\alpha} S(s-u) B(X(u,x)) dW(u) \right|^p ds$$

$$\leq k\mathbb{E} \int_0^1 \left(\int_0^s (s-u)^{-2\alpha} \|S(s-u)\|_{HS}^2 \|B(X(u,x))\|_{HS}^2 du \right)^{p/2} ds$$

$$\leq c^p k 2^{p/2} \mathbb{E} \int_0^1 \left(\int_0^s (s-u)^{-2\alpha} K^2(s-u)(1+|X(u,x)|^2) du \right)^{p/2} ds.$$

By the Young inequality,

$$\mathbb{E} \int_0^1 |Y(s,x)|^p ds \leq kc^p 2^{p/2} \left(\int_0^1 t^{-2\alpha} K^2(t) dt \right)^{p/2}$$

$$\times \quad \mathbb{E} \int_0^1 (1+|X(u,x)|^2)^{p/2} du$$

$$\leq \quad k_1(1+|x|^p),$$

for some constant $k_1 > 0$. Moreover, by (5.2.15) there exists $k_2 > 0$ such that

$$\mathbb{E} \int_0^1 |F(X(s,x))|^p ds \leq k_2(1+|x|^p), \quad x \in H.$$

Now let $|x| \leq r$. Since the mapping γ defined by (6.1.7) is one–to–one, one has $X(1,x) \notin K(r)$ if and only if either $|F(X(\cdot,x))|_p \geq r$ or $|Y(\cdot,x)|_p \leq \frac{\pi r}{\sin \alpha \pi}$. It follows that

$$\mathbb{P}(X(1,x) \notin K(r)) \leq \mathbb{P}(|F(X(\cdot,x))|_p > r) + \mathbb{P}\left(|Y(\cdot,x)|_p > \frac{\pi r}{\sin \alpha \pi}\right)$$

and, by the Chebyshev inequality,

$$\mathbb{P}(X(1,x) \notin K(r)) \leq r^{-p} \mathbb{E}(|F(X(\cdot,x))|_p^p)$$

$$+ \quad r^{-p} \frac{\sin^p \alpha \pi}{\pi} \mathbb{E}(|Y(\cdot,x)|_p^p)$$

$$\leq \quad r^{-p} \left(\pi^{-p} k_1 + k_2\right) (1+|x|^p),$$

and the lemma follows.   ∎

Proof of Theorem 6.1.2 —   For any $t > 1$, we have by the Markov property, noting $X(t, x) = X(t)$,

$$\mathbb{P}(X(t) \in K(r)) = \mathbb{E}(P_1(X(t-1), K(r)))$$

$$\geq \mathbb{E}\left\{ P_1(X(t-1), K(r)) \chi_{|X(t-1)| \leq r_1} \right\}.$$

By Lemma 6.1.5, for $r > r_1 > 0$

$$\mathbb{P}(X(t) \in K(r)) \geq \left(1 - cr^{-p}(1 + r_1^p)\right) \mathbb{P}(|X(t-1)| \leq r_1).$$

Consequently

$$\frac{1}{T} \int_1^{T+1} \mathbb{P}(X(t) \in K(r))dt \geq (1 - c(r^{-p}(1 + r_1^p)))$$

$$\times \frac{1}{T} \int_0^T \mathbb{P}(|X(t)| \leq r_1)dt.$$

Taking first r_1 and then $r > r_1$ sufficiently large we see that the family

$$\frac{1}{T} \int_1^{T+1} P_t(x, \cdot)dt, \ T \geq 1,$$

is tight. This finishes the proof.   ∎

Property (6.1.3) is implied by various kinds of conditions implying boundedness of moments . If, for instance, for some $x \in H$, $T_0 > 0$ and $p > 0$

$$\sup_{t \geq T_0} \mathbb{E}|X(t, x)|^p < +\infty \qquad (6.1.8)$$

then, by the Chebyshev inequality, the condition (6.1.3) is satisfied.

The final proposition gives sufficient conditions for boundedness of the second moment of the solutions of nonlinear equations. In its proof we use $\| \cdot \|^2$ as Liapunov function.

Proposition 6.1.6 *Assume that A, F and B satisfy conditions of Theorem 6.2.1. Assume in addition that*

(i) there exist constants $a > 0, b, c$ such that

$$\langle Ax + F(x + y), x \rangle \leq -a|x|^2 + b|y|^2 + c, \ x \in D(A), \ y \in H,$$

(ii) we have $\displaystyle\int_0^{+\infty} \sup_{x \in H} \|S(t)B(x)\|_{HS}^2 \, dt = k < +\infty.$

Then for arbitrary $x \in H$

$$\sup_{t \geq 0} \mathbb{E}|X(t, x)|^2 < +\infty.$$

Proof —   Write

$$Z(t) = \int_0^t S(t - s)B(X(s))dW(s),$$

$$Y(t) = X(t) - Z(t), \ t \geq 0,$$

where $X(t) = X(t, x)$, $t \geq 0$. It follows from (ii) that

$$\sup_{t \geq 0} \mathbb{E}|Z(t)|^2 \ = \ \sup_{t \geq 0} \mathbb{E} \int_0^t \|S(t - s)B(X(s))\|_{HS}^2 \, ds \tag{6.1.9}$$

$$\leq \ k < +\infty.$$

Obviously

$$Y(t) = S(t)x + \int_0^t S(t - s)F(X(s))ds, \ t \geq 0.$$

Let $Y_\lambda(t) = \lambda(\lambda - A)^{-1}Y(t)$ and $F_\lambda(x) = \lambda(\lambda - A)^{-1}F(x)$, where $\lambda > 0$ is sufficiently large. Since $Y_\lambda(t) \in D(A)$ for any $t \geq 0$ and

$$Y_\lambda(t) = \lambda(\lambda - A)^{-1}S(t)x + \int_0^t S(t - s)F_\lambda(X(s))ds$$

then Y_λ satisfy the following equations:

$$\frac{d}{dt}Y_\lambda(t) = AY_\lambda(t) + F_\lambda(X(t)) = AY_\lambda(t) + F(Y_\lambda(t) + Z(t)) + \delta_\lambda(t),$$

where $\delta_\lambda(t) = F_\lambda(X(t)) - F(Y_\lambda(t) + Z(t)) \to 0$ as $\lambda \to +\infty$. We have

$$\frac{1}{2}\frac{d}{dt}|Y_\lambda(t)|^2 \ = \ \langle AY_\lambda(t) + F(Y_\lambda(t) + Z(t)) + \delta_\lambda(t), Y_\lambda(t) \rangle$$

$$\leq \ -a|Y_\lambda(t)|^2 + b|Z(t)|^2 + c + |Y_\lambda(t)||\delta_\lambda(t)|$$

$$\leq \ -\frac{a}{2}|Y_\lambda(t)|^2 + b|Z(t)|^2 + c + \frac{2}{a}|\delta_\lambda(t)|^2.$$

By a well known comparison theorem

$$|Y_\lambda(t)|^2 \le 2e^{-at/2}|x|^2 + 2\int_0^t e^{-a(t-s)/2}(b|Z(s)|^2 + c + \frac{2}{a}|\delta_\lambda(s)|^2)ds.$$

Letting λ tend to infinity we obtain

$$|Y(t)|^2 \le 2e^{-at/2}|x|^2 + 2\int_0^t e^{-a(t-s)/2}(b|Z(s)|^2 + c)ds.$$

Finally, by (6.1.9) we find

$$\mathbb{E}\left(|Y(t)|^2\right) \le 2e^{-at/2}|x|^2 + 2\int_0^t e^{-a(t-s)/2}(bk + c)ds$$

$$\le 2 + \frac{4}{a}(bk + c). \qquad \blacksquare$$

We will not go further into Liapunov type sufficient conditions implying (6.1.3), but we restrict ourselves to two important special cases which will be referred to later: to linear systems and dissipative systems.

6.2   Linear systems

We are here concerned with the linear equation, introduced in §5.2,

$$\left.\begin{array}{l} dX(t) = AX(t)dt + BdW(t), \ t \ge 0, \\[2mm] X(0) = \xi, \end{array}\right\} \qquad (6.2.1)$$

under the following assumptions.

Hypothesis 6.1 *(i) A is the infinitesimal generator of a strongly continuous semigroup $S(t)$, $t \ge 0$, on H.*
(ii) B is a linear continuous mapping from U into H.
(iii) For any $t > 0$ the linear operator Q_t

$$Q_t x = \int_0^t S(s)QS^*(s)x dt, \ \ x \in H, \qquad (6.2.2)$$

where $Q = BB^*$, is of trace class.

Obviously Hypothesis 6.1 implies Hypothesis 5.1 so that there exists a unique mild solution of (6.2.1),

$$X(t,\xi) = S(t)\xi + \int_0^t S(t-s)BdW(s), \ t > 0.$$

6.2.1 A description of invariant measures

As is easily checked for any $t \geq 0$ and for any $x \in H$, $X(t,x)$ is a Gaussian random variable $\mathcal{N}(S(t)x, Q_t)$, that is with mean $S(t)x$ and covariance Q_t. The transition semigroup corresponding to (6.2.1) is given by

$$R_t\varphi(x) = \int_H \varphi(y)\mathcal{N}(S(t)x, Q_t)(dy), \text{ for all } \varphi \in C_b(H). \quad (6.2.3)$$

Clearly a probability measure on $(H, \mathcal{B}(H))$ is invariant if

$$\int_H R_t\varphi(x)\mu(dx) = \int_H \varphi(x)\mu(dx), \text{ for all } \varphi \in C_b(H), \text{ and } t > 0.$$

Given $h \in H$ we set $\varphi_h(x) = e^{i\langle h, x \rangle}$, $x \in H$. Then $R_t\varphi_h$ coincides with the characteristic functional of the measure $\mathcal{N}(S(t)x, Q_t)$ and we have

$$R_t\varphi_h(x) = e^{i\langle h, S(t)x \rangle - \frac{1}{2}\langle Q_t h, h \rangle}, \ x \in H.$$

It follows that μ is an invariant measure for (6.2.1) if and only if its characteristic functional $\hat{\mu}$ is given by

$$\hat{\mu}(\lambda) = \hat{\mu}(S^\star(t)\lambda)e^{-\frac{1}{2}\langle Q_t \lambda, \lambda \rangle}, t \geq 0, \lambda \in H. \quad (6.2.4)$$

The following theorem is from J. Zabczyk [164]. The finite dimensional version is due to J. Snyders and M. Zakai. [143]

Theorem 6.2.1 *Assume Hypothesis 6.1. Then the following conditions are equivalent.*

(i) There exists an invariant measure for problem (6.2.1).

(ii) $\sup\limits_{t\geq0}$ Tr $[Q_t] < +\infty.$

(iii) There exists a trace–class operator P in the cone $\Sigma^+(H)$, of all symmetric nonnegative operators on H, satisfying the equation

$$2\langle PA^\star x, x \rangle + \langle Qx, x \rangle = 0, \text{ for all } x \in D(A^\star). \quad (6.2.5)$$

If any of the conditions (i), (ii), (iii) holds then any invariant measure for (6.2.1) is of the form

$$\nu * \mathcal{N}(0, Q_\infty),$$

where ν is an invariant measure for the deterministic system $z' = Az$ and $P = Q_\infty$ is the minimal nonnegative solution to (6.2.5) given by the formula

$$Q_\infty x = \int_0^{+\infty} S(t) Q S^*(t) x \, dt, \ x \in H.$$

Proof—   (i)$\Rightarrow$ (ii) Let us assume that μ is an invariant measure for (6.2.1), and let $\widehat{\mu}$ be its characteristic functional given by (6.2.4). It follows that

$$\langle Q_t \lambda, \lambda \rangle \leq 2 \log \left(\frac{1}{\operatorname{Re} \widehat{\mu}(\lambda)} \right), \text{ for all } \lambda \in H.$$

Moreover, by the Bochner theorem, see e.g. G. Da Prato and J. Zabczyk [44, page 48], there exists a trace–class operator $S_0 \in \Sigma^+(H)$ such that

$$\lambda \in H, \ \langle S_0 \lambda, \lambda \rangle \leq 1 \ \Rightarrow \ \operatorname{Re} \widehat{\mu}(\lambda) \geq \frac{1}{2}.$$

Thus the following implication holds.

$$\lambda \in H, \ \langle S_0 \lambda, \lambda \rangle \leq 1 \Rightarrow \langle Q_t \lambda, \lambda \rangle \leq 2 \log 2,$$

which yields

$$0 \leq Q_t \leq 2 \log 2 \, S_0.$$

So

$$\operatorname{Tr}[Q_t] \leq 2 \log 2 \operatorname{Tr}[S_0].$$

(iii)$\Rightarrow$(ii) Let $P \in \Sigma^+(H)$ be a solution of (6.2.5) and let $x \in D(A^*)$. Then we have

$$\frac{d}{dt} \langle P S^*(t) x, S^*(t) x \rangle = -\langle Q S^*(t) x, S^*(t) x \rangle.$$

By integrating this identity between 0 and t we get

$$\langle Px, x \rangle = \int_0^t \langle P S^*(s) x, S^*(s) x \rangle ds + \langle Q_t x, x \rangle, \ x \in H, \qquad (6.2.6)$$

which implies Tr $Q_t \leq$ Tr P.

(ii)$\Rightarrow$(iii) If (ii) holds then there exists a trace–class operator $Q_\infty \in \Sigma^+(H)$ such that

$$Q_\infty x = \int_0^{+\infty} S(s)QS^\star(s)x ds = \lim_{t\uparrow+\infty} \int_0^t S(s)QS^\star(s)x ds, \ x \in H.$$

On the other hand, if $x \in D(A^\star)$ we have

$$2\langle Q_t A^\star x, x\rangle = \int_0^t \frac{d}{ds}\langle QS^\star(s)x, S^\star(s)x\rangle ds \qquad (6.2.7)$$
$$= \langle QS^\star(t)x, S^\star(t)x\rangle - \langle Qx, x\rangle.$$

Since $\int_0^\infty \langle QS^\star(s)x, S^\star(s)x\rangle ds < +\infty$, there exists a sequence $t_n \uparrow +\infty$ such that

$$\lim_{n\to\infty}\langle QS^\star(t_n)x, S^\star(t_n)\rangle = 0.$$

Thus, setting in (6.2.7) $t = t_n$ and letting n tend to infinity we find

$$2\langle Q_\infty A^\star x, x\rangle = -\langle Qx, x\rangle,$$

and (iii) is proved.

(ii)$\Rightarrow$(i) If (ii) holds then the linear operator

$$Q_\infty x = \int_0^{+\infty} S(t)QS^\star(t)x dt, \ x \in H,$$

is of trace class. Let us prove that $\mu = \mathcal{N}(0, Q_\infty)$ is an invariant measure. For this it suffices to show that (6.2.4) holds. We have in fact

$$\widehat{\mu}(\lambda) = e^{-\frac{1}{2}\langle Q_\infty \lambda, \lambda\rangle},$$

which implies

$$\widehat{\mu}(S^\star(t)\lambda) = e^{-\frac{1}{2}\langle S(t)Q_\infty S^\star(t)\lambda, \lambda\rangle} = e^{-\frac{1}{2}\langle Q_t \lambda, \lambda\rangle} e^{-\frac{1}{2}\langle Q_\infty \lambda, \lambda\rangle}, \qquad (6.2.8)$$

so (i) is proved.

We want to show now the last part of the theorem. Let μ be an invariant measure for (6.2.1) and let $Q_\infty x = \int_0^\infty S(s)QS^\star(s)x ds, x \in H$. Then, letting t tend to $+\infty$ in (6.2.8), we have

$$\widehat{\mu}(\lambda) = e^{-\frac{1}{2}\langle Q_\infty \lambda, \lambda\rangle}\widehat{\psi}(\lambda), \lambda \in H, \qquad (6.2.9)$$

where
$$\widehat{\psi}(\lambda) = \lim_{t\uparrow+\infty} \widehat{\mu}(S^\star(t)\lambda), \lambda \in H.$$

It remains only to prove that $\widehat{\psi}(\lambda)$ is the characteristic function of a probability measure ν, which is invariant for $S(\cdot)$ since

$$\widehat{\nu}(S(s)\lambda) = \lim_{t\uparrow+\infty} \widehat{\mu}(S^\star(t+s)\lambda) = \widehat{\nu}(\lambda), \lambda \in H.$$

In fact by Bochner's theorem, given $\varepsilon > 0$ there exists a positive operator S of trace class, such that

$$\text{Re } \widehat{\mu}(\lambda)(x) \geq 1 - \varepsilon \quad \text{if } \langle Sx, x \rangle \leq 1.$$

Thus, if $x \in H$ is such that $\langle Sx, x \rangle \leq 1$, it follows that

$$\text{Re } \widehat{\psi}(\lambda)(x) = \text{ Re } \widehat{\mu}(\lambda)(x)e^{\frac{1}{2}\langle Q_\infty \lambda, \lambda \rangle} \geq 1 - \varepsilon.$$

So, using Bochner's theorem once again, there exists a probability measure ν in $H, \mathcal{B}(H)$ such that $\widehat{\psi}(\cdot) = \widehat{\nu}(\cdot)$ as required. In conclusion by (6.2.9) we have $\mu = \mathcal{N}(0, Q_\infty) \star \nu$. ∎

Remark 6.2.2 Assume that there exists an invariant measure for (6.2.1). Then $Q_\infty x = \int_0^{+\infty} S(t)QS^\star(t)xdt, \ x \in H$, is the *minimal solution* of (6.2.5). In fact let $P \in \Sigma^+(H)$ be a solution of (6.2.5); then, by (6.2.7),

$$\langle Px, x \rangle \geq \langle Q_t x, x \rangle, \text{for all } t > 0.$$

Now, letting t tend to infinity, we get $P \geq Q_\infty$. Equation (6.2.5) is sometimes called the *Liapunov equation* for equation (6.2.1).

The following characterization shows that for linear equations boundedness in probability of a solution implies existence of an invariant measure, however, compare Remark 6.1.1.

Theorem 6.2.3 *Assume Hypothesis 6.1. The following conditions are equivalent.*

   (i) There exists an invariant measure for problem (6.2.1).
   (ii) There exists a solution $X(t)$, $t \geq 0$, of (6.2.1), bounded in probability.
   (iii) The solution $X(t,0)$, $t \geq 0$, of (6.2.1) is bounded in probability.

Proof — (i)⇒(ii) If μ is an invariant measure for (6.2.1), and X is a solution such that $\mathcal{L}(X(0)) = \mu$, then $\mathcal{L}(X(t)) = \mu$ for $t \geq 0$ and the boundedness in probability of X follows.

(ii)⇒(iii) Let

$$X(t) = S(t)\xi + X(t,0),\ t \geq 0,$$

be a solution bounded in probability. Denote by μ_t, μ_t^0, ν_t the distributions of $X(t)$, $X(t,0)$ and $S(t)\xi$ respectively. Then

$$\mu_t = \nu_t * \mu_t^0,\ t \geq 0.$$

We claim that for arbitrary $R > 0$

$$\mathbb{P}(|X(t)| \geq R) \geq \frac{1}{2}\mathbb{P}(|X(t,0)| \geq R),\ t > 0. \tag{6.2.10}$$

For $B = B(0, R)$ one has

$$\mu_t(B) = \int_H \mu_t^0(B - x)\nu_t(dx),$$

and therefore, for some $\hat{x} \in H$, $\mu_t^0(B - \hat{x}) \geq \mu_t(B)$. It is clear that

$$B \supset (B - \hat{x}) \cap (-B + \hat{x}),$$

and therefore

$$
\begin{aligned}
\mu_t^0(B) &\geq \mu_t^0((B - \hat{x}) \cap (-B + \hat{x})) \\[4pt]
&\geq 1 - \mu_t^0((B - \hat{x})^c \cup (-B + \hat{x})^c) \\[4pt]
&\geq 1 - \mu_t^0((B - \hat{x})^c) - \mu_t^0((-B + \hat{x})^c) \\[4pt]
&\geq -1 + \mu_t^0(B - \hat{x}) + \mu_t^0(-B + \hat{x}).
\end{aligned}
$$

Since the measure μ_t^0 is symmetric and

$$\mu_t^0(B - \hat{x}) \geq \mu_t(B),$$

we finally have

$$
\begin{aligned}
1 - \mathbb{P}(|X(t,0)| \geq R) &\geq -1 + 2\mu_t(B) \\[4pt]
&\geq -1 + 2(1 - \mathbb{P}(|X(t,0)| \geq R)),
\end{aligned}
$$

and thus inequality (6.2.10) follows. Consequently (ii)⇒(iii)– To show that(iii)⇒(i) it is enough to prove, compare Theorem 6.2.1, that boundedness in probability of $X(t,0)$, $t \geq 0$, implies

$$\sup_{t>0} \mathbb{E}\left(|X(t,0)|^2\right) = \sup_{t\geq 0} \text{Tr } Q_t < +\infty.$$

It follows from Fernique's theorem, see e.g. G. Da Prato and J. Zabczyk [44], that there exist constants $R > 0$, $\gamma \in]0,1[$, $\lambda > 0$, and $C > 0$ such that if $\mu_t^0\left(B(0,R)\right) \geq \gamma$ then

$$\int_H e^{\lambda|z|^2} \, \mu_t^0(dz) \leq C,$$

and in particular

$$\mathbb{E}\left(|X(t,0)|^2\right) \leq \frac{C}{\lambda}.$$

This proves the required implication.   ∎

6.2.2   Invariant measures and recurrence

The concept of a recurrent set for a Markov process was introduced in §3.4. Here we will comment on recurrence for solutions to (6.2.1). If dim $H < +\infty$ and the semigroup R_t ,$t \geq 0$, is irreducible at some $t > 0$, then all solutions to (6.2.1) are recurrent with respect to all open nonempty sets if and only if all eigenvalues of A, with the exception of at most two, have negative real parts and the remaining eigenvalues either are pure imaginary of multiplicity 1 or are equal to 0, see H. Dym [58], R. Erickson [62] and J. Zabczyk [165]. There is also a simple connection between existence of invariant measures and recurrence. If there exists an invariant measure for an irreducible semigroup R_t, $t \geq 0$, corresponding to (6.2.1) then any solution to (6.2.1) is recurrent with respect to all open sets of H. In the infinite dimensional case such a characterization is not known. However, some results in this direction are possible, see J. Zabczyk [164]. In particular we have the following

Proposition 6.2.4 *(i) Assume that the transition semigroup R_t, $t \geq 0$, corresponding to (6.2.1) is irreducible at some $t > 0$ and has an*

invariant measure. If $\lim_{t \to +\infty} S(t)x = 0$ *then any solution to* (6.2.1) *is recurrent with respect to all open sets of H.*

(ii) There exists an equation of the form (6.2.1) such that the corresponding semigroup R_t, $t \geq 0$, *is irreducible for all* $t > 0$, *and has an invariant measure, but a solution* $X(t,x)$, $t \geq 0$, *to (6.2.1) is not recurrent with respect to a ball.*

Proof — (i) If μ is an invariant measure then μ is unique, Gaussian with its support equal to H. Moreover for arbitrary $x \in H$ $P_t(x, \cdot) \to \mu$ weakly as $t \to +\infty$. Consequently, for any open, nonempty set $\Gamma \subset H$

$$\liminf_{t \to +\infty} P_t(x, \Gamma) = \mu(\Gamma) > 0,$$

and it is enough to apply Proposition 3.4.5.

(ii) Let us consider the transition semigroup R_t, $t \geq 0$, corresponding to the linear equation

$$dZ(t) = AZ(t)dt + BdW(t), \quad Z(0) = x.$$

Fix $R > 0$. It is not difficult to find $\hat{x} \in H$ such that

$$\|S(t)\hat{x}\| \geq t + R, \ t \geq 0.$$

Then
$$P_t(\hat{x}, B(0, R)) \leq P_t(0, B^c(0, t)), \ t \geq 0.$$

By Chebyshev's inequality

$$P_t(0, B^c(0, t)) \leq \frac{1}{t^2} \, \mathbb{E}\left(|X(0, t)|^2\right), \ t > 0.$$

It easily follows now that

$$\int_0^{+\infty} P_t(\hat{x}, B(0, R))dt \leq C \int_1^{+\infty} \frac{dt}{t^2} < +\infty. \tag{6.2.11}$$

Assume that with probability 1 the set

$$\{t > 0 : |X(t, \hat{x})| \leq \frac{R}{2}\}$$

is unbounded and define

$$\tau_R^x = \inf\{t \geq 0 : |X(t, \hat{x})| \leq R\}.$$

Then

$$\inf_{\|x\|\leq R/2} \mathbb{E}(\tau_R^x) = a > 0.$$

Therefore the expected time the process $X(t, \widehat{x})$ spends in $B(0, R)$ after each visit to B(0,R/2) and before reaching $\partial B(0, R)$ is at least a. Since by (6.2.11) the total time spent by $X(\cdot, \widehat{x})$ in $B(0, R)$ is finite we easily obtain the contradiction.   ∎

6.3   Dissipative systems

Dissipative systems were introduced in §5.4. Here we study their long–time behaviour. We start from general comments on asymptotic properties of solutions to the deterministic problem

$$\left.\begin{array}{l} y'(t) = Ay(t) + F(y(t)), \ t \geq 0, \\[2mm] y(0) = x \in E, \end{array}\right\} \tag{6.3.1}$$

under the hypothesis of Proposition 5.5.4, with $\lambda = 0$, which implies that $f = A + F$ is dissipative. We denote by $y(\cdot, x)$ the strong solution to (6.3.1).

Dissipativity of f implies an important contraction property for the solutions of (6.3.1). In fact let $y, z \in E$. Then, by Proposition 5.5.4, it follows that

$$\frac{d^-}{dt}\|y(t, x) - y(t, z)\| = \langle f(y(t, x)) - f(y(t, z)), x_{t,x,z}^* \rangle, \tag{6.3.2}$$

for some $x_{t,x,z}^* \in \partial\|y(t, x) - y(t, z)\|$. Consequently

$$\|y(t, x) - y(t, z)\| \leq \|x - z\|, \ \text{for all } t \geq 0, \text{ and } x, z \in E. \tag{6.3.3}$$

Assume now in addition that f is strongly dissipative and let $\omega > 0$ be such that $f + \omega I$ is dissipative. Arguing as before, we get the estimate

$$\|y(t, x) - y(t, z)\| \leq e^{-\omega t}\|x - z\|, \ \text{for all } t \geq 0, \text{ and } x, z \in E. \tag{6.3.4}$$

Consequently, if f is m–dissipative and strongly dissipative, then there exists a unique $\bar{x} \in D(A)$ such that $f(\bar{x}) = 0$, that is $\bar{x}$ is a unique equilibrium point for f. Moreover, from (6.3.4) it follows that

$$\lim_{t \to +\infty} y(t, x) = \bar{x}.$$

Consider now the corresponding transition semigroup

$$P_t \varphi(x) = \varphi(y(t, x)), \ x \in E, \ \varphi \in B_b(E).$$

Then

$$\lim_{t \to +\infty} P_t^* \delta_x = \delta_{\bar{x}}, \text{ weakly for all } x \in E,$$

and therefore

$$\lim_{t \to +\infty} P_t^* \nu = \delta_{\bar{x}}, \text{ weakly for all } x \in E,$$

for arbitrary probability measure ν. This proves the following result.

Proposition 6.3.1 *If f is strongly m–dissipative, then there exists a unique invariant measure for the system 6.3.1 which is moreover strongly mixing.*

6.3.1 General noise

A stochastic version of Proposition 6.3.1 is given by the following theorem.

Theorem 6.3.2 *Assume that A, B and F satisfy Hypothesis 5.2. Assume in addition that there exists $\omega > 0$ such that*

$$2\langle A_n(x - y) + F(x) - F(y), x - y \rangle + \|B(x) - B(y)\|_2^2$$
$$\leq -\omega|x - y|^2, \text{ for all } x, y \in H, \ n \in \mathbb{N},$$
(6.3.5)

where $A_n = \lambda A(n - A)^{-1}$ are the Yosida approximations of A. Then there exists exactly one invariant measure μ for (6.3.1), it is strongly mixing and for arbitrary $\nu \in \mathcal{M}_1(H)$, $P_t^* \nu \to \mu$ weakly as $t \to +\infty$. Moreover there exists $C > 0$ such that, for any bounded Lipschitz continuous function φ, all $t > 0$ and all $x \in H$,

$$|P_t \varphi(x) - \langle \varphi, \mu \rangle| \leq C(1 + |x|)e^{-\omega t/2}\|\varphi\|_{\text{Lip}}.$$

Proof —   It is useful to introduce another cylindrical Wiener process $V(t)$, $t \geq 0$, independent of $W(t)$, $t \geq 0$, and to define

$$\left.\begin{array}{l} \overline{W}(t) = \left\{ \begin{array}{ll} W(t) & \text{if } t \geq 0, \\ V(-t) & \text{if } t \leq 0, \end{array} \right. \\[2ex] \overline{\mathcal{F}}_t = \sigma(\overline{W}(s), s \leq t), \ t \in \mathbb{R}. \end{array}\right\} \tag{6.3.6}$$

Next, for any $s \in \mathbb{R}$ and $x \in H$, we consider the regularized equation

$$\left.\begin{array}{l} dX_n = (A_n X_n + F(X_n))dt + B(X_n)d\overline{W}(t), t \geq s, \\[2ex] X_n(s) = x. \end{array}\right\} \tag{6.3.7}$$

which has a strong solution $X_n(t) = X_n(t, s, x)$, $t \geq s$. It is easy to see that as $n \to \infty$, $X_n(t, s, x)$ converges, in mean square, to the solution $X(t) = X(t, s, x)$, $t \geq s$ of the equation

$$\left\{\begin{array}{l} dX = (AX + F(X))dt + B(X)d\overline{W}(t), \ t \geq s, \\[2ex] X(s) = x. \end{array}\right. \tag{6.3.8}$$

We now proceed in two steps.

Step 1—   A priori estimate

We apply Ito's lemma to the process $|X_n(t)|^2$, $t \geq s$. Then

$$\begin{aligned} d|X_n(t)|^2 &= \{2\langle AX_n(t) + F(X_n(t)), X_n(t)\rangle + \|B(X_n(t))\|_2^2\}dt \\ &\quad + 2\langle X_n(t), B(X_n(t))dW(t)\rangle. \end{aligned}$$

Consequently, integrating in $[s, t]$, and taking expectations, we have

$$\mathbb{E}\left(|X_n(t)|^2\right) = |x|^2$$

$$+ \mathbb{E} \int_s^t \{2\langle AX_n(\sigma) + F(X_n(\sigma)), X_n(\sigma)\rangle + \|B(X_n(\sigma))\|_2^2\}d\sigma.$$

It follows that

$$\frac{d}{dt}\mathbb{E}\left(|X_n(t)|^2\right) = \mathbb{E}(2\langle AX_n(t) + F(X_n(t)), X_n(t)\rangle + \|B(X_n(t))\|_2^2)$$

$$\leq -\omega\,\mathbb{E}\left(|X_n(t)|^2\right) + 2|F(0)|\,\mathbb{E}(|X_n(t)|)$$

$$+ \|B(0)\|_2^2 + 2\|B(0)\|_2\mathbb{E}(|B(X_n(t))|_2)$$

$$\leq -\frac{\omega}{2}\mathbb{E}\left(|X_n(t)|^2\right) + C_1,$$

for a suitable constant $C_1 > 0$. So there exists a constant $C_2 > 0$ such that

$$\mathbb{E}|X_n(t)|^2 \leq e^{-\omega(t-s)/2}(|x|^2 + C_2),\ t \geq s.$$

As n tends to infinity we find

$$\mathbb{E}|X(t)|^2 \leq e^{-\omega(t-s)/2}(|x|^2 + C_1) \leq C_3(1 + |x|^2),\ t \geq s, \qquad (6.3.9)$$

for some constant $C_3 > 0$.

Step 2 — For $\delta > \gamma > 0$ define

$$Z(t) = X(t, -\gamma, x) - X(t, -\delta, x),\ t \geq -\gamma.$$

Using the Ito lemma and proceeding as before we find the estimate

$$\mathbb{E}|Z(t)|^2 \leq e^{-\omega(t+\gamma)}\mathbb{E}|X(-\gamma, -\delta, x)|^2,\ t > -\gamma$$

from which, recalling (6.3.9),

$$\mathbb{E}|X(0, -\gamma, x) - X(0, -\delta, x)|^2 \leq C_3(1 + |x|^2)e^{-\omega\gamma},\ \delta > \gamma. \qquad (6.3.10)$$

It follows that the sequence of random variables $\{X(0, -\gamma, x)\}_{\gamma \geq 0}$, satisfies the Cauchy condition in $L^2(\Omega; H)$, as $\gamma \to +\infty$, and therefore it is convergent to a random variable $\eta \in L^2(\Omega, H)$, which is independent of x. We claim that the law of η is the invariant measure with the required properties. To see this it is enough to remark that

$$\mathcal{L}(X(t, 0, x)) = \mathcal{L}(X(0, -t, x)) \to \mathcal{L}(\eta) = \mu,\ \text{weakly as } t \to +\infty,$$

which is equivalent to

$$P_t^* \delta_x \to \mu \text{ weakly as } t \to +\infty.$$

Finally, let $\varphi \in \text{Lip } (H)$, then by (6.3.10)

$$
\begin{aligned}
|P_t\varphi(x) - P_s\varphi(x)|^2 &= |\mathbb{E}(\varphi(X(0,-t,x))) - \mathbb{E}(\varphi(X(0,-s,x)))|^2 \\
&\leq \|\varphi\|_{\text{Lip}}^2 \, \mathbb{E}|X(0,-t,x) - X(0,-s,x)|^2 \\
&\leq C_3(1 + |x|^2)\|\varphi\|_{\text{Lip}}^2 \, e^{-\omega t}.
\end{aligned}
$$

As $s \to +\infty$ we find

$$|P_t\varphi(x) - \langle \varphi, \mu \rangle|^2 \leq C_3(1 + |x|^2)\|\varphi\|_{\text{Lip}}^2 \, e^{-\omega t}.$$

This finishes the proof.   ■

6.3.2   Additive noise

Now we consider the setting of §5.5.2. We want to prove the existence of an invariant measure for the problem

$$
\left.
\begin{aligned}
dX &= (AX + F(X))dt + BdW(t), \\
X(0) &= x,
\end{aligned}
\right\}
\qquad (6.3.11)
$$

in the space H or in the Banach space K continuously and densely embedded in H. As before we set

$$W_A(t) = \int_0^t S(t-s)BdW(s).$$

We shall assume, besides Hypotheses 5.4 and 5.5, the following.

Hypothesis 6.2 *(i) There exists $\omega_1, \omega_2 \in \mathbb{R}$ such that $\omega = \omega_1 + \omega_2 > 0$ and the operators $A + \omega_1 I$, $F + \omega_2 I$ are dissipative on H.*
(ii) We have

$$\sup_{t \geq 0} \mathbb{E}\left(|W_A(t)|_H + |F(W_A(t))|_H\right) < +\infty.$$

We now prove

Theorem 6.3.3 *Assume that Hypotheses 5.4, 5.5 and 6.2 hold. Then there exists exactly one invariant measure μ for (6.3.11), it is strongly mixing and for arbitrary $\nu \in \mathcal{M}_1(H)$, $P_t^* \nu \to \mu$ weakly as $t \to +\infty$. Moreover there exists $C > 0$ such that, for any bounded Lipschitz continuous function φ, all $t > 0$ and all $x \in H$*

$$|P_t\varphi(x) - \langle \varphi, \mu \rangle| \leq C(1 + |x|)e^{-\omega t/2} \|\varphi\|_{\text{Lip}}.$$

Proof— Proceeding as in the previous theorem we consider problem (6.3.11) for $t \geq s$, $s \in \mathbb{R}$.

$$\left. \begin{array}{l} dX = (AX + F(X))dt + Bd\overline{W}(t),\ t \geq s \\ \\ X(s) = x, \end{array} \right\} \tag{6.3.12}$$

By Theorem 5.5.8 we know that there exists a unique generalized solution of (6.3.12), which we shall denote by $X(t, s, x)$, $t \geq s$.

We show now that there exists $c_1 > 0$ such that

$$\mathbb{E}(|X(t, -\lambda, x)|) \leq C + |x|, \text{ for all } \lambda > 0, \text{ for all } x \in H, \text{ for all } t > -\lambda. \tag{6.3.13}$$

For any $\lambda > 0$ define

$$W_{A,\lambda}(t) = \int_{-\lambda}^{t} S(t - s)Bd\overline{W}(s),$$

where $\overline{W}(\cdot)$ is defined in (6.3.6). We first remark that

$$Z_\lambda(t) = X(t, -\lambda, x) - W_{A,\lambda}(t),\ t \geq -\lambda,$$

is the mild solution of the problem

$$\left. \begin{array}{l} \dfrac{d}{dt}Z = AZ + F(Z + W_{A,\lambda}), \\ \\ Z(-\lambda) = x,\ t \geq -\lambda. \end{array} \right\}$$

It follows, denoting by $x_{\lambda,t}^{\star}$ an element from the subdifferential of $|Z_\lambda(t)|$, that

$$
\begin{aligned}
\frac{d^-}{dt}|Z_\lambda(t)| &= \langle AZ_\lambda(t) + F(Z_\lambda(t) + W_{A,\lambda}(t)) \\
&\quad - F(W_{A,\lambda}(t)), x_{\lambda,t}^{\star} \rangle + \langle F(Z_\lambda(t)), x_{\lambda,t}^{\star} \rangle \qquad (6.3.14) \\
&\leq -\omega|Z_\lambda(t)| + |F(W_{A,\lambda}(t))|.
\end{aligned}
$$

Consequently

$$
|Z_\lambda(t)| \leq e^{-\omega(t+\lambda)}|x| + \int_{-\lambda}^{t} e^{-\omega(t-s)}|F(W_{A,\lambda}(s))|ds, \ t \geq -\lambda,
$$

and

$$
\mathbb{E}|Z_\lambda(t)| \leq |x| + \frac{1}{\omega} \sup_{s\in[-\lambda,t]} \mathbb{E}|F(W_{A,\lambda}(s))|, \ t \geq -\lambda.
$$

Taking into account the definition of the process Z_λ and Hypothesis 6.2–(ii) one arrives at (6.3.13).

In a similar way one shows that for all $-\lambda \in \,]-\gamma, t[$

$$
\begin{aligned}
\mathbb{E}|X(t,-\lambda,x) - X(t,-\gamma,x)| &\leq e^{-\omega(t+\lambda)}\mathbb{E}|X(-\lambda,-\gamma,x) - x| \\
&\leq e^{-\omega(t+\lambda)}(2|x| + C).
\end{aligned}
$$

$$(6.3.15)$$

Therefore there exists a random variable ζ, the same for all $x \in H$, such that

$$
\lim_{\lambda \to -\infty} \mathbb{E}|X(0,-\lambda,x) - \zeta| = 0. \qquad (6.3.16)
$$

We claim that the law $\mu = \mathcal{L}(\zeta)$ is the unique invariant measure for $P_t, \ t \geq 0$. To see this it is enough to remark that, by (6.3.16), for arbitrary $x \in H$,

$$
P_t(x,\cdot) = \mathcal{L}(X(t,x))
$$

$$
= \mathcal{L}(X(0,-t,x)) \to \mu, \ \text{weakly as } t \to +\infty.
$$

Finally, let φ be a bounded Lipschitz function on H, then, by (6.3.15), for $s \geq t \geq 0$

$$
\begin{aligned}
|P_t\varphi(x) - P_s\varphi(x)| &= |\mathbb{E}(\varphi(X(t,0,x)) - \varphi(X(s,0,x)))| \\
&= |\mathbb{E}(\varphi(X(0,-t,x)) - \varphi(X(0,-s,x)))| \\
&\leq \|\varphi\|_{\text{Lip}} \, \mathbb{E}|X(0,-t,x)) - X(0,-s,x)| \\
&\leq \|\varphi\|_{\text{Lip}} \, e^{-\omega t}(2|x| + C),
\end{aligned}
$$

and the estimate follows. ∎

Remark 6.3.4 Notice that assumptions of Theorem 6.3.3 are satisfied in particular if
(i) $A + \omega_1$ is dissipative,
(ii) F is Lipschitz continuous with constant ω_2 such that $\omega_1 + \omega_2 = \omega > 0$,
(iii) there exists an invariant measure for the linear system (6.3.11) with $B = 0$.
See also Remark 5.5.7.

The following theorem is a variation of Theorem 6.3.3. It implies, however, only the existence of an invariant measure.

Theorem 6.3.5 *Assume that the operators $S_K(t)$, $t > 0$, generated by A_K are compact and that for an $\omega > 0$*

$$
\|S_K(t)\| \leq e^{-\omega t}, \; t > 0.
$$

If in addition to Hypothesis 5.6

$$
\sup_{t \geq 0} \mathbb{E}\left[\|W_A(t)\| + a\left(\|W_A(t)\|\right)\right] < +\infty
$$

then there exists an invariant measure for the solution of (5.5.18) on K.

Proof — Arguing as in the proof of Theorem 6.3.3 one obtains that

$$\|X(t) - W_A(t)\| \le e^{-\omega t}\|x\| + \int_0^t e^{\omega(t-s)}a\left(\|W_A(s)\|\right)ds$$

and therefore

$$\sup_{t\ge 0}\mathbb{E}\|X(t)\| \le \sup_{t\ge 0}\mathbb{E}\left[\|W_A(t)\| + \frac{1}{\omega}a\left(\|W_A(t)\|\right)\right] < +\infty.$$

Similarly as in §6.1 we will show that for arbitrary $x \in K$ the laws $\mathcal{L}(X(t,x))$, $t \ge 1$, form a tight family of measures on K. This is certainly enough for the existence of an invariant measure for (5.5.18).

For $\varepsilon \in [0,1]$ define an operator G^ε from $C([0,1];K)$ into K by the formula

$$G^\varepsilon\varphi = \int_\varepsilon^1 S_K(u)\varphi(u)du, \quad \varphi \in C([0,1];K).$$

Since for $\varepsilon \in \,]0,1]$

$$G^\varepsilon\varphi = S_K(\varepsilon)\int_\varepsilon^1 S_K(u-\varepsilon)\varphi(u)du, \quad \varphi \in C([0,1];K),$$

and $S_K(\varepsilon)$ is a compact operator, it is clear that G^ε is a compact operator for $\varepsilon \in \,]0,1]$. However,

$$\|G^0\varphi - G^\varepsilon\varphi\| \le \sup_{u\in\,]0,1]}\|\varphi(u)\|\int_0^\varepsilon \|S_K(v)\|dv$$

and consequently o the operator G^0 is also compact.

Let us remark that for $t \ge 1$

$$X(t+1,x) = S_K(1)X(t,x)$$

$$+ \int_0^1 S_K(u)F(X(t+1-u,x))du + \int_t^{t+1} S_K(t+(1-u))dW(u)$$

$$= S_K(1)X(t,x) + \int_0^1 S_K(u)F(X(t+(1-u),x))du + W_{A,t}(1)$$

$$(6.3.17)$$

where
$$W_{A,t}(u) = \int_t^{t+u} S(t+u-r)BdW(r), \ u \geq 0.$$

However, compare the beginning of the proof,

$$\|X(t+u,x) - W_{A,t}(u)\| \leq e^{-\omega u}\|X(t,x)\|$$

$$+ \int_0^u e^{-\omega(u-s)} a\left(\|W_{A,t}(s)\|\right) ds, \ u \geq 0,$$

and therefore

$$\sup_{u \in [0,1]} \|X(t+u,x)\| \leq \|X(t,x)\|$$

$$+ \sup_{u \in [0,1]} \|W_{A,t}(u)\| + \sup_{u \in [0,1]} a\left(\|W_{A,t}(u)\|\right).$$

But $\sup_{t \geq 0} \mathbb{E}\|X(t,x)\| < +\infty$ and consequently for arbitrary $\varepsilon > 0$ there exists $C > 0$ such that

$$\mathbb{P}(\|X(t,x)\| \geq C) < \varepsilon, \ \text{for all } t \geq 0.$$

Since the process $W_A(\cdot)$ has K–continuous trajectories and all the processes $W_{A,t}(\cdot)$ have the same distributions as $W_A(\cdot)$, the constant $C > 0$ can be chosen to give also

$$\mathbb{P}\left(\sup_{u \in [0,1]} \|W_{A,t}(u)\| \geq C\right) < \varepsilon,$$

$$\mathbb{P}\left(\sup_{u \in [0,1]} a\left(\|W_{A,t}(u)\|\right) \geq C\right) < \varepsilon.$$

But then

$$\mathbb{P}\left(\sup_{u \in [0,1]} \|X(t+u,x)\| \geq 3C\right) \leq \mathbb{P}(\|X(t,x)\| \geq C)$$

$$+\mathbb{P}\left(\sup_{u \in [0,1]} \|W_{A,t}(u)\| \geq C\right) + \mathbb{P}\left(\sup_{u \in [0,1]} a\left(\|W_{A,t}(u)\|\right) \geq C\right) \leq 3\varepsilon,$$

for all $t \geq 0$. Taking into account this estimate, the identity (6.3.17), andthe compactness of the operators $S_K(1)$ and G^0, we easily obtain that the family $\mathcal{L}(X(t + 1, x))$, $t \geq 0$, is tight, as required.   ∎

By Theorem 5.5.11 and Theorem 6.3.5 we get the following result.

Theorem 6.3.6 *If in addition to the assumptions of Theorem 6.3.5 there exists $\omega \in \mathbb{R}$ such that the mapping $F - \omega$ is dissipative in H then there exists an invariant measure for equation (6.3.11) which in addition is concentrated on K.*

6.4 Genuinely dissipative systems

We assume here that Hypotheses 5.4, 5.5 and 6.2 hold and moreover the following.

Hypothesis 6.3 *(i) $S(\cdot)$ is of negative type.*
(ii) There exist $r > 1$ and $C_r > 0$ such that

$$\langle F(x) - F(y), x - y \rangle \leq -C_r |x - y|^{1+r}.$$

Hypothesis 6.3 implies not only that our system is strongly dissipative, but in addition that the nonlinear part is superlinear. In finite dimensions such systems have been considered by S. Cerrai [18]. Under Hypothesis 6.3 we will be able to prove that $P_t\varphi(x) \to \int_H \varphi(x)\mu(dx)$ for all $\varphi \in \text{Lip}(H)$ uniformly on x. Moreover, we show that the convergence of $P_t\varphi(x)$ holds for all $\varphi \in B_b(H)$, as $t \to +\infty$, under the additional hypothesis following.

Hypothesis 6.4 *For any $\varphi \in B_b(H)$ and $t > 0$ there exists $K_t > 0$ such that $P_t\varphi$ is Lipschitz continuous and*

$$|P_t\varphi(x) - P_t\varphi(y)| \leq K_t |x - y| \, \|\varphi\|_0,$$

for all $x, y \in H$.

Theorem 6.4.1 *Assume that Hypotheses 5.4, 5.5, 6.2 and 6.3 hold and let μ be the invariant measure for (6.3.11) given by Theorem*

6.3.3. *Then there exists $C_1 > 0$ such that, for any bounded Lipschitz continuous function φ, all $t > 0$ and all $x \in H$*

$$|P_t\varphi(x) - \langle \varphi, \mu \rangle| \leq C_1 e^{-\omega t/2} \|\varphi\|_{\text{Lip}}. \tag{6.4.1}$$

If in addition Hypothesis 6.4 holds, then there exists $C_2 > 0$ such that

$$|P_t\varphi(x) - \langle \varphi, \mu \rangle| \leq C_2 e^{-\omega t/2} \|\varphi\|_0,$$

for all $\varphi \in B_b(H)$.

Proof— Proceeding as in the proof of Theorem 6.3.3 we consider the unique generalized solution of (6.3.12), which we shall denote by $X(t, s, x)$.

We set

$$Z_{\lambda,\gamma}(t) = X(t, -\lambda, x) - X(t, -\gamma, x), \ 0 < \lambda < \gamma.$$

Then $Z_{\lambda,\gamma}$ is the mild solution of the problem

$$\left. \begin{array}{l} \dfrac{d}{dt} Z_{\lambda,\gamma} = A Z_{\lambda,\gamma} + F(X(t, -\lambda, x)) - F(X(t, -\gamma, x)), \\[2mm] Z_{\lambda,\gamma}(-\lambda) = x - X(-\lambda, -\gamma, x), \ t \geq -\lambda \end{array} \right\}$$

It follows, denoting by $x^\star_{\lambda,\gamma,t}$ an element from the subdifferential of $|Z_{\lambda,\gamma}(t)|$, that

$$\frac{d^-}{dt}|Z_{\lambda,\gamma}(t)| \leq \langle A Z_{\lambda,\gamma}(t) + F(X(t, -\lambda, x)) - F(X(t, -\gamma, x)), x^\star_{\lambda,\gamma,t}\rangle$$

$$\leq -\omega|Z_{\lambda,\gamma}(t)| - C_r|Z_{\lambda,\gamma}(t)|^r.$$

$$\tag{6.4.2}$$

By a well known comparison theorem it follows that

$$|Z_{\lambda,\gamma}(t)| \leq \eta(t, x),$$

where η is the solution to the initial value problem

$$\left. \begin{array}{l} \eta'(t) = -\omega\eta(t) - C_r|\eta(t)|^r, \ t \geq -\lambda, \\[2mm] \eta(0) = |x - X(-\lambda, -\gamma, x)|. \end{array} \right\}$$

By proceeding as in S. Cerrai [18], one can show that there exists a constant $C_2(t)$, independent of x, such that

$$\mathbb{E}|X(t, -\lambda, x) - X(t, -\gamma, x)| \leq e^{-\omega(t+\lambda)}\mathbb{E}|X(-\lambda, -\gamma, x) - x|$$

$$\leq C_3 e^{-\omega(t+\lambda)}, \ t \geq -\lambda.$$

$$(6.4.3)$$

Let now ζ be the random variable such that (6.3.16) holds, and let φ be a bounded Lipschitz function on H, then, by (6.4.3) we easily find (6.4.1).

Let us prove the last statement of the theorem. Let $\psi \in B_b(H)$ and let $\varepsilon > 0$. Then, setting $\varphi = P_\varepsilon \psi$ in (6.4.1), we have, by Hypothesis 6.4, and by the invariance of μ,

$$|P_{t+\varepsilon}\psi(x) - \langle P_\varepsilon\psi, \mu\rangle| \leq C_1 e^{-\omega t/2}\|P_\varepsilon\psi\|_{\mathrm{Lip}}$$

$$\leq C_1 K_\varepsilon e^{-\omega t/2}\|\psi\|_0.$$

It follows that

$$|P_t\psi(x) - \langle \psi, \mu\rangle| \leq K_\varepsilon C_1 e^{-\omega(t-\varepsilon)/2}\|\psi\|_0,$$

for all $\psi \in C_b(H)$ and for all $t > \varepsilon$. Since

$$|P_t\psi(x) - \langle \psi, \mu\rangle| \leq 2\|\psi\|_0,$$

the proof is complete.   ∎

We conclude this section by providing a sufficient condition for to Hypothesis 6.4 hold. We follow here G. Da Prato, D. Elworthy and J. Zabczyk [37].

Theorem 6.4.2 *Assume that Hypotheses 5.4, 5.5 hold, and that $B = I$. Then Hypothesis 6.4 is fulfilled.*

Proof — By Theorem 5.5.8 the equation

$$\left.\begin{array}{l} dX = (AX + F(X))dt + dW(t), \\[2mm] X(0) = x, \end{array}\right\} \qquad (6.4.4)$$

has a unique generalized solution $X(\cdot, x)$. Moreover $X(t, x)$ is the L^2 limit of the solution $X_\alpha(t, x)$ of the problem

$$\left. \begin{array}{l} dX_\alpha = (AX_\alpha + F_\alpha(X_\alpha))dt + dW(t), \\[2mm] X(0) = x, \end{array} \right\} \qquad (6.4.5)$$

where F_α are the Yosida approximations of F. Since F_α are Lipschitz continuous we can apply Theorem 7.1.1 in next chapter, and conclude that for any $T > 0$ there exists a constant C_T, independent of α, such that

$$|P_t^\alpha \varphi(x) - P_t^\alpha \varphi(y)| \le \frac{C_T}{\sqrt{t}} \, \|\varphi\|_0 |x - y|, x, y \in H, \qquad (6.4.6)$$

for any $\varphi \in B_b(H)$, where

$$P_t^\alpha \varphi(x) = \mathbb{E}\varphi(X_\alpha(t, x)), \ t > 0, \alpha > 0, \ \varphi \in B_b(H).$$

Letting α tend to 0 gives the conclusion. ∎

Remark 6.4.3 Differential stochastic equations in $\mathbb{R}^n$ with polynomial drift have been studied by S. Kusuoka and D. W. Stroock [98], when the drift term is the gradient of a potential plus a Lipschitz continuous function.

6.5 Dissipative systems in Banach spaces

We are concerned with existence and uniqueness of invariant measures for the problem

$$\left. \begin{array}{l} dX = (AX + F(X))dt + dW(t), \\[2mm] X(0) = x, \end{array} \right\} \qquad (6.5.1)$$

in a Banach space K (norm $\|\cdot\|$) continuously and densely embedded in H.

Theorem 6.5.1 *Assume that Hypothesis 5.7 holds with $\omega + \eta = -\delta < 0$ and that*

$$\sup_{t \geq 0} \mathbb{E}\left(\|W_A(t)\|\right) < +\infty.$$

Then there exists exactly one invariant measure μ for (6.5.1), it is strongly mixing and for arbitrary $\nu \in \mathcal{M}_1(K)$, $P_t^* \nu \to \mu$ weakly as $t \to +\infty$. Moreover there exists $C > 0$ such that, for any bounded Lipschitz continuous function φ, all $t > 0$ and all $x \in E$,

$$|P_t\varphi(x) - \langle\varphi,\mu\rangle| \leq (C + 2\|x\|)e^{-\delta t/2}\|\varphi\|_{\text{Lip}}.$$

Proof — As in the proof of Theorem 6.3.3 we consider problem (6.5.1) for $t \geq -\lambda$, $\lambda > 0$,

$$\left. \begin{array}{l} dX = (AX + F(X))dt + Bd\overline{W}(t), t \geq -\lambda, \\[2mm] X(-\lambda) = x, \end{array} \right\} \tag{6.5.2}$$

By Theorem 5.5.13 we know that there exists a unique generalized solution of (6.5.2), which we shall denote by $X(t, -\lambda, x)$, $t \geq -\lambda$.

We show now that there exists $c_1 > 0$ such that

$$\mathbb{E}(\|X(t, -\lambda, x)\|) \leq C + \|x\|, \text{ for all } \lambda > 0, \text{ for all } x \in H, \text{for all } t > -\lambda. \tag{6.5.3}$$

We first remark that

$$Z_\lambda(t) = X(t, -\lambda, x) - W_{A,\lambda}(t), t \geq -\lambda,$$

is the mild solution of the problem

$$\left. \begin{array}{l} \dfrac{d}{dt}Z = AZ + F(Z + W_{A,\lambda}), \\[2mm] Z(-\lambda) = x, \ t \geq -\lambda \end{array} \right\}.$$

It follows, recalling that F is dissipative, that there exists $x_{\lambda,t}^* \in \partial\|Z_\lambda(t)\|$ such that

$$\begin{aligned} \frac{d^-}{dt}\|Z_\lambda(t)\| &= \langle AZ_\lambda(t) + F(Z_\lambda(t) + W_{A,\lambda}(t)), x_{\lambda,t}^*\rangle \\[2mm] &\quad - \langle F(W_{A,\lambda}(t)), x_{\lambda,t}^*\rangle + \langle F(Z_\lambda(t)), x_{\lambda,t}^*\rangle \\[2mm] &\leq -\omega\|Z_\lambda(t)\| + \|F(W_{A,\lambda}(t))\|. \end{aligned} \tag{6.5.4}$$

Consequently

$$\|Z_\lambda(t)\| \leq e^{-\omega(t+\lambda)}\|x\| + \int_{-\lambda}^{t} e^{-\omega(t-s)}\|F(W_{A,\lambda}(s))\|ds, \ t \geq -\lambda,$$

and

$$\mathbb{E}\|Z_\lambda(t)\| \leq \|x\| + \frac{1}{\omega} \sup_{s \in [-\lambda, t]} \mathbb{E}\|F(W_{A,\lambda}(s))\|, \ t \geq -\lambda.$$

Taking into account the hypothesis imposed one arrives at (6.3.13). Now the proof is completely similar to that of Theorem 6.3.3. ∎

Chapter 7

Uniqueness of invariant measures

We are still concerned with invariant measures of the transition semi-group P_t, $t \geq 0$, associated with problem (5.3.1). As we know from Theorem 4.2.1, uniqueness of invariant measure is a consequence of regularity. Moreover regularity of P_t, $t \geq 0$, follows from the strong Feller property and irreducibility for some $t > 0$, see Proposition 4.1.1. We will present here methods, applicable in several important cases, which imply the strong Feller property and irreducibility for P_t, $t \geq 0$. We do not present the most general results because this would complicate the formulation, but restrict to consider typical situations. In several applications in Part III, some modifications of the arguments used here will be needed. We start from the strong Feller property.

7.1 Strong Feller property for non–degenerate diffusions

We are concerned with the problem

$$
\left.
\begin{aligned}
dX(t) &= (AX + F(X))dt + B(X)dW(t), \\
X(0) &= x \in H.
\end{aligned}
\right\}
\tag{7.1.1}
$$

In all this section we will assume that $U = H$. We will need the following hypothesis, stronger than Hypothesis 5.1.

Hypothesis 7.1 *(i) A is the infinitesimal generator of a strongly continuous semigroup $S(t)$, $t \geq 0$, on H.*
(ii) F is a mapping from H into H and there exists a constant $c_0 > 0$ such that

$$|F(x)| \leq c_0(1 + |x|), \quad x \in H,$$

$$|F(x) - F(y)| \leq c_0|x - y|, \quad x, y \in H.$$

(iii) B is a strongly continuous mapping from H into $L(H)$ and there exists a constant $K > 0$ such that

$$\|B(x)\| \leq (1 + |x|), \quad x \in H,$$

$$\|B(x) - B(y)\| \leq K|x - y|, \quad t > 0, \ x, y \in H.$$

Moreover, for all $z \in H$, $B(z)$ is invertible and

$$\|B^{-1}(z)\| \leq K, \text{ for all } z \in H.$$

(iv) We have $S(t) \in L_2(H)$ for all $t > 0$, and

$$\int_0^1 \|S(t)\|_{HS}^2 \, dt < +\infty.$$

We shall denote, as usual, by $X(\cdot, x)$ the mild solution to (7.1.1) and by P_t, $t > 0$, the corresponding transition semigroup . The following theorem is taken from S. Peszat and J. Zabczyk [124].

Theorem 7.1.1 *Assume that Hypothesis 7.1 holds. Then for any $T > 0$ there exists a constant $C_T > 0$ such that for all $\psi \in B_b(H)$ and $t \in [0, T]$*

$$|P_t\psi(x) - P_t\psi(y)| \leq \frac{C_T}{\sqrt{t}} \|\psi\|_0 \, |x - y|, x, y \in H. \tag{7.1.2}$$

In particular P_t, $t \geq 0$, is a strong Feller semigroup for all $t > 0$.

Proof — The proof will consist of two parts. In the first part we prove the desired estimate under the additional hypothesis that F, B and ψ are regular. In the second part we show how to dispense with that assumption.

Part 1 . Here we assume that $\psi \in C_b^2(H)$, $F \in C_b^2(H; H)$ and $B \in C_b^2(H; L(H))$.
We have divided this part into a sequence of lemmas.

Lemma 7.1.2 *Assume that F, B and ψ are twice differentiable functions with bounded and continuous derivatives up to the second order. Then*

$$\psi(X(t, x)) = P_t \psi(x)$$

$$+ \int_0^t \langle D_x P_{t-s} \psi(X(s, x)), B(X(s, x)) dW(s) \rangle, \; \mathbb{P}\text{–a.s..} \tag{7.1.3}$$

Proof — Let $\{e_n\}$ be a complete orthonormal system in H. For each n let $X_n(\cdot, x)$ be the solution of the problem

$$\left. \begin{aligned} dX_n(t) &= (A_n X_n + F(X_n)) dt + B(X_n) Q_n dW(t), \\[2mm] X_n(0) &= x, \end{aligned} \right\}$$

where $A_n = nA(n - A)^{-1}$ is the Yosida approximation of A and Q_n is the orthogonal projection of H onto

$$\text{lin}\{e_1, ..., e_n\}.$$

It follows from Theorem 5.4.2 that the function

$$v_n(t, x) = \mathbb{E}(\psi(X_n(t, x))), \; (t, x) \in \,]0, +\infty[\,\times H,$$

is a strict solution to the Kolmogorov equation

$$\left. \begin{aligned} \frac{d}{dt} v^n(t, x) &= \frac{1}{2} \, \text{Tr} \, [B^*(x) Q_n B(x) v_{xx}^n(t, x)] \\[2mm] &\quad + \langle A_n x + F(x), v_x^n(t, x) \rangle, \\[2mm] v^n(0, x) &= \psi(x), \; x \in H, \; t \geq 0. \end{aligned} \right\}$$

Applying the Ito formula, see e.g. G. Da Prato and J. Zabczyk [44], to the process $v^n(t - s, X_n(s, x))$, $s \in [0, t]$, we obtain that

$$\psi(X_n(t, x)) = v^n(t, x)$$

$$+ \int_0^t \langle v_x^n(t - s, X_n(s, x)), B(X_n(s, x))Q_n dW(s) \rangle, \ \mathbb{P}\text{–a.s.}.$$

Letting $n \to +\infty$ gives the desired result.   ∎

Now we extend the finite–dimensional Elworthy formula, see D. Elworthy [61].

Lemma 7.1.3 *Assume that F, B and ψ are twice differentiable functions with bounded and continuous derivatives up to the second order. Then the directional derivatives $\langle D_x P_t \psi(x), h \rangle$ are given by*

$$\langle D_x P_t \psi(x), h \rangle$$

$$= \frac{1}{t} \mathbb{E} \left\{ \psi(X(t, x)) \int_0^t \langle B^{-1}(X(s, x)) X_x(s, x) h, dW(s) \rangle \right\}, \ \mathbb{P}\text{–a.s.}$$

$$(7.1.4)$$

Proof —   Fix $h \in H$ and set $v(t, x) = P_t \psi(x), t \geq 0, \ x \in H$. Multiplying both sides of (7.1.3) by

$$\int_0^t \langle B^{-1}(X(s, x)) X_x(s, x) h, dW(s) \rangle,$$

and taking the expectation we get

$$\mathbb{E} \left(\psi(X(t, x)) \int_0^t \langle B^{-1}(X(s, x)) X_x(s, x) h, dW(s) \rangle \right)$$

$$= \mathbb{E} \int_0^t \langle B^*(X(s, x)) v_x(t - s, X(s, x)), B^{-1}(X(s, x)) X_x(s, x) h \rangle ds$$

$$= \int_0^t \langle D_x \mathbb{E}(D_x P_{t-s} \psi(X(s, x))), h \rangle \, ds$$

$$= \int_0^t \langle v_x(t - s, X(s, x)), h \rangle \, ds = \int_0^t \langle D_x P_t \psi(x), h \rangle \, ds$$

$$= t \langle D_x P_t \psi(x), h \rangle,$$

which yields (7.1.4). ∎

Lemma 7.1.4 *Assume that F, B and ψ are twice differentiable functions with bounded and continuous derivatives up to the second order. Then the estimate (7.1.2) holds true.*

Proof — Fix $T > 0$, then by (7.1.4) we have

$$|\langle D_x P_t \psi(x), h \rangle|^2 \leq \frac{1}{t^2} \|\psi\|_0^2 \, \mathbb{E} \left\{ \int_0^t |B^{-1}(X(s,x)) X_x(s,x) h|^2 ds \right\}$$

$$\leq \frac{K^2}{t^2} \|\psi\|_0^2 \, \mathbb{E} \int_0^t |X_x(s,x) h|^2 ds,$$

and the conclusion follows from (5.4.2). ∎

Part 2 —

We first show that if (7.1.2) holds for $\psi \in C_b^2(H)$ then it holds for all $\psi \in B_b(H)$. This follows from the lemma

Lemma 7.1.5 *Let P_t, $t \geq 0$, be a Markov semigroup on $B_b(H)$ and let $c > 0$ and $t > 0$ be fixed. Then the following conditions are equivalent.*

(i) for all $\varphi \in C_b^2(H)$, for all $x, y \in H$, $|P_t\varphi(x) - P_t\varphi(y)| \leq c\|\varphi\|_0 |x - y|$.

(ii) for all $\varphi \in B_b(H)$, for all $x, y \in H$, $|P_t\varphi(x) - P_t\varphi(y)| \leq c\|\varphi\|_0 |x - y|$.

(iii) for all $x, y \in H$, $\mathrm{Var}\,(P_t(x, \cdot) - P_t(y, \cdot)) \leq c\|\varphi\|_0 |x - y|$.

Proof — Let

$$\mathcal{K}_1 = \{\varphi \in C_b(H) : \|\varphi\|_0 \leq 1\},$$

and

$$\mathcal{K}_2 = \{\varphi \in C_b^2(H) : \|\varphi\|_0 \leq 1\}.$$

Since each bounded continuous function on H may be approximated pointwise by functions of $C_b^2(H)$ we have

$$\sup_{\varphi \in \mathcal{K}_1} |P_t\varphi(x) - P_t\varphi(y)| = \sup_{\varphi \in \mathcal{K}_2} |P_t\varphi(x) - P_t\varphi(y)|,$$

for all $x, y \in H$. As a simple consequence of the Hahn decomposition theorem, we have

$$\sup_{\varphi \in \mathcal{K}_1} |P_t\varphi(x) - P_t\varphi(y)| = \text{Var} \ (P_t(x, \cdot) - P_t(y, \cdot)).$$

Therefore (i) implies (iii). However, if (iii) holds then for all $\varphi \in B_b(H)$

$$|P_t\varphi(x) - P_t\varphi(y)| \ \leq \ \left| \int_H \varphi(z)(P_t(x, dz) - P_t(y, dz)) \right|$$

$$\leq \ \|\varphi\|_0 \ \text{Var} \ (P_t(x, \cdot) - P_t(y, \cdot))$$

$$\leq \ \|\varphi\|_0 |x - y|,$$

and we see that (ii) is true. Obviously (ii) implies (i) and the proof of the lemma is complete.   ∎

Therefore, to prove the theorem in the general case, we can assume that $\varphi \in C_b^2(H)$. We will now construct approximations F_n of F and B_n of B with the following properties.

(i) F_n, B_n, $n \in \mathbb{N}$, are twice Fréchet differentiable with bounded and continuous derivatives.

(ii) F_n, B_n, $n \in \mathbb{N}$, satisfy the Lipschitz condition with the constant c.

(iii) for all $\widetilde{K} \geq K, \exists \ n_0 \in \mathbb{N}$ such that the operators $B_n(z), z \in H$, are invertible and

$$\sup_{z \in H} \|B_n^{-1}(z)\| \leq \widetilde{K}, \ n \geq n_0.$$

(iv) $\lim_{n\to\infty} |F(z) - F_n(z)| = 0$ and $\lim_{n\to\infty} \|B(z) - B_n(z)\| = 0$.

Before giving the construction we will show how to finish the proof of the theorem.

Let $X_n(\cdot, x)$ be the solution of the equation

$$\left. \begin{array}{l} dX_n(t) = (A_nX_n + F(X_n))dt + B(X_n)Q_ndW(t), \\[2mm] X_n(0) = x, \end{array} \right\}$$

and let $P_{n,t}$, $t \geq 0$, be the corresponding transition semigroup. Fix $t > 0$, $\psi \in C_b^2(H)$ and a number $\tilde{c}_t$ greater than c_t. Lemma 7.1.5 applied to $P_{n,t}$, $t \geq 0$, shows that

$$|P_{n,t}\psi(x) - P_{n,t}\psi(y)| \leq \tilde{c}_t \|\psi\|_0 |x - y|, x, y \in H.$$

However, for all $t > 0$ and $x \in H$ there exists a sub–sequence $\{X_{n_j}\}$ such that $X_{n_j}(t,x) \to X(t,x)$ a.s. as $j \to +\infty$. Since ψ is a bounded continuous function we have

$$P_{n,t}\psi(x) = \mathbb{E}\left(\psi(X_{n_j}(t,x))\right) \to \mathbb{E}(\psi(X(t,x))),$$

as $j \to +\infty$, and we have

$$|P_t\psi(x) - P_t\psi(y)| \leq c_t \|\psi\|_0 |x - y|, \ x, y \in H,$$

which proves the theorem.

We return now to the construction of the sequences $\{F_n\}$ and $\{B_n\}$. We take a sequence of non–negative, twice differentiable functions $\{\rho_n\}$ such that

$$\text{supp} \{\rho_n\} \subset \left\{\xi \in \mathbb{R}^n : |\xi|_{\mathbb{R}^n} \leq \frac{1}{n}\right\}$$

and

$$\int_{\mathbb{R}^n} \rho_n(\xi) d\xi = 1.$$

Identifying $\mathbb{R}^n$ with $\text{lin}\{e_1, ..., e_n\}$ we define

$$F_n(x) = \int_{\mathbb{R}^n} \rho_n(\xi - Q_n x) F\left(\sum_{i=1}^n \xi_i e_i\right) d\xi,$$

$$B_n(x) = \int_{\mathbb{R}^n} \rho_n(\xi - Q_n x) B\left(\sum_{i=1}^n \xi_i e_i\right) d\xi.$$

Observe that F_n and B_n are twice Fréchet differentiable functions with bounded and continuous derivatives. Moreover, for all $x, y \in H$

and $n \in \mathbb{N}$

$$|F_n(x) - F_n(y)|$$

$$= \left| \int_{\mathbb{R}^n} \rho_n(\xi) \left[F \left(\sum_{i=1}^{n} \xi_i e_i + Q_n x \right) - F \left(\sum_{i=1}^{n} \xi_i e_i + Q_n y \right) \right] d\xi \right|$$

$$\leq c |Q_n(x - y)| \int_{\mathbb{R}^n} \rho_n(\xi) \leq c|x - y|.$$

In the same manner we can see that

$$|B_n(x) - B_n(y)| \leq c|x - y|.$$

We next prove that, for sufficiently large n, the operators B_n, $n \in \mathbb{N}$, are invertible and that

$$\limsup_{n \to +\infty} \left[\sup_{z \in H} \|B_n^{-1}(z)\| \right] \leq K.$$

To this end observe that the operators $B(Q_n z)$, $z \in H, n \in \mathbb{N}$, are invertible,

$$\sup_{n \in \mathbb{N}} \sup_{z \in H} \|B^{-1} Q_n z\| \leq \sup_{z \in H} \|B^{-1}(z)\| \leq K$$

and

$$\sup_{z \in H} \|B(Q_n z) - B_n(z)\|$$

$$= \sup_{z \in H} \left\| \int_{\mathbb{R}^n} \rho_n(\xi) \left[B \left(\sum_{i=1}^{n} \xi_i e_i + Q_n x \right) - B \left(\sum_{i=1}^{n} \xi_i e_i + Q_n y \right) \right] d\xi \right\|$$

$$\leq c \int_{\mathbb{R}^n} \rho_n(\xi) \left| \sum_{i=1}^{n} \xi_i e_i \right| d\xi \leq \frac{c}{n}.$$

Consequently for sufficiently large n the operators B_n, $n \in \mathbb{N}$, are invertible and

$$\limsup_{n \to +\infty} \left[\sup_{z \in H} \| B_n^{-1}(z) \| \right]$$

$$= \limsup_{n \to +\infty} \left[\sup_{z \in H} \| [B(Q_n z) - (B(Q_n z) - B(z))]^{-1} \| \right]$$

$$\leq \limsup_{n \to +\infty} \left[\sup_{z \in H} \| [B^{-1}(Q_n z)\| \sum_{j=0}^{\infty} \| B^{-1}(Q_n z)(B(Q_n z) - B(z)) \|^j \right]$$

$$\leq K \limsup_{n \to +\infty} \sum_{j=0}^{\infty} (Kcn^{-1})^j = K.$$

It is also easy to check that (iv) holds. ∎

Remark 7.1.6 The strong Feller property for equation (7.1.1) with f only bounded and continuous has been obtained by D. Gątarek and B. Goldys [77] by using the concept of martingale solutions, see G. Da Prato and J. Zabczyk [44].

7.2 Strong Feller property for degenerate diffusion

The invertibility Hypothesis 7.1(iii) may not be satified but, nevertheless the semigroup P_t, $t \geq 0$, corresponding to (5.2.1) can be a strong Feller one. We will examine this possibility for equations with constant diffusion coefficient:

$$\left. \begin{aligned} dX(t) &= (AX + F(X))dt + BdW(t), \\ X(0) &= x. \end{aligned} \right\} \tag{7.2.1}$$

We again assume a stronger hypothesis than Hypothesis 5.1.

Hypothesis 7.2 *(i) A is the infinitesimal generator of a strongly continuous semigroup $S(t)$, $t \geq 0$, on H.*

(ii) F is a mapping from H into H and there exists a constant $c_0 > 0$ such that

$$|F(x)| \leq c_0(1 + |x|), \quad x \in H,$$

$$|F(x) - F(y)| \leq c_0|x - y|, \quad x, y \in H.$$

(iii) B is a linear continuous mapping from U into H.

(iv) For any $t > 0$ the linear operator Q_t,

$$Q_t x = \int_0^t S(s)QS^*(s)x \, ds, \quad x \in H, \tag{7.2.2}$$

where $Q = BB^*$, is of trace class.

In some cases we shall assume further

Hypothesis 7.3 Image $S(t) \subset$ Image $Q_t^{1/2}$.

Hypothesis 7.3 is equivalent to the fact that the deterministic system

$$\xi'(t) = A\xi(t) + Bu(t)$$

is *null controllable*, see e.g. G. Da Prato and J. Zabczyk [44, §B.3]. If Hypothesis 7.3 holds we denote by $\Gamma(t)$ the linear bounded operator

$$\Gamma(t) = Q_t^{-1/2}S(t), \ t > 0.$$

We start with the linear case,

$$\left. \begin{array}{l} dZ(t) = AZdt + BdW(t), \\[2mm] Z(0) = x. \end{array} \right\} \tag{7.2.3}$$

We remark that, for all $t > 0$, $X(t)$ is a Gaussian random variable, $\mathcal{N}(S(t)x, Q_t)$, that is with mean $S(t)x$ and covariance Q_t.

We shall denote by P_t, $t > 0$, and R_t, $t > 0$, the transition semigroups corresponding to problems (7.2.1) and (7.2.3) respectively:

$$P_t\varphi(x) = \mathbb{E}\varphi(X(t,x)), \ t \geq 0, \ \varphi \in C_b(H),$$

$$R_t\varphi(x) = \mathbb{E}\varphi(Z(t,x)), \ t \geq 0, \ \varphi \in C_b(H).$$

We have the following basic result.

Theorem 7.2.1 *Let R_t, $t \geq 0$, be the semigroup corresponding to (7.2.1), and let $t > 0$. Then the following conditions are equivalent.*
(i) R_t, $t \geq 0$, is t–regular.
(ii) R_t, $t \geq 0$, is a strong Feller semigroup at time t.
(iii) Hypothesis 7.3 holds.

Proof — Since transition probabilities are Gaussian

$$R_t(x, \cdot) = \mathcal{N}(S(t)x, Q_t),$$

the equivalence (i)$\Longleftrightarrow$(iii) is an immediate consequence of the Cameron–Martin formula.

To show that (ii) $\Longleftrightarrow$ (iii) assume that (ii) holds. If (iii) is not true then for some $\bar{x} \in H, S(t)\bar{x} \neq \operatorname{Im} Q_t^{1/2}$. From the linearity $S(t)\left(\frac{1}{n}\bar{x}\right) \neq \operatorname{Image} Q_t^{1/2}$, $n \in \mathbb{N}$. Since all measures

$$R_t\left(\frac{\bar{x}}{n}, \cdot\right)$$

are singular with respect to $R_t(0, \cdot)$ there exists a set $\Gamma \subset \mathcal{B}(H)$ such that

$$R_t\left(\frac{\bar{x}}{n}, \Gamma\right) = 0, \quad n \in \mathbb{N},$$

and $R_t(0, \Gamma) = 1$. But this implies that the function $R_t \chi_\Gamma$ is not continuous at 0. So (ii) $\Longrightarrow$(iii).

To show that (iii) $\Longrightarrow$(ii), we first recall that, by the Cameron–Martin formula:

$$\frac{d\mathcal{N}(S(t)x, Q_t)}{d\mathcal{N}(0, Q_t)}(z) = d(t, x, z),$$

where

$$d(t, x, z) = \exp\left(-\frac{1}{2}|Q_t^{-1/2}S(t)x|^2 + \langle Q_t^{-1/2}z, Q_t^{-1/2}x\rangle\right)$$
$$= \exp\left(-\frac{1}{2}|\Gamma(t)x|^2 + \langle\Gamma(t)x, Q_t^{-1/2}z\rangle\right).$$

Setting $v(t, x) = R_t\varphi(x)$ it follows that

$$v(t,x) = \int_H \varphi(z)\mathcal{N}(S(t)x, Q_t)(dz)$$

$$= \int_H \varphi(z)d(t,x,z)\mathcal{N}(0, Q_t)(dz).$$

Set

$$f_N(x,z) = \exp\left(-\frac{1}{2}\sum_{k=1}^N \frac{|\langle S(t)x, e_k(t)\rangle|^2}{\lambda_k(t)}\right)$$

$$\times \exp\left(\sum_{k=1}^N \frac{\langle S(t)x, e_k(t)\rangle \langle z, e_k(t)\rangle}{\lambda_k(t)}\right),$$

where $\{e_k(t)\}$ and $\{\lambda_k(t)\}$ are eigensequences relative to Q_t. We recall that

$$\lim_{N\to\infty} f_N(\cdot, z) = d(t, \cdot, z),$$

for any $t \geq 0, z \in H$, in $L^p(H, \mathcal{B}(H), \mathcal{N}(0, Q_t))$ and almost surely. If $h \in H$, we have,

$$\langle \frac{d}{dx}f_N(x,z), h\rangle = -d(t,x,z)\sum_{k=1}^N \frac{\langle S(t)x, e_k(t)\rangle \langle S(t)h, e_k(t)\rangle}{\lambda_k(t)}$$

$$+d(t,x,z)\sum_{k=1}^N \frac{\langle S(t)h, e_k(t)\rangle \langle z, e_k(t)\rangle}{\lambda_k(t)},$$

and it is easy to check that

$$\lim_{N\to\infty} \langle \frac{d}{dx}f_N(x,z), h\rangle =$$

$$d(t,x,z)\left(-\langle \Gamma(t)x, \Gamma(t)h\rangle + \langle \Gamma(t)\xi, Q_t^{-1/2}z\rangle\right),$$

in $L^p(H, \mathcal{B}(H), \mathcal{N}(0, Q_t))$.

By the dominate convergence theorem, it follows that

$$\langle v_x(t,x), h \rangle$$

$$= \int_H \varphi(z) d(t,x,z) \langle Q_t^{-1/2} z - \Gamma(t)x, \Gamma(t)h \rangle \mathcal{N}(0, Q_t)(dz)$$

$$= \int_H \varphi(z) \langle Q_t^{-1/2} z - \Gamma(t)x, \Gamma(t)h \rangle \mathcal{N}(S(t)x, Q_t)(dz).$$

Consequently the directional derivative of $v(t,x)$ at x in direction ξ does exist and is given by the formula

$$\langle v_x(t,x), \xi \rangle = \int_H \langle \Gamma(t)\xi, Q_t^{-1/2} z \rangle \varphi(S(t)x + z) \mathcal{N}(0, Q_t)(dz). \quad (7.2.4)$$

It follows from (7.2.4) that for arbitrary $x, y \in H$ and some $s \in [0,1]$

$$|v(t,x) - v(t,y)| = |\langle v_x(t, x + s(y-x)), x - y \rangle|$$

$$\leq \int_H |\langle \Gamma(t)(x-y), Q_t^{-1/2} z \rangle| \, |\varphi(S(t)(x + s(y-x)) + z)| \mathcal{N}(0, Q_t)(dz)$$

$$\leq \|\varphi\|_0 \int_H |\langle \Gamma(t)(x-y), Q_t^{-1/2} z \rangle| \, \mathcal{N}(0, Q_t)(dz)$$

$$\leq \|\varphi\|_0 \left(\int_H |\langle \Gamma(t)(x-y), Q_t^{-1/2} z \rangle|^2 \, \mathcal{N}(0, Q_t)(dz) \right)^{1/2}$$

$$\leq \|\varphi\|_0 \|\Gamma(t)\| \, |(x-y)|.$$

This proves (ii). ∎

Remark 7.2.2 For general Markovian semigroups regularity is a consequence of the strong Feller property and irreducibility, but for the Gaussian case irreducibility is irrelevant. Take for instance $H = L^2(0,1)$ and $S(t)$, $t \geq 0$, the shift semigroup:

$$S(t)x(\xi) = \begin{cases} x(\xi + t) & \text{if } \xi \in [0, 1-t], \\ 0 & \text{if } \xi \in \,]1-t, 1]. \end{cases}$$

Then Image $S(t) = \{0\}$ for $t \geq 1$ and condition (iii) is satisfied for an arbitrary operator Q of trace class. If $Q = 0$ then certainly the transition semigroup P_t, $t \geq 0$, is not irreducible, but it is t–regular for $t \geq 1$. It is, however, an easy consequence of (iii) that t–regularity for some $t > 0$ implies regularity for arbitrary $t > 0$.

Before going to the nonlinear case let us remark that, although t–regularity is a sufficient condition for the uniqueness of invariant measure it is not a necessary one. This is clearly shown by the following special case of Theorem 6.2.1.

Proposition 7.2.3 *Assume that there exists an invariant measure for (7.2.3). Then any invariant measure for (7.2.3) is of the form*

$$\nu * \mathcal{N}(0, \overline{P}),$$

where ν is invariant for $S(t)$, $t \geq 0$, and

$$\overline{P}x = \int_0^{+\infty} S(t)QS^*(t)x\,dt, \quad x \in H.$$

We continue now our study of the strong Feller property for equation (7.2.1).

Theorem 7.2.4 *Assume that Hypotheses 7.2 and 7.3 hold, and that $\|\Gamma(\cdot)\|$ belongs to $L^1_{loc}(0, +\infty)$, where $\Gamma(t) = Q_t^{-1/2}S(t)$. Then for arbitrary $\varphi \in B_b(H)$ and $x, y \in H$*

$$|P_t\varphi(x) - P_t\varphi(y)| \leq \xi(t)|x - y|, \quad t > 0, \qquad (7.2.5)$$

where $\xi(\cdot)$ is the solution to the integral equation,

$$\xi(t) = \|\Gamma(t)\| \, \|\varphi\|_0 + \|F\|_0 \int_0^t \|\Gamma(t - s)\|\xi(s)ds. \qquad (7.2.6)$$

In particular P_t, $t \geq 0$, is a strong Feller semigroup.

Proof — Denote by $D(\mathcal{A}_0)$ the set of all φ such that
 (i) $\varphi \in C_b^2(H)$,
 (ii) $\varphi_x(x) \in D(A^*)$ for all $x \in H$ and the mapping

$$H \to \mathbb{R}, x \to \langle x, A^*\varphi_x \rangle,$$

belongs to $C_b(H)$.

(iii) $\varphi_{xx} \in C_b(H, L_1(H))$.

Now define a linear operator $\mathcal{A}_0$ on $C_b(H)$ by setting

$$
\mathcal{A}_0\varphi \;=\; \frac{1}{2} \operatorname{Tr}\left[Q\varphi_{xx}\right] + \langle x, A^*\varphi_x \rangle,
$$

$$
\left.
\begin{aligned}
D(\mathcal{A}_0) \;=\; & \{\varphi \in C_b^2(H) : \varphi_{xx} \in L_1(H), \; \sup_{x \in H} \|\varphi_{xx}\|_{L_1(H)} < +\infty, \\
& \text{and there exists } \psi \in C_b^2(H) \text{ such that } \varphi(x) = \psi(A^{-1}x) \\
& \text{and the mapping } x \to \langle x, \psi_x(A^{-1}x) \rangle \text{ is in } C_b(H)\}.
\end{aligned}
\right\}
$$

It has been shown in S. Cerrai and F. Gozzi [20] that the set $D(\mathcal{A}_0)$ is dense in $C_b(H)$ in the following sense:

For arbitrary $\varphi \in C_b^1(H)$ there exists a sequence $\{\varphi_n\} \subset D(\mathcal{A}_0)$ such that

(i) $\|\varphi_n\|_0 \leq 2\|\varphi\|_0$, $n \in \mathbf{N}$,

(ii) $\varphi_n \to \varphi$ uniformly on any compact subset of H.

Moreover $R_t(D(\mathcal{A}_0)) \subset D(\mathcal{A}_0)$ and $\mathcal{A}_0 R_t = R_t \mathcal{A}_0$ on $D(\mathcal{A}_0)$, for arbitrary $t \geq 0$.

Assume first that F is twice differentiable with bounded and continuous derivatives up to the second order and that $\varphi \in D(\mathcal{A}_0)$. Then the function

$$
v(t, x) = P_t\varphi(x), \; t \geq 0, \; x \in H,
$$

is the strict solution of the Kolmogorov equation:

$$
\begin{aligned}
v_t(t, x) \;=\; & \frac{1}{2} \operatorname{Tr}\left[Q v_{xx}(t, x)\right] + \langle x, A^* v_x(t, x) \rangle \\
& + \langle F(x), v_x(t, x) \rangle, \; t \geq 0,
\end{aligned}
\tag{7.2.7}
$$

which can be written equivalently as

$$
\left.
\begin{aligned}
\frac{dv}{dt} &= \mathcal{A}_0 v(t) + \mathcal{F}v(t), \\
v(0) &= \varphi, \; t \geq 0.
\end{aligned}
\right\}
\tag{7.2.8}
$$

In (7.2.8) $\mathcal{A}_0$ is the operator defined above and

$$
\mathcal{F}\psi(x) = \langle F(x), \psi_x \rangle, \; \psi \in C_b^1(H), \; x \in H.
$$

For fixed $t > 0$ consider the function $R_{t-s}(v(s))$, $s \in [0, T]$. Then, for $s \in [0, T]$,

$$\frac{d}{ds}[R_{t-s}(v(s))] = -\mathcal{A}_0[R_{t-s}(v(s))] + R_{t-s}\left(\frac{dv}{dt}(s)\right)$$

$$= -\mathcal{A}_0[R_{t-s}(v(s))] + R_{t-s}[\mathcal{A}_0 v(s) + \mathcal{F}v(s)].$$

Since the operators $\mathcal{A}_0$ and R_s commute we obtain that

$$\frac{d}{ds}[R_{t-s}(v(s))] = R_{t-s}\mathcal{F}, \quad s \in [0, t]. \qquad (7.2.9)$$

Integrating (7.2.9) over the interval $[0, t]$ one obtains that

$$v(t) = R_t\varphi + \int_0^t R_{t-s}(\mathcal{F}v(s))ds, \quad t \geq 0. \qquad (7.2.10)$$

The equation (7.2.10) is called a *mild* Kolmogorov equation. It follows that, for arbitrary $\psi \in C_b(H)$ and $t \geq 0$, $R_t\psi$ is a function bounded and continuous together with its first derivative and

$$|D_x R_t\psi| \leq \|\Gamma(t)\| \, \|\psi\|_0. \qquad (7.2.11)$$

Note that

$$\int_0^t \|D_x R_{t-s}\| \, \|\mathcal{F}v(s)\|_0 ds \leq \int_0^t \|\Gamma(t-s)\| \, \|\mathcal{F}v(s)\|_0 ds$$

$$\leq \int_0^t \|\Gamma(t-s)\| \, \|F\|_0 \|D_x v(s)\|_0 ds.$$

Under our conditions on φ and F the function $\|D_x v(s)\|_0$, $s \geq 0$, is locally bounded. Taking into account (7.2.11) and taking derivatives of both sides of (7.2.10), we arrive at the inequality

$$\|D_x v(t)\|_0 \leq \|\Gamma(t)\| \, \|\varphi\|_0 + \|F\|_0 \int_0^t \|\Gamma(t-s)\| \, \|D_x v(s)\|_0 ds, \quad t \geq 0,$$

valid for all $t > 0$.

Since $\|D_x v(\cdot)\|_0$ is locally bounded by a generalization of Gronwall's lemma,

$$\|D_x v(t)\|_0 \leq \xi(t), \quad t > 0, \qquad (7.2.12)$$

and (7.2.5) follows for all $\varphi \in C_b(H)$. The generalization to arbitrary Borel funtions φ and to bounded, Lipschitz continuous F can be done again as in Theorem 5.2.3. ∎

7.3 Irreducibility for non–degenerate diffusions

Irreducibility, as a rule, requires some kind of non–degeneracy of the diffusion term. We divide the material into two sections. In the present section we consider the case when the mappings B have bounded inverses; next section will be devoted to the case when B is constant, but not necessarily invertible. The main result of this section is the following.

Theorem 7.3.1 *Assume besides Hypothesis 5.1 that F is bounded and the mappings $B(x)$, $x \in H$ are invertible, and*

$$\sup_{x \in H} \left(\|B(x)\| + \|B^{-1}(x)\| \right) < +\infty.$$

Then the transition semigroup P_t, $t \geq 0$, corresponding to problem (7.1.1) is irreducible.

Proof — We first prove, in the following lemma, that irreducibility of P_t, $t \geq 0$, is equivalent to irreducibility of the system

$$\left. \begin{array}{l} dZ = (AZ + \varphi(t))dt + B(Z(t))dW(t),\ t \geq 0, \\[2mm] Z(0) = x, \end{array} \right\} \tag{7.3.1}$$

where φ is any bounded and adapted process. Then, by choosing φ in a suitable way, we prove that system (7.3.1) is irreducible.

Lemma 7.3.2 *Assume, besides the hypotheses on A, B of Theorem 7.3.1, that the process φ is bounded. Then the laws in $(H, \mathcal{B}(H))$ of $X(t, x)$ and $Z(t, x)$ are equivalent.*

Proof — Let $\alpha(t) = B^{-1}(Z(t))[\varphi(t) - F(Z(t))]$. Then α is bounded and adapted, so that the equation

$$\left. \begin{array}{l} dM(t) = -\langle \alpha(t), dW(t) \rangle M(t),\ t \geq 0, \\[2mm] M(0) = 1, \end{array} \right\}$$

has a unique solution $M(\cdot)$ and $\mathbb{E}(M(t)) = 1$, $t \geq 0$. Now fix $T > 0$ and set

$$d\mathbb{P}^*(\omega) = M(T)d\mathbb{P}(\omega).$$

Then $\mathbb{P}^*$ is a probability on $(\Omega, \mathcal{F})$ and therefore, by the Girsanov theorem (see e.g. G. Da Prato and J. Zabczyk [44]), the process

$$W^*(t) = W(t) + \int_0^t \alpha(s)ds, \ t \in [0, T],$$

is a cylindrical Wiener process in H defined on the probability space $(\Omega, \mathcal{F}, \mathbb{P}^*)$. Note that $\mathbb{P}$ and $\mathbb{P}^*$ are equivalent. Now

$$Z(t) = S(t)x + \int_0^t S(t-s)\varphi(s)ds + \int_0^t S(t-s)B(Z(s))dW(s)$$

$$= S(t)x + \int_0^t S(t-s)F(Z(s))ds + \int_0^t S(t-s)B(Z(s))dW^*(s).$$

Consequently Z is the solution of (5.1.1) on the probability space $(\Omega, \mathcal{F}, \mathbb{P}^*)$. Since the law of the solution does not depend on the particular choice of the probability space, for every Borel set $\Gamma \subset H$ we have $\mathbb{P}(X(t,x) \in \Gamma) = \mathbb{P}^*(Z(t,x) \in \Gamma)$. Since $\mathbb{P}$ and $\mathbb{P}^*$ are equivalent the laws of $X(t,x)$ and $Z(t)$ are also equivalent which is our claim. ∎

We can now proceed to the proof of Theorem 7.3.1. We fix $t_0 > 0$, $r > 0$, and $x, a \in H$. We will choose a bounded process φ such that, if Z is the corresponding solution of (7.3.1), we have

$$\mathbb{P}(|Z(t_0) - a| < r) > 0, \qquad (7.3.2)$$

and we will apply Lemma 7.3.2.

Let us consider an auxiliary system

$$dY = AYdt + B(Y)dW(t), \ Y(0) = x,$$

which clearly has a unique solution Y. Let $t_1 \in [0, t_0[$ be a moment to be chosen later. Now we choose φ by setting

$$\varphi(t) = \begin{cases} 0 \text{ if } t \in [0, t_1[, \\ \\ f(t, Y(t_1)) \text{ if } t \in [t_1, t_0], \end{cases}$$

where $f : [t_1, t_0] \times H \to H$ is bounded, $f(t, \cdot)$ is Lipschitz, and moreover

$$f(s, y) = \begin{cases} 0 & \text{if } |y| \geq 2R, \ s \in [t_1, t_0], \\[2mm] \frac{1}{t_0 - t_1} S(s - t_1)(\tilde{a} - y) - A\tilde{a} & \text{if } |y| \leq R, \ s \in [t_1, t_0], \end{cases}$$

where $R > 0$ and $\tilde{a} \in D(A)$ will also be chosen later. We remark that, as is easily checked, the following implication holds:

$$y \in H, \ |y| \leq R \Rightarrow S(t_0 - t_1)y + \int_{t_1}^{t_0} S(t_0 - s)f(s, y)ds = \tilde{a}. \quad (7.3.3)$$

We set moreover

$$Z(t_0) = I_1 + I_2,$$

where

$$I_1 = S(t_0 - t_1)Y(t_1) + \int_{t_1}^{t_0} S(t_0 - s)f(s, Y(t_1))ds,$$

and

$$I_2 = \int_{t_1}^{t_0} S(t_0 - s)B(Z(s))dW(s).$$

Now we choose $\tilde{a} \in D(A)$ such that

$$|a - \tilde{a}| \leq \frac{r}{3},$$

and claim that

$$\mathbb{P}(|Z(t_0) - a| < r) = \mathbb{P}(|(I_1 - \tilde{a}) + (I_2 + \tilde{a} - a)| < r)$$

$$\geq \mathbb{P}\left(I_1 = \tilde{a} \ \& \ |I_2| \leq \tfrac{r}{3}\right).$$

In fact if $I_1 = \tilde{a}$ and $|I_2| \leq \frac{r}{3}$ we have

$$|(I_1 - \tilde{a}) + (I_2 + \tilde{a} - a)| \leq |I_2| + |\tilde{a} - a| \leq \frac{2}{3}r.$$

Thus the claim is proved and we have

$$\mathbb{P}(|Z(t_0) - a| < r) \geq \mathbb{P}(I_1 = \tilde{a}) - \mathbb{P}\left(|I_2| \geq \frac{r}{3}\right). \quad (7.3.4)$$

We now choose $t_1 \in [0, t_0[$ such that

$$\mathbb{P}\left(|I_2| \geq \frac{r}{3}\right) \leq \frac{1}{4}, \qquad (7.3.5)$$

and $R > 0$ such that

$$\mathbb{P}(|Y(t_1)| \leq R) \geq \frac{3}{4}. \qquad (7.3.6)$$

By (7.3.3) it follows that if $|Y(t_1)| \leq R$ we have $I_1 = \tilde{a}$. Consequently, by (7.3.6)

$$\mathbb{P}(I_1 = \tilde{a}) \geq \mathbb{P}(|Y(t_1)| \leq R) \geq \frac{3}{4}. \qquad (7.3.7)$$

By (7.3.4), (7.3.5), and (7.3.7) it follows finally that

$$\mathbb{P}(|z(t_0) - a| < r) \geq \mathbb{P}(I_1 = \tilde{a}) - \mathbb{P}(I_2 \geq \frac{r}{3})$$

$$\geq \frac{3}{4} - \frac{1}{4} = \frac{1}{2}.$$

The proof is complete   ∎

Remark 7.3.3 Irreducibility for equation (7.1.1) for a much larger class of f including polynomial nonlinearities, has been obtained by D. Gątarek and B. Goldys [77] by using the concept of martingale solutions, see G. Da Prato and J. Zabczyk [44].

7.4   Irreducibility for equations with additive noise

Here we are concerned with nonlinear systems of the form

$$\left.\begin{array}{l} dX = (AX + F(X))dt + Q^{1/2}dW(t), \\[2mm] X(0) = x. \end{array}\right\} \qquad (7.4.1)$$

We assume that $U = H$, that $B = Q^{1/2}$ and that the deterministic system

$$\left.\begin{array}{l} y'(t) = Ay(t) + F(y(t)) + Q^{1/2}u(t), \\[2mm] y(0) = x, \end{array}\right\} \qquad (7.4.2)$$

is approximately controllable. We denote by $y(\cdot, x; u)$ the solution of (7.4.2). We recall that the system (7.4.2) is *approximately controllable* in time $T > 0$ if, for arbitrary $x, z \in H$ and $\varepsilon > 0$, there exists $u \in L^2(0, T; H)$ such that $|y(T, x; u) - \overset{\cdot\cdot}{z}| \le \varepsilon$. Note that Hypothesis 5.1(iii) implies continuity of the solution $Z(\cdot)$ of the linear equation

$$
\left.
\begin{aligned}
dZ &= AZ dt + Q^{1/2} dW(t), \\
Z(0) &= 0.
\end{aligned}
\right\}
\tag{7.4.3}
$$

Theorem 7.4.1 *Assume that the operator A generates a C_0-semigroup $S(t)$, $t \ge 0$, F is a locally Lipschitz mapping, and that the deterministic system (7.4.2) is approximately controllable in time $T > 0$. Then the transition semigroup P_t, $t \ge 0$, corresponding to problem (7.4.1) is irreducible in time T.*

Proof — We first remark that the law $\mathcal{L}(Z(\cdot))$ of the process Z is a Gaussian measure on $L^2(0, T; H)$ with a covariance operator $\mathcal{Q}$ given by

$$
\mathcal{Q}\varphi(t) = \int_0^T g(t, s)\varphi(s)ds, \ \varphi \in L^2(0, T: H);
$$

where

$$
g(t, s)x = \int_0^{t \wedge s} S(t - r)QS^*(s - r)sdr, \ x \in H.
$$

Moreover

$$
\text{Image } \mathcal{Q}^{1/2} = \Big\{ v \in L^2(0, T; H) : v(t) = \int_0^t S(t - s)Q^{1/2}u(s)ds,
$$

$$
t \in [0, T], \ u \in L^2(0, T; H) \Big\}.
$$

Since $\mathcal{L}(Z(\cdot))$ is concentrated on $C_0 = \{u \in C([0, T]; H) : u(0) = 0\}$, the closure of Image $\mathcal{Q}^{1/2}$ in the C_0-norm is the support of $\mathcal{L}(Z(\cdot))$, see G. Da Prato and J. Zabczyk [44, page 41].

Now fix x and z in H, and $\varepsilon > 0$. We want to prove that

$$
\mathbb{P}(|X(T, x) - z| < \varepsilon) > 0.
\tag{7.4.4}
$$

We first remark that, by the approximate controllability assumption, there exists $u \in L^2(0, T; H)$ such that

$$|y(T, x; u) - z| < \varepsilon.$$

Take R such that

$$|y(s, x; u)| < \frac{R}{2}, \quad \text{for all } s \in [0, T],$$

and let L_R be the Lipschitz constant of F restricted to the ball $\{z \in H : |z| \leq R\}$. Fix $\varepsilon \in \,]0, 1[$. Since the support of $\mathcal{L}(Z(\cdot))$ is the closure of Image $Q^{1/2}$, we have, with positive probability,

$$\sup_{s \in [0,T]} \left| Z(s) - \int_0^s S(s - \sigma) Q^{1/2} u(\sigma) d\sigma \right| \leq \varepsilon \frac{R}{2} e^{-L_R M T}, \quad (7.4.5)$$

where $M = \sup_{t \in [0,T]} \|S(t)\|$. We claim that if (7.4.5) holds, then we have

$$\sup_{s \in [0,T]} |X(s, x)| \leq R \qquad (7.4.6)$$

and

$$\sup_{s \in [0,T]} |X(s, x) - y(s, x; u)| \leq \varepsilon \frac{R}{2}. \qquad (7.4.7)$$

This will imply the result.

Assume by contradiction that (7.4.5) holds but

$$\sup_{s \in [0,T]} |X(s, x)| > R. \qquad (7.4.8)$$

Let

$$T_1 = \inf\{t \geq 0 : |X(t, x)| > R\}.$$

Then $T_1 < T$ and for all $t \in [0, T_1]$

$$|X(t, x)| \leq R, \ |y(t, x; u)| \leq \frac{R}{2}.$$

Note that for $t \in [0, T]$

$$X(t, x) - y(t, x; u) = \int_0^t S(t - s)(F(X(s, x) - F(y(s, x; u)))) ds$$

$$+ \ Z(t) - \int_0^t S(t - s) Q^{1/2} u(s) ds.$$

Consequently

$$|X(t,x) - y(t,x;u)| \leq L_R M \int_0^t |X(s,x) - y(s,x;u)| ds$$

$$+ \sup_{s \in [0,t]} \left| Z(s) - \int_0^s S(s-\sigma) Q^{1/2} u(\sigma) d\sigma \right|.$$

By Gronwall's lemma, for all $t \leq T_1$ we have

$$|X(t,x) - y(t,x;u)| \leq e^{L_R M T} \sup_{s \in [0,t]} \left| Z(s) - \int_0^s S(s-\sigma) Q^{1/2} u(\sigma) d\sigma \right|.$$

In particular, for all $t \leq T_1$

$$|X(t,x)| \leq \frac{R}{2} + e^{M L_R T} \frac{R}{2} e^{-M L_R T} \varepsilon < R.$$

On the other hand, $|X(T_1, x)| \geq R$, a contradiction. Thus $T_1 = T$. By similar calculations, for all $t \leq T$

$$|X(t,x) - y(t,x;u)| \leq \varepsilon \frac{R}{2}.$$

This finishes the proof. ■

Theorem 7.4.2 *Assume that the operator A generates a C_0-semigroup $S(t)$, $t \geq 0$, F is a locally Lipschitz mapping, and that $\text{Ker}Q = \{0\}$. Then the transition semigroup P_t, $t \geq 0$, corresponding to problem (7.4.1) is irreducible in time T.*

Proof — It is enough to show that the deterministic system (7.4.2) is approximately controllable in time $T > 0$. Let $\varphi \in C([0,T]; H)$ be a function such that $\varphi(0) = x$ and $\varphi(T) = z$ and let

$$\psi(t) = \varphi(t) - S(t)x - \int_0^t S(t-s)F(\varphi(s))ds, t \in [0,T].$$

Since Image $Q^{1/2}$ is dense in H there exists a sequence $\{u_n\} \subset C([0,T]; H)$ such that the sequence

$$\delta_n(t) = \psi(t) - \int_0^t S(t-s)Q^{1/2} u_n(s)ds, t \in [0,T],$$

tends to 0 uniformly on $[0, T]$. Let $y_n = y(\cdot, x; u_n)$, then

$$\varphi(t) - y_n(t) = \delta_n(t) + \int_0^t S(t - s)(F(y(s)) - F(y_n(s)))ds.$$

Therefore, for a constant L and all $n \in \mathbb{N}$

$$|z - y_n(T)| \leq |\delta_n(T)| + LM \int_0^T |y(s) - y_n(s)|ds,$$

it follows that $\lim_{n \to \infty} y_n(T) = z$. This proves approximate controllability. ∎

Remark 7.4.3 Theorem 7.4.2, under Lipschitz conditions, was proved by B. Maslowski [116, Prop. 2.11].

We finish this section with a result, see J. Zabczyk [164], on the interplay between irreducibility and uniqueness of invariant measure, showing difference between finite dimensional and infinite dimensional systems. The proposition below shows also that irreducibility alone is not sufficient for uniqueness of invariant measures.

Proposition 7.4.4 *Let R_t, $t \geq 0$, denote the transition semigroup corresponding to equation (7.2.3).*

(i) If $\dim H < \infty$ and R_t, $t \geq 0$, is irreducible at some $t > 0$ then there exists at most one invariant measure for R_t, $t \geq 0$.

(ii) There exist an infinite dimensional Hilbert space H and an irreducible, at all $t > 0$, semigroup R_t, $t \geq 0$, for which there exist at least two different invariant measures.

Proof — (i) For sufficiently large $t > 0$ (in fact for all $t > 0$), Image $Q_t = H$ and therefore also Image $Q_\infty = H$. Consequently by a control theoretic argument, see e. g. G. Da Prato and J. Zabczyk [44, Appendix D.1], existence of an invariant measure for the system implies $\lim_{t \to +\infty} \|S(t)\| = 0$. By Proposition 7.2.3 uniqueness follows.

(ii) Define $H = L^2(0, +\infty)$, $U = \mathbb{R}$, and

$$S(t)x(\theta) = e^t x(\theta + t), \ t \geq 0, \ \theta \geq 0,$$

$$Bu = bu, \ u \in \mathbb{R},$$

where $b(\theta) = e^{-\theta^2}$, $\theta \geq 0$. We first show that the semigroup R_t, $t \geq 0$, is irreducible. Assume in contradiction that there exists $t > 0$ such that

$$\overline{\text{Image } Q_t^{1/2}} = \overline{\text{Image } Q_t} \neq H.$$

Then there exists an element $x_0 \in H$ different of 0 such that

$$\langle Q_t^{1/2} z, x_0 \rangle = 0 \text{ for all } z \in H,$$

and consequently $Q_t^{1/2} x_0 = 0$. Since

$$|Q_t^{1/2} x_0|^2 = \int_0^t |\langle S(s)b, x_0 \rangle|^2 \, ds$$

and

$$\langle S(s)b, x_0 \rangle = \int_0^{+\infty} e^s e^{-(\theta+s)^2} x_0(\theta) d\theta,$$

we have, for almost all $s \in]0, t[$

$$\psi(s) = \int_0^{+\infty} e^{-2\theta s} \left(e^{-\theta^2} x_0(\theta) \right) \, d\theta = 0.$$

From elementary properties of the Laplace transform we have $x_0 = 0$. This proves irreducibility of R_t, $t \geq 0$, for all $t > 0$. ∎

Note that

$$\text{Tr } Q_\infty = \int_0^{+\infty} |S(t)b|^2 dt$$

$$= \int_0^{+\infty} \int_0^{+\infty} e^t e^{-2t\theta} e^{-t^2} e^{-\theta^2} dt \, d\theta < +\infty,$$

and therefore $\mathcal{N}(0, Q_\infty)$ is an invariant measure for R_t, $t \geq 0$. However, if $y(\theta) = e^{-\theta}$, $\theta \geq 0$, then $S(t)y = y$ for all $t \geq 0$ and by Proposition 7.2.3 the measure $\mathcal{N}(\delta_{\{y\}}, Q_\infty)$ is also invariant.

Chapter 8

Densities of invariant measures

This chapter is devoted to the existence and regularity of the densities of an invariant measure with repect to a properly chosen Gaussian reference measure. Conditions are given under which transition functions for linear and nonlinear equations are absolutely continuous with respect to the reference measure.

It is shown also that the density exists and satisfies an adjoint equation. Under additional conditions the density belongs to some Sobolev space. We examine also the so–called gradient systems for which an explicit formula for the density exists.

8.1   Introduction

We consider a stochastic equation on a separable Hilbert space H,

$$\left. \begin{array}{l} dX = (AX + F(X))dt + \sqrt{Q}dW, \\[2mm] X(0) = x \in H, \end{array} \right\} \tag{8.1.1}$$

having densities with respect to the invariant measure of the linear version

$$\left. \begin{array}{l} dZ = AZdt + \sqrt{Q}\, dW, \\[2mm] Z(0) = x \in H. \end{array} \right\} \tag{8.1.2}$$

147

We will assume

Hypothesis 8.1 *(i) A is the infinitesimal generator of a strongly continuous semigroup $S(t)$, $t \geq 0$, on H of negative type.*
(ii) F is a Lipschitz continuous and bounded mapping from H into H.
(iii) Q is a bounded symmetric nonnegative operator on H.
(iv) W is a cylindrical Wiener process on H.
(v) We have

$$\int_0^{+\infty} \text{Tr}\, [S(t)QS^*(t)]\, dt < +\infty.$$

Under Hypothesis 8.1 equations (8.1.1) and (8.1.2) have unique mild solutions $X(t,x)$, $t \geq 0$, and $Z(t,x)$, $t \geq 0$, respectively, in virtue of Theorem 5.3.1. We shall denote by $P_t, t > 0$, and by $R_t, t > 0$, the corresponding transition semigroups,

$$P_t\varphi(x) = \mathbb{E}[\varphi(X(t,x))],\ t > 0, x \in H,\ \varphi \in B_b(H),$$

$$R_t\varphi(x) = \mathbb{E}[\varphi(Z(t,x))],\ t > 0, x \in H,\ \varphi \in B_b(H).$$

By Theorem 6.2.1 there exists a unique invariant measure μ for the semigroup $R_t, t > 0$, being a Gaussian measure, $\mu = \mathcal{N}(0, Q_\infty)$, with mean vector 0 and covariance operator Q_∞:

$$Q_\infty x = \int_0^\infty S(s)QS^*(s)x ds,\ x \in H.$$

Since μ is an invariant measure for the semigroup $R_t, t > 0$, we can extend R_t, bya standard argument, to a contraction semigroup in $L^2(H,\mu)$, which we still denote by R_t, $t \geq 0$. We have in fact

$$|R_t\varphi(x)|^2 = (\mathbb{E}[\varphi(Z(t,x))])^2 \leq \mathbb{E}\left[\varphi^2(Z(t,x))\right] = R_t(\varphi^2)(x),$$

so that

$$\int_H |R_t\varphi(x)|^2 \mu(dx) \leq \int_H R_t(\varphi^2)(x)\mu(dx) = \int_H \varphi^2(x)\mu(dx).$$

In Theorem 8.4.4 we give conditions under which there exists an invariant measure μ_F for the solution of (8.1.1) *absolutely continuous* with respect to μ with density $\frac{d\mu_F}{d\mu}$,

$$\frac{d\mu_F}{d\mu} \in L^2(H,\mu).$$

Under additional sets of conditions, see Theorems 8.7.5, 8.8.1 and 8.8.2, we show that

$$\frac{d\mu_F}{d\mu} \in W_Q^{1,2}(H,\mu),$$

or even

$$\frac{d\mu_F}{d\mu} \in W_Q^{2,2}(H,\mu),$$

where $W_Q^{1,2}(H,\mu)$ and $W_Q^{2,2}(H,\mu)$ are appropriately defined Sobolev spaces of functions on H.

For generalizations to L^p and $W^{1,p}$ of the results on regularity of the densities, see A. Chojnowska–Michalik and B. Goldys [26].

Regularity of invariant measures for equations of the form

$$dX = \left(-\frac{1}{2}X + F(X)\right) dt + Q^{1/2}dW_t, \quad X(0) = x \in H, \quad (8.1.3)$$

on $H = L^2(0,1)$, was an object of paper by I. Shigekawa [139] and R. V. Vintschger [153]. These papers treates the case of the operator Q defined by the formula

$$Q\varphi(\xi) = \int_0^1 (\xi \wedge \eta)\varphi(\eta)d\eta, \quad \varphi \in H, \quad \xi \in [0,1], \quad (8.1.4)$$

and the F nonlinearity having values in Image $Q^{1/2}$. In particular in I. Shigekawa [139] the mapping was assumed to be bounded. We remark that the cases covered here are different and motivated by applications.

In §8.6 we investigate the so–called *gradient systems* for which explicit formulae for the density of an invariant measure can be derived.

8.2   Sobolev spaces

We start by recalling definitions and basic properties of some function spaces. We assume Hypothesis 8.1 and set $\mu = \mathcal{N}(0, Q_\infty)$. We denote by $\{e_n\}$ a complete orthonormal basis of eigenvectors of Q_∞ and by $\{q_n\}$ the sequence of its positive eigenvalues.

We are also given a self–adjoint operator R with the same set of eigenvectors $\{e_n\}$ and with positive eigenvalues $\{r_n\}$. Let $\mathcal{P}$ be the set of all functions φ on H of the form

$$\varphi(x) = p(\langle x, e_1 \rangle, \cdots, \langle x, e_n \rangle), \ x \in H,$$

where p is a polynomial in n variables, $n \in \mathbb{N}$.

The spaces $W_R^{1,2}(H, \mu)$ and $W_R^{2,2}(H, \mu)$ are defined as the closures of $\mathcal{P}$ with respect to the norms

$$\|\varphi\|^2_{W_R^{1,2}(H,\mu)} = \|\varphi\|^2_{L^2(H,\mu)} + \int_H |R^{1/2} D_x \varphi(x)|^2 \mu(dx)$$

and

$$\|\varphi\|^2_{W_R^{2,2}(H,\mu)} = \|\varphi\|^2_{W_R^{1,2}(H,\mu)} + \int_H \|R^{1/2} D_{xx} \varphi(x) R^{1/2}\|^2_{HS} \, \mu(dx),$$

respectively. Here $\| \cdot \|_{HS}$ denotes the Hilbert–Schmidt norm of an operator.

If $R = I$ we set $W_R^{1,2}(H, \mu) = W^{1,2}(H, \mu)$ and $W_R^{2,2}(H, \mu) = W^{2,2}(H \ \mu)$.

Let

$$H_n(\xi) = \frac{(-1)^n}{\sqrt{n!}} \, e^{\frac{\xi^2}{2}} \, \frac{d^n}{d\xi^n}(e^{-\frac{\xi^2}{2}}), n \in \mathbb{N}, \ \xi \in \mathbb{R},$$

be Hermite polynomials forming, as is well known, a complete orthonormal basis on $L^2(\mathbb{R}, \mathcal{N}(0,1))$.

Denote by Γ the set of all mappings

$$\gamma : \mathbb{N} \to \mathbb{N} \cup \{0\}, n \to \gamma_n,$$

such that

$$|\gamma| = \sum_{n=1}^{\infty} \gamma_n < +\infty.$$

For any $\gamma \in \Gamma$ we define

$$H_\gamma(x) = \prod_{n=1}^{\infty} H_{\gamma_n} \left(\frac{x_n}{\sqrt{q_n}} \right), \ x_n = \langle x, e_n \rangle. \tag{8.2.1}$$

Then it is well known that $\{H_\gamma : \gamma \in \Gamma\}$ is an orthonormal complete system on $L^2(H, \mu)$. For any $\varphi \in L^2(H, \mu)$ we set

$$\varphi_\gamma = \langle \varphi, H_\gamma \rangle_{L^2(H,\mu)}.$$

Proposition 8.2.1 *(i) For any $\varphi \in W_R^{1,2}(H, \mu)$*

$$\int_H |R^{1/2} D_x \varphi|^2 \mu(dx) = \sum_{\gamma \in \Gamma} \left\langle \frac{r}{q}, \gamma \right\rangle |\varphi_\gamma|^2. \qquad (8.2.2)$$

(ii) For any $\varphi \in W_R^{2,2}(H, \mu)$

$$\int_H \|R^{1/2} D_{xx} \varphi(x) R^{1/2}\|_{HS}^2 \, \mu(dx) = \sum_{\gamma \in \Gamma} \left[\left| \left\langle \frac{r}{q}, \gamma \right\rangle \right|^2 - \left\langle \frac{r^2}{q^2}, \gamma \right\rangle \right] |\varphi_\gamma|^2. \qquad (8.2.3)$$

In the above formulae we have used the notation

$$\left\langle \frac{r}{q}, \gamma \right\rangle = \sum_{n=1}^\infty \frac{r_n}{q_n} \gamma_n, \quad \left\langle \frac{r^2}{q^2}, \gamma \right\rangle = \sum_{n=1}^\infty \frac{r_n^2}{q_n^2} \gamma_n.$$

Proof — Recalling that $H_n'(x) = \sqrt{n} \, H_{n-1}(x)$, $n \in \mathbb{N}, x \in \mathbb{R}$, we have:

$$D_x \varphi(x) = \sum_{n=1}^\infty \left[\sum_{\gamma \in \Gamma, \gamma_n > 0} \varphi_\gamma \sqrt{\frac{\gamma_n}{q_n}} H_{\gamma_n - 1} \left(\frac{x_n}{\sqrt{q_n}} \right) \prod_{h \neq n} H_{\gamma_h} \left(\frac{x_h}{\sqrt{q_h}} \right) \right] e_n,$$

and so

$$\int_H |R^{1/2} D_x \varphi|^2 \mu(dx) = \sum_{n=1}^\infty \sum_{\gamma \in \Gamma, \gamma_n > 0} |\varphi_\gamma|^2 \frac{r_n \gamma_n}{q_n},$$

and (8.2.2) follows. To prove (8.2.3) we remark that

$$D_{x_n}^2 \varphi = \sum_{\gamma \in \Gamma, \gamma_n > 0, \gamma_{n-1} > 0} \varphi_\gamma \frac{\gamma_n}{q_n} \sqrt{\gamma_n(\gamma_n - 1)} H_{\gamma_n - 2} \left(\frac{x_n}{\sqrt{q_n}} \right)$$

$$\times \prod_{h \neq n} H_{\gamma_h - 1} \left(\frac{x_h}{\sqrt{q_h}} \right),$$

and, if $m \neq n$,

$$
D_{x_n} D_{x_m} = \sum_{\gamma \in \Gamma, \gamma_n > 0, \gamma_m > 0} \varphi_\gamma \sqrt{\frac{\gamma_n \gamma_m}{q_n q_m}} H_{\gamma_n - 1}\left(\frac{x_n}{\sqrt{q_n}}\right) H_{\gamma_m - 1}\left(\frac{x_m}{\sqrt{q_m}}\right)
$$

$$
\times \prod_{h \neq n, m} H_{\gamma_h - 1}\left(\frac{x_h}{\sqrt{q_h}}\right).
$$

It follows that

$$
\int_H r_n^2 \left|D_{x_n}^2 \varphi\right|^2 \mu(dx) = \sum_{\gamma \in \Gamma, \gamma_n > 0, \gamma_{n-1} > 0} |\varphi_\gamma|^2 \frac{r_n^2}{q_n^2} \gamma_n (\gamma_n - 1),
$$

and

$$
\int_H r_n r_m \left|D_{x_n} D_{x_m} \varphi\right|^2 \mu(dx) = \sum_{\gamma \in \Gamma, \gamma_n > 0, \gamma_m > 0} |\varphi_\gamma|^2 \frac{r_n r_m \gamma_n \gamma_m}{q_n q_m}
$$

which yields (8.2.3).   ■

The following proposition was proved in G. Da Prato, P. Malliavin and D. Nualart [40] and independently in S. Peszat [123]

Proposition 8.2.2 *For an arbitrary Gaussian measure μ on a Hilbert space H the embedding*

$$
W_R^{1,2}(H, \mu) \subset L^2(H, \mu)
$$

is compact if and only if

$$
\lim_{n \to \infty} \frac{r_n}{q_n} = +\infty. \tag{8.2.4}
$$

In particular $W^{1,2}(H, \mu)$ is compactly embedded in $L^2(H, \mu)$.

Proof — Note that (8.2.4) holds if and only if for any ordering $\gamma(k)$, $k \in \mathbb{N}$, of Γ, $\lim_{k \to \infty}\langle \frac{r}{q_0}, \gamma(k)\rangle = +\infty$. This easily implies the required result.   ■

Remark 8.2.3 Note that the embedding

$$W^{2,2}(H,\mu) \subset W^{1,2}(H,\mu),$$

is not compact. To see this define

$$\varphi_m(x) = \langle x, e_m \rangle, \quad m \in \mathbb{N}, x \in H.$$

Then

$$D_x\varphi_m(x) = e_m, \quad D_x^2\varphi_m(x) = 0.$$

Therefore

$$\|\varphi_m\|_{L^2(H,\mu)} = q_m^{1/2}$$

and

$$\|\varphi_m\|_{W^{1,2}(H,\mu)} = \|\varphi_m\|_{W^{2,2}(H,\mu)} = (1 + q_m)^{1/2}.$$

So the sequence $\{\varphi_m\}$ is bounded in $W^{2,2}(H,\mu)$, converges to 0 in $L^2(H,\mu)$ and does not contain any subs–equence convergent to 0 in $W^{1,2}(H,\mu)$.

8.3 Properties of the semigroup R_t, $t > 0$, on $L^2(H,\mu)$

For our study of the transition semigroups R_t, $t \geq 0$, and P_t, $t \geq 0$ we will need, besides Hypothesis 8.1, the following.

Hypothesis 8.2 *(i) Image $S(t) \subset$ Image $Q_t^{1/2}$, $t > 0$.*

(ii) Setting $\Gamma(t) = Q_t^{-1/2}S(t)$, the function $\|\Gamma(t)\|$, $t > 0$, is Laplace transformable, i.e. for sufficiently large λ, $\int_0^{+\infty} e^{-\lambda t}\|\Gamma(t)\|dt$ is finite.

Note that if Hypotheses 8.1 and 8.2 hold, then the assumptions of Theorem 7.2.4 are fulfilled, so that R_t, $t \geq 0$ is a strong Feller semigroup.

Remark 8.3.1 Note that if Q coincides with the identity operator I, then Hypothesis 8.2 is satisfied, see G. Da Prato and J. Zabczyk [44, Corollary 9.22]. On the other hand if $H = \mathbb{R}^d$, $d \in \mathbb{N}$, then Hypotheses 8.1 and 8.2 hold if and only if Q is invertible, see T. Seidman [137].

We introduce, following [26], the linear operator in $L^2(H,\mu)$:

$$
\left.
\begin{aligned}
\mathcal{A}_0\varphi &= \frac{1}{2}\operatorname{Tr}[Q\varphi_{xx}] + \langle x, A^*\varphi_x\rangle, \\[2mm]
D(\mathcal{A}_0) &= \Big\{\varphi(x) = f(\langle x, a_1\rangle, ..., \langle x, a_n\rangle) : f \in C_b^2(\mathbb{R}^n), \\[2mm]
& \qquad a_1, ..., a_n \in D(A^*),\ n \in \mathbb{N}\Big\}.
\end{aligned}
\right\}
$$

It is easy to see that $D(\mathcal{A}_0)$ is invariant with respect to the semigroup $R_t, t \geq 0$, and is dense in $L^2(H,\mu)$.

The following result is proved in G. Da Prato and J. Zabczyk [47].

Theorem 8.3.2 *Assume Hypothesis 8.1.*

(i) $\mathcal{A}_0$ is closable in $L^2(H,\mu)$ and its closure $\mathcal{A}$ is the infinitesimal generator of a C_0-semigroup $e^{t\mathcal{A}}, t \geq 0$, in $L^2(H,\mu)$ identical with $R_t,\ t \geq 0$:

$$
e^{t\mathcal{A}}\varphi(x) = \mathbb{E}\left[\varphi\left(S(t)x + \int_0^t S(t-s)dW(s)\right)\right] = R_t\varphi(x),
$$

$$
t \geq 0,\ \varphi \in L^2(H,\mu).
$$

$$(8.3.1)$$

(ii) If moreover Hypothesis 8.2(i) holds, then for all $t > 0$ and $\varphi \in L^2(H,\mu)$ we have $R_t\varphi \in W^{1,2}(H,\mu)$, and

$$
\|D_x R_t\varphi\|_{L^2(H,\mu)} \leq \|\Gamma(t)\|\ \|\varphi\|_{L^2(H,\mu)}, \qquad (8.3.2)
$$

and for any $t > 0$, R_t is a compact operator on $L^2(H,\mu)$.

(iii) If, in addition, Hypothesis 8.2(ii) holds, then

$$
D(\mathcal{A}) \subset W^{1,2}(H,\mu).
$$

Proof — (i) If $\varphi \in D(\mathcal{A}_0)$ then, for arbitrary $x \in H$

$$
R_t\varphi(x) = \varphi(x) + \int_0^t R_s(\mathcal{A}_0\varphi)(x)ds.
$$

Therefore $D(\mathcal{A}_0) \subset D(\mathcal{A})$ where $\mathcal{A}$ is the infinitesimal generator of the semigroup $R_t,\ t \geq 0$. From Theorem 1.9 in E. B. Davies [49], it follows that $D(\mathcal{A}_0)$ is dense in the graph norm of $D(\mathcal{A})$ and therefore the closure of $\mathcal{A}_0$ is $\mathcal{A}$, and (i) follows.

Now we prove (ii). Let $\varphi \in C_b(H)$. Since

$$R_t\varphi(x) = \int_H \varphi(y)\mathcal{N}(S(t)x, Q_t)(dy),$$

it follows by the Cameron–Martin formula that, for any $h \in H$,

$$\langle D_x R_t\varphi(x), h\rangle = \int_H \langle \Gamma(t)h, Q_t^{-1/2}y\rangle \varphi(y)\mathcal{N}(S(t)x, Q_t)(dy).$$

Consequently

$$|\langle D_x R_t\varphi(x), h\rangle| \leq \int_H |\langle \Gamma(t)h, Q_t^{-1/2}y\rangle|\, |\varphi(S(t)x + y)|\mathcal{N}(0, Q_t)(dy)$$

and by the Schwarz inequality

$$\begin{aligned}|\langle D_x R_t\varphi(x), h\rangle|^2 &\leq \int_H |\langle \Gamma(t)h, Q_t^{-1/2}y\rangle|^2\mathcal{N}(0, Q_t)(dy) \\ &\times \int_H |\varphi(S(t)x + y)|^2\mathcal{N}(0, Q_t)(dy) \\ &\leq \|\Gamma(t)\|^2|h|^2(R_t\varphi^2)(x).\end{aligned}$$

Thus, integrating both sides with respect to μ, we arrive at (8.3.2). The compactness of R_t follows from Proposition 8.2.2.

Finally as far as (iii) is concerned, it is a simple consequence of (8.3.2) and the formula

$$D_x(\lambda - \mathcal{A})^{-1}\varphi = \int_0^{+\infty} e^{-\lambda t}D_x e^{t\mathcal{A}}\varphi dt,$$

which yields

$$\|D_x(\lambda - \mathcal{A})^{-1}\varphi\|_{L^2(H,\mu)} \leq \int_0^{+\infty} e^{-\lambda t}\|\Gamma(t)\|\, dt\, \|\varphi\|_{L^2(H,\mu)}.$$

The proof is complete. ∎

8.4   Existence and absolute continuity of the invariant measure of P_t, $t > 0$, with respect to μ

Now we are going to show that the semigroup P_t can be extended to a C_0–semigroup on $L^2(H,\mu)$. We will need the following result which is a direct extension of Theorem 3.5 of E. B. Davies [49].

Proposition 8.4.1 *Let $\mathcal{A}$ be the infinitesimal generator of a C_0– semigroup on a Hilbert space $\mathcal{H}$, and let $\mathcal{C}$ be a closed operator on $\mathcal{H}$ such that*

$$D(\mathcal{C}) \supset \bigcup_{t>0} e^{t\mathcal{A}}(\mathcal{H}), \quad \int_0^1 \|\mathcal{C}e^{t\mathcal{A}}\|dt < +\infty.$$

Then $D(\mathcal{C}) \supset D(\mathcal{A})$ and for an arbitrary bounded operator $\mathcal{F}$ on $\mathcal{H}$ the operator

$$\mathcal{L} = \mathcal{A} + \mathcal{F}\mathcal{C}, \quad D(\mathcal{L}) = D(\mathcal{A}), \tag{8.4.1}$$

is the infinitesimal generator of a C_0–semigroup $e^{t\mathcal{L}}$, $t \geq 0$, on $\mathcal{H}$ and we have

$$e^{t\mathcal{L}}\varphi = \sum_{n=0}^{\infty} T_n(t)\varphi, \ t > 0, \ \varphi \in \mathcal{H}, \tag{8.4.2}$$

where

$$T_0(t)\varphi = e^{t\mathcal{A}}\varphi, \quad T_n(t)\varphi = \int_0^t e^{(t-s)\mathcal{A}}\mathcal{F}\mathcal{C}T_{n-1}(s)\varphi ds, \quad n \in \mathbb{N}. \tag{8.4.3}$$

Remark 8.4.2 Theorem 3.5 in E. B. Davies [49] covers the case of $\mathcal{F}$ being the identity operator, however, its proof is valid in the present more general situation.

Now we can prove the result

Theorem 8.4.3 *Assume that Hypotheses 8.1 and 8.2 hold. Define*

$$\mathcal{B}\varphi = \langle F(\cdot), D_x\varphi \rangle, \ \varphi \in W^{1,2}(H,\mu),$$

and

$$\mathcal{L} = \mathcal{A} + \mathcal{B}, \quad D(\mathcal{L}) = D(\mathcal{A}),$$

where $\mathcal{A}$ *is the infinitesimal generator of the semigroup* $R_t, t > 0$, *on* $L^2(H,\mu)$. *Then the operator* $\mathcal{L}$ *generates a* C_0–*semigroup* $e^{t\mathcal{L}}$, $t \geq 0$, *in* $L^2(H,\mu)$ *which is an extension of* P_t *to* $L^2(H,\mu)$. *Moreover*
(i) $D(\mathcal{L}) \subset W^{1,2}(H,\mu)$,
(ii) *Operators* $e^{t\mathcal{L}}$ *are compact for all* $t > 0$.

Proof — Note that the operator $\mathcal{C}$,

$$\mathcal{C}\varphi = D_x\varphi, \ D(\mathcal{C}) = W^{1,2}(H,\mu),$$

is closed on $L^2(H,\mu)$ and the operator $\mathcal{F}$,

$$\mathcal{F}\varphi = \langle F(x),\varphi\rangle, \ D(\mathcal{F}) = L^2(H,\mu),$$

is bounded. Moreover, by (8.3.2) we have

$$\|\mathcal{C}e^{t\mathcal{A}}\varphi\|_{L^2(H,\mu)} \leq \|\Gamma(t)\| \ \|\varphi\|_{L^2(H,\mu)}.$$

Therefore the assumptions of Proposition 8.4.1 are satisfied and the operator $\mathcal{L}$ generates a C_0–semigroup $e^{t\mathcal{L}}$, $t \geq 0$, in $L^2(H,\mu)$.

To show that $e^{t\mathcal{L}}$, $t \geq 0$, is the required extension, consider $\varphi \in D(\mathcal{A}_0)$ and define $u(t,x) = P_t\varphi(x)$, $t > 0$, $x \in H$. Assume, for a moment, that $F \in C_b^2(H;H)$. Then the function u is the unique strict solution of the associated Kolmogorov equation

$$\left.\begin{array}{l} u_t = \dfrac{1}{2}\operatorname{Tr}[Qu_{xx}] + \langle Ax + F(x), u_x\rangle, \\[2mm] u(0,x) = \varphi(x). \end{array}\right\} \qquad (8.4.4)$$

In particular the $L^2(H,\mu)$–valued function $u(t,\cdot), t \geq 0$, is a solution of the problem

$$u'(t) = \mathcal{L}u(t), \ t > 0, \ \ u(0) = \varphi,$$

and therefore $u(t,\cdot) = e^{t\mathcal{L}}\varphi$ for $t > 0$. If F is only a Lipschitz continuous function one can find a uniformly bounded sequence of mappings $\{F_n\} \subset C_b^2(H)$ such that $\lim_{n\to\infty} F_n(x) = F(x)$, $x \in H$. It easily follows from (8.4.2) and the definition of P_t that for the corresponding semigroups $e^{t\mathcal{L}_n}$, $t > 0$, P_t^n, $t > 0$, and $\varphi \in D(\mathcal{A}_0)$, we will have

$$e^{t\mathcal{L}}\varphi = \lim_{n\to\infty} e^{t\mathcal{L}_n}\varphi = \lim_{n\to\infty} P_t^n\varphi = P_t\varphi.$$

So $P_t\varphi = e^{t\mathcal{L}}\varphi, t > 0$ for $\varphi \in D(\mathcal{A}_0)$ and therefore for all $\varphi \in L^2(H,\mu)$.

Property (i) follows directly from the definition of $\mathcal{L}$. To show (ii) it is enough to remark that

$$\sum_{n=0}^{\infty} \|D_x T_n(t)\| < +\infty$$

for $t > 0$, where T_n is defined in (8.4.3), and to use (8.4.2). ∎

We are now in a position to prove the existence of an invariant measure μ_F for (8.1.1) and that it is absolutely continuous with respect to μ.

Theorem 8.4.4 *Assume that Hypotheses 8.1 and 8.2 hold. Then there exists $\psi \in D(\mathcal{L}^*), \psi \geq 0$, such that*

$$\mathcal{L}^*\psi = 0, \quad \int_H \psi(x)\,\mu(dx) = 1.$$

Moreover ψ is the density of an invariant measure μ_F for (8.1.1). Finally, if ImageQ is dense in H, the invariant measure is unique.

Proof — Since the operators $e^{t\mathcal{L}}, t > 0$, are compact on $L^2(H,\mu)$, the same is true for their adjoints $e^{t\mathcal{L}^*}, t > 0$. Since 1 belongs to the spectrum $\sigma(e^{t\mathcal{L}})$ of $e^{t\mathcal{L}}$, 1 is also an eigenvalue for $e^{t\mathcal{L}^*}$. Repeating the argument from the proof of Theorem 7.8 in E. B. Davies [49], one obtains that the set

$$G = \{\psi \in L^2(H,\mu) : e^{t\mathcal{L}^*}\psi = \psi \,\forall\, t \geq 0\}$$

$$= \{\psi \in L^2(H,\mu) : \mathcal{L}^*\psi = 0\},$$

is linear and of finite, positive dimension. It is easy to show that elements of G are densities of possibly signed invariant measures for $e^{t\mathcal{L}}, t \geq 0$. That there are nonnegative and nonzero elements in $\mathcal{G}$ follows from Lemma 3.4 in I. Shigekawa [139].

To prove uniqueness, it is enough to show that P_t, $t > 0$ on $B_b(H)$ is a strong Feller and irreducible semigroup; however this is so by Theorems 7.1.1 and 7.2.4. ∎

8.5 Locally Lipschitz nonlinearities

We assume here Hypothesis 8.1(i)(iii)(iv)(v) but, instead of 8.1(ii), we assume that $F : H \to H$ is locally Lipschitz continuous and that problem (8.1.1) has a unique mild solution $X(t,\cdot)$, $t \geq 0$. We want to extend Theorem 8.4.4 to this more general situation. For this we define the mappings F_n, $n \in \mathbb{N}$,

$$F_n(x) = \begin{cases} F(x), & |x| \leq n, \\ F\left(n\frac{x}{|x|}\right), & |x| \geq n, \end{cases} \tag{8.5.1}$$

which are Lipschitz continuous and bounded. Then we consider, besides equation (8.1.1), the following equation corresponding to F_n :

$$\left. \begin{array}{l} dX_n = (AX_n + F_n(X_n))dt + \sqrt{Q}dW, \\[2mm] X_n(0) = x \in H. \end{array} \right\} \tag{8.5.2}$$

Let $P_t(x,\cdot)$, $P_t^n(x,\cdot)$, $t > 0, n \in \mathbb{N}$, $x \in H$ BE the transition probabilities corresponding to solutions of (8.1.1) and (8.5.2) respectively.

We first prove a result on absolute continuity of transition probabilities.

Proposition 8.5.1 *Assume that Hypothesis 8.1(i)(iii)(iv)(v) hold, that F is locally Lipschitz continuous and that problem (8.1.1) has a unique mild solution $X(t,\cdot)$, $t \geq 0$. If for some $t > 0$ and all $n \in \mathbb{N}$ the measures $P_t^n(x,\cdot)$ are absolutely continuous with respect to a measure ν, then the measure $P_t(x,\cdot)$ is absolutely continuous with respect to ν.*

Proof — Let $\Gamma \subset H$ be a Borel set such that $\nu(\Gamma) = 0$. Then

$$\begin{aligned} P_t(x,\Gamma) &= \mathbb{P}(X(t,x) \in \Gamma) = \mathbb{P}(X(t,x) \in \Gamma, \ \tau_n^x > t) \\[2mm] &\quad + \mathbb{P}(X(t,x) \in \Gamma, \ \tau_n^x \leq t), \end{aligned}$$

where

$$\tau_n^x = \inf\{s \geq 0 : |X(s,x)| > n\}.$$

Since for all $s \leq \tau_n^x$, $X(s,x) = X_n(s,x)$, we have

$$P_t(x,\Gamma) \leq \mathbb{P}(X_n(t,x) \in \Gamma,\ \tau_n^x > t) + \mathbb{P}(\tau_n^x \leq t)$$

$$\leq P_t^n(x,\Gamma) + \mathbb{P}(\tau_n^x \leq t).$$

But $P_t^n(x,\Gamma) = 0$ and $\mathbb{P}\left(\lim_{n \to +\infty} \tau_n^x = +\infty\right) = 1$, and consequently $P_t(x,\Gamma) = 0$. ∎

Now we prove the main result of this section under the following assumptions.

Hypothesis 8.3 *(i) Hypotheses 8.1(i)(iii)(iv)(v) and 8.2 hold and Image Q is dense in H.*

(ii) F is locally Lipschitz continuous and problem (8.1.1) has a unique mild solution.

(iii) Problem (8.1.1) has a unique invariant measure μ_F.

Theorem 8.5.2 *Assume that Hypothesis 8.3 holds. Then the invariant measure μ_F to problem (8.1.1) is absolutely continuous with respect to the invariant measure μ to problem (8.1.2).*

Proof — From Theorem 8.4.4 for any $n \in \mathbb{N}$ problem (8.5.2) has a unique invariant measure $\mu_{F_n} \ll \mu$. Then the conclusion follows from Proposition 8.5.1. ∎

8.6   Gradient systems

Before investigating the problem of regularity of the density $\frac{d\mu_F}{d\mu}$ we devote this section to an important class of stochastic systems, called *gradient systems*, for which an explicit expression for the density exists.

Again let H (norm $|\cdot|$) be a separable Hilbert space and E (norm $\|\cdot\|$) a separable Banach space densely and continuously embedded into H.

Gradient processes are solutions X on E of the equation

$$\left.\begin{array}{l} dX(t) = (AX(t) + U'(X(t)))dt + dW(t),\ t \geq 0, \\ X(0) = x \in E, \end{array}\right\} \qquad (8.6.1)$$

under the following hypothesis.

Hypothesis 8.4 *(i) A is a self-adjoint negative definite operator on H such that A^{-1} is of trace class.*

(ii) For any $t \geq 0$, E is an invariant subspace of $S(t) = e^{tA}$. Moreover for any $x \in E$ the mapping $S(\cdot)x$ is continuous on $]0, +\infty[$, with the topology of E.

(iii) The Ornstein–Uhlenbeck process $W_A(t) = \int_0^t S(t-s)dW(s)$, W being the cylindrical Wiener process on H, has a version concentrated on E.

(iv) $U : E \to \mathbb{R}$ is continuous and there exists the directional derivative $DU(x; h)$ of U at any point $x \in E$ and at any direction $h \in E$. Moreover there exists a mapping $F : E \to H$ such that

$$DU(x; h) = \langle F(x), h \rangle.$$

(v) U is bounded from above, and $F : E \to H$ is a locally Lipschitz mapping: for arbitrary $r > 0$ there exists $k_r > 0$ such that

$$|F(x) - F(y)|_H \leq K_r \|x - y\|_E.$$

(vi) For arbitrary $x \in E$ there exists an E–continuous solution of the integral equation

$$X(t) = S(t) + \int_0^t S(t-s)F(X(s))ds + W_A(t)x,\ t \geq 0. \qquad (8.6.2)$$

Example 8.6.1 Define E as the space $C[0, \pi]$ of continuous functions on $[0, \pi]$ and $H = L^2(0, \pi)$. Let

$$Ax = \frac{d^2 x}{d\xi^2},\ \forall\, x \in H^2(0, \pi) \cap H_0^1(0, \pi).$$

Let U be of the form

$$U(x) = \int_0^\pi p(x(\xi))d\xi, \qquad (8.6.3)$$

where p is a polynomial of even order with negative leading coefficient. Then, recalling Theorem 5.5.11, it is not difficult to see that Hypothesis 8.4 holds.

Equations of the form (8.6.1) were studied in particular by R. Marcus [112], B. Maslowski [116], T. Funaki [75], P. L. Chow [27] and by J. Zabczyk [168]. Here we follow J. Zabczyk [168].

It turns out that, under Hypothesis 8.4, there exists an invariant measure μ_F for (8.6.1) on E that it is absolutely continuous with respect to the measure μ which in the present situation is of the form

$$\mu = \mathcal{N}(0, Q_\infty) = \mathcal{N}\left(0, -\frac{1}{2}A^{-1}\right).$$

Before proving this result let us introduce some notation.

Let $\{e_n\}$ and $\{\lambda_n\}$ be the sequences of all eigenvectors and all eigenvalues of A. We assume that

$$0 > -\lambda_1 \geq -\lambda_2 \geq \cdots.$$

We denote by H_n the closed subspace of H spanned by $\{e_1, ..., e_n\}$ and we set

$$\Pi_n(x) = \sum_{j=1}^n \langle x, e_j \rangle e_j, \; x \in H, \; n \in \mathbb{N}.$$

We assume moreover that the cylindrical process $W(t)$ is such that

$$\langle W(t), x \rangle = \sum_{j=1}^\infty \beta_j(t)\langle x, e_j \rangle, \; \forall \; x \in H,$$

where $\{\beta_j\}$ is a sequence of mutually independent, standard, real Wiener processes. We have the following lemma.

Lemma 8.6.2 *For arbitrary* $n \in \mathbb{N}$, $H_n \subset E$ *and we have*

$$\lim_{n \to +\infty} \|x - \Pi_n(x)\|_E = 0,$$

for μ*–almost all* $x \in E$.

Proof — Since the measure μ is concentrated on E, E contains the reproducing kernel of μ which is

$$\text{Image } Q_{\infty}^{1/2} = D((-A)^{1/2}).$$

Thus $D((-A)^{1/2}) \subset E$ and so $\{e_k\} \subset E$. Moreover the random variables $\xi_j \in L^2(\mu, E)$ given by

$$\xi_j(x) = \sqrt{2\lambda_j}\langle x, e_j \rangle, x \in E,$$

are Gaussian, normalized and independent and therefore the sequence

$$\sum_{j=1}^{n} \xi_j(x)\frac{e_j}{\sqrt{2\lambda_j}}, \ x \in E, \ n \in \mathbb{N},$$

converges μ–almost surely in E norm to a random variable ζ such that $\mathcal{L}(\zeta) = \mu$, see G. Da Prato and J. Zabczyk [44, §2.2.3]. Since E is continuously embedded in H, the limit has to be $\zeta(x) = x$. ∎

We can now prove the result

Theorem 8.6.3 *Under Hypothesis 8.4 the measure*

$$\mu_F(dx) = e^{2U(x)}\mu(dx), \ x \in E, \tag{8.6.4}$$

is invariant for (8.6.1) and the corresponding transition semigroup P_t, $t \geq 0$, is μ_F–symmetric

$$\int_E \varphi(x)P_t\psi(x)\mu_F(dx) = \int_E \psi(x)P_t\varphi(x)\mu_F(dx), \tag{8.6.5}$$

for all $\varphi, \psi \in B_b(E)$.

For some comments on symmetric semigroups see the end of §2.2.

Proof — It is enough to show the identity (8.6.5). Define

$$W_n(t) = \sum_{j=1}^{n}\beta_j(t)e_j,$$

and

$$Z_n(t) = \int_0^t S(t-s)dW_n(s) = \sum_{j=1}^{n}\int_0^t e^{-\lambda_j(t-s)}d\beta_j(s)e_j.$$

It is clear that the processes Z_n are E–continuous and satisfy the equation

$$dZ_n = AZ_n dt + dW_n, \quad Z_n(0) = 0.$$

It follows from Hypothesis 8.4(vi) that on an arbitrary finite interval $[0, T]$

$$\lim_{n \to +\infty} Z_n(\cdot) = Z(\cdot),$$

uniformly in E–norm. Let P_t^n, $t \geq 0$, be the transition semigroup corresponding to the solution $X_n(t, x)$ of the equation

$$X_n(t) = S(t) + \int_0^t S(t - s) \Pi_n F(\Pi_n X_n(s)) ds + Z_n(t) x, \; t \geq 0.$$

It is well known that the theorem is true in finite dimensions. Consequently, if μ^n are the projections of μ on $H_n = \Pi_n H$ and $U_n(x) = U(\Pi_n(x))$, $x \in H$, $n \in \mathbb{N}$, then setting

$$\mu_{F_n}^n(dx) = k_n e^{U_n x} \mu(dx), \; x \in E,$$

one has

$$\int_E \varphi(x) P_t^n \psi(x) e^{U(\Pi_n x)} \mu^n(dx) = \int_E \psi(x) P_t^n \varphi(x) e^{U(\Pi_n x)} \mu^n(dx). \tag{8.6.6}$$

But

$$\int_E \varphi(x) P_t^n \psi(x) e^{U(\Pi_n x)} \mu^n(dx) = \int_E \varphi(\Pi_n x) P_t^n \psi(\Pi_n x) e^{U(\Pi_n x)} \mu^n(dx). \tag{8.6.7}$$

Moreover one can check, see J. Zabczyk [168], that for any sequence $\{x_n\} \subset E$ convergent to x in E and for any function $\psi \in C_b(H)$ one has

$$\lim_{n \to +\infty} P_t^n \psi(x_n) = P_t \psi(x). \tag{8.6.8}$$

By (8.6.7) and (8.6.8) we can pass to the limit in (8.6.6) provided $\varphi, \psi \in C_b(H)$. Therefore

$$\int_E \varphi(x) P_t \psi(x) e^{U(x)} \mu(dx) = \int_E \psi(x) P_t \varphi(x) e^{U(x)} \mu(dx), \; \forall \; \varphi, \psi \in C_b(H). \tag{8.6.9}$$

By a monotone class argument (8.6.9) follows for arbitrary $\varphi, \psi \in B_b(H)$. ∎

8.7 Regularity of the density when $\mathcal{L}$ is variational

Theorem 8.4.4 implicitly states that the density $\frac{d\mu_F}{d\mu}$ is in the domain $D(\mathcal{L}^*)$. Therefore, one way of obtaining information about the regularity of $\frac{d\mu_F}{d\mu}$ is to characterize elements of $D(\mathcal{L}^*)$. In general this seems to be a difficult task. In the present section we restrict ourselves to systems such that the infinitesimal generator of the semigroup P_t, $t \geq 0$, is variational. We recall that $\mathcal{L}$ is said to be *variational* if there exists another Hilbert space $\mathcal{V}$ continuously and densely embedded in $\mathcal{H} =: L^2(H, \mu)$ and a continuous bilinear form

$$a : \mathcal{V} \times \mathcal{V} \to \mathbb{R}, \ (\varphi, \psi) \to a(\varphi, \psi),$$

such that

$$\langle \mathcal{L}\varphi, \psi \rangle_{L^2(H, \mu)} = -a(\varphi, \psi), \ \forall \ \varphi \in D(\mathcal{L}), \ \forall \ \psi \in \mathcal{V}.$$

If the operator $\mathcal{L}$ is variational then obviously

$$D(\mathcal{L}) \subset \mathcal{V},$$

and since $\mathcal{L}^*$ is also variational we also have

$$D(\mathcal{L}^*) \subset \mathcal{V}.$$

We assume here that

Hypothesis 8.5 *(i)* Image$Q_\infty \subset D(A)$.
(ii) $Q = I$.

Remark 8.7.1 If $S(\cdot)$ is analytic, and there exists $\omega \in \mathbb{R}$ such that $\|S(t)\| \leq e^{\omega t}$ for all $t \geq 0$, then Hypothesis 8.5 is fulfilled, see G. Da Prato [35, Proposition 2.5].

Remark 8.7.2 More general hypotheses were considered in V. I. Bogachev, M. Röckner and B. Schmuland [12], see also M. Fuhrman [73].

Following [12], we introduce the space $\mathcal{E}$ of all exponential functions:

$$\mathcal{E} = \{\varphi \in C_b(H): \ \varphi(x) = e^{i\langle h, x\rangle}, \ x \in H, \ h \in D(A^*)\}.$$

Clearly $\mathcal{E}$ is linearly dense in $L^2(\mu, H)$ and in $D(\mathcal{L})$. Moreover if $\varphi(x) = e^{i\langle h, x\rangle}$, $x \in H$, we have

$$\mathcal{L}\varphi(x) = -\frac{1}{2}|h|^2 + i\langle x, A^*h\rangle. \tag{8.7.1}$$

We now set $\mathcal{V} = W^{1,2}(\mu, H)$ and introduce the bilinear form in $\mathcal{V} \times \mathcal{V}$,

$$a(\varphi, \psi) = \int_H \langle D_x\varphi(x), AQ_\infty D_x\psi(x)\rangle \mu(dx), \ \varphi, \psi \in W^{1,2}(\mu, H). \tag{8.7.2}$$

Under Hypothesis 8.5, a is clearly well defined and continuous. Let us check, following M. Fuhrman [73], that it is coercive.

Proposition 8.7.3 *We have*

$$a(\varphi, \varphi) = -\frac{1}{2} \|\varphi\|_{W^{1,2}(\mu, H)}^2, \ \forall \ \varphi \in W^{1,2}(\mu, H). \tag{8.7.3}$$

Proof — We first remark that for any $x \in D(A^*)$ the Liapunov equation holds:

$$-|x|^2 = \langle Q_\infty x, A^*x\rangle + \langle Q_\infty A^*x, x\rangle.$$

By Hypothesis 8.5(i) this implies

$$AQ_\infty + (AQ_\infty)^* = -I. \tag{8.7.4}$$

We can now prove the statement. We have

$$
\begin{aligned}
a(\varphi, \varphi) &= \frac{1}{2}\int_H \langle D_x\varphi(x), AQ_\infty D_x\varphi(x)\rangle \mu(dx) \\[2mm]
&= \frac{1}{2}\int_H \langle AQ_\infty D_x\varphi(x), D_x\varphi(x)\rangle \mu(dx) \\[2mm]
&= \frac{1}{2}\int_H \langle D_x\varphi(x), (AQ_\infty + (AQ_\infty)^*)D_x\varphi(x)\rangle \mu(dx) \\[2mm]
&= -\frac{1}{2}\|\varphi\|_{W^{1,2}(\mu, H)}^2, \ \varphi \in W^{1,2}(\mu, H).
\end{aligned}
$$

■

We finally check that a is the Dirichlet form associated with the infinitesimal generator $\mathcal{L}$ of P_t, $t \geq 0$.

Proposition 8.7.4 *We have*

$$D(\mathcal{L}) = \{\varphi \in W^{1,2}(\mu, H) : a(\cdot, \varphi) \text{ is continuous in } L^2(\mu, H)\}, \tag{8.7.5}$$

and

$$a(\varphi, \psi) = \langle \mathcal{L}\varphi, \psi \rangle, \ \forall \ \varphi \in D(\mathcal{L}), \ \forall \ \psi \in W^{1,2}(\mu, H). \tag{8.7.6}$$

Proof — Since $\mathcal{E}$ is linearly dense in $D(\mathcal{L})$ it is enough to prove the identity in (8.7.6) for all $\varphi, \psi \in \mathcal{E}$. For this purpose we notice that

$$\int_H e^{i\langle h,x \rangle} \langle k, x \rangle \mu(dx) = i \langle Q_\infty^{1/2} h, Q_\infty^{1/2} k \rangle \, e^{-\frac{1}{2}|Q_\infty^{1/2} h|^2}, \ h, k \in H. \tag{8.7.7}$$

This follows by setting

$$F(\varepsilon) = \int_H e^{i\langle h,x \rangle + \varepsilon \langle k,x \rangle} \mu(dx) = e^{-\frac{1}{2}|Q_\infty^{1/2}(h - i\varepsilon k)|^2}, \ \varepsilon > 0,$$

and by computing $F'(0)$.

Let now $\varphi(x) = e^{i\langle \alpha,x \rangle}$, $\psi(x) = e^{i\langle \beta,x \rangle}$, $\alpha, \beta \in D(A^*)$. By (8.7.1) and (8.7.7), we have

$$\langle \mathcal{L}\varphi, \psi \rangle_{L^2(\mu,H)} = \frac{1}{2}|\alpha|^2 \int_H e^{i\langle \alpha+\beta,x \rangle} \mu(dx)$$

$$+ i \int_H \langle x, A^*\alpha \rangle e^{i\langle \alpha+\beta,x \rangle} \mu(dx)$$

$$= -\frac{1}{2}|\alpha|^2 + \langle Q_\infty A^*\alpha + AQ_\infty \alpha, \alpha + \beta \rangle e^{-\frac{1}{2}|Q_\infty^{1/2}(\alpha+\beta)|^2}$$

$$= -\frac{1}{2}\langle \alpha, \beta \rangle \, e^{-\frac{1}{2}|Q_\infty^{1/2}(\alpha+\beta)|^2} = a(\varphi, \psi).$$

■

We can now prove the result

Theorem 8.7.5 *Assume that Hypotheses 8.1, 8.2 and 8.5 hold. Then* $D(\mathcal{L}^*) \subset W_Q^{1,2}(H,\mu)$, *and there exists an invariant measure* μ_F *such that*

$$\frac{d\mu_F}{d\mu} \in W_Q^{1,2}(H,\mu).$$

Proof — It is enough to notice that $D(\mathcal{A}) = D(\mathcal{L})$ and that the adjoint $\mathcal{L}^*$ is the variational operator in $\mathcal{V}$ corresponding to the bilinear form

$$a^*(\varphi,\psi) = a(\psi,\varphi), \quad \varphi,\psi \in \mathcal{V}.$$

This implies $D(\mathcal{L}^*) \subset W^{1,2}(H,\mu)$. ∎

8.8 Further regularity results in the diagonal case

We consider here the case when the operators A and Q are symmetric and diagonal with respect to a given orthonormal basis. We assume here

Hypothesis 8.6 *(i) The operator A is negative definite with eigenvalues and eigenvectors equal respectively to* $-\alpha_n$, $e_n, n \in \mathbb{N}$.

(ii) There exists a sequence $\{q_n\}$ of positive numbers such that $Qe_n = q_n e_n, n \in \mathbb{N}$.

Let us remark that regularity of the transition densities was studied, in the present situation, by B. Gaveau and J. M. Moulinier [78] and I. Simao [141]. Their results are not, at least directly, applicable to invariant measures. Their methods are based on Girsanov's theorem, ours are more analytical.

Theorem 8.8.1 *Assume that Hypotheses 8.1, 8.2 and 8.6 hold and moreover that* $\text{Image}F(x) \subset \text{Image}Q^{1/2}$ *for all* $x \in H$, *and* $Q^{-1/2}F$ *is bounded. Then* $D(\mathcal{L}^*) \subset W_Q^{1,2}(H,\mu)$, *and there exists an invariant measure* μ_F *such that*

$$\frac{d\mu_F}{d\mu} \in W_Q^{1,2}(H,\mu).$$

Proof — Set $\mathcal{V} = W_Q^{1,2}(H,\mu)$ and let a be the bilinear form on $\mathcal{V} \times \mathcal{V}$,

$$a(\varphi,\psi) = \int_H [\langle Q^{1/2}D_x\varphi, Q^{1/2}D_x\psi\rangle \tag{8.8.1}$$
$$+ \langle Q^{-1/2}F(x), Q^{1/2}D_x\varphi\rangle\psi]\,\mu(dx).$$

Let, in addition, $\mathcal{A}$ be the infinitesimal generator of the semigroup $R_t, t \geq 0$, on $L^2(H;\mu)$. Recalling the identity $H_n''(x) - xH_n'(x) = -nH_n(x), n \in \mathbb{N}, x \in \mathbb{R}$, we have, for $\gamma \in \Gamma$, see also §8.1,

$$\mathcal{A}H_\gamma = \sum_{n=1}^\infty \left(\frac{q_n}{2}\frac{\partial^2 H_\gamma}{\partial x_n^2} - \alpha_n x_n \frac{\partial H_\gamma}{\partial x_n} \right)$$

$$= \sum_{n=1}^\infty \left(\alpha_n H_{\gamma_n}'' \left(\sqrt{\frac{2\alpha_n}{q_n}}x_n \right) - \alpha_n \sqrt{\frac{2\alpha_n}{q_n}} x_n H_{\gamma_n}' \left(\sqrt{\frac{2\alpha_n}{q_n}}x_n \right) \right)$$

$$\times \prod_{n \neq h} H_{\gamma_h} \left(\sqrt{\frac{2\alpha_h}{q_h}}x_h \right)$$

$$= -\sum_{n=1}^\infty \alpha_n \gamma_n H_\gamma(x).$$

Thus H_γ is an eigenvector for $\mathcal{A}$ and

$$\mathcal{A}H_\gamma = -\sum_{n=1}^\infty \alpha_n \gamma_n H_\gamma = -\langle \alpha, \gamma \rangle H_\gamma. \tag{8.8.2}$$

Consequently

$$\mathcal{A}\varphi = -\sum_{\gamma \in \Gamma} \langle \alpha, \gamma \rangle \varphi_\gamma H_\gamma,$$

$$D(\mathcal{A}) = \{\varphi \in L^2(H,\mu) : \sum_{\gamma \in \Gamma} |\langle \alpha, \gamma \rangle|^2 |\varphi_\gamma|^2 < +\infty\},$$

and

$$D((-\mathcal{A})^{1/2}) = \{\varphi \in L^2(H,\mu) : \sum_{\gamma \in \Gamma} |\langle \alpha, \gamma \rangle| |\varphi_\gamma|^2 < +\infty\}.$$

It easily follows from Proposition 8.2.1 that

$$W_Q^{1,2}(H,\mu) = D((-A)^{1/2}),$$

and

$$D(A) \subset W_Q^{2,2}(H,\mu).$$

Moreover

$$a(\varphi,\psi) = \langle A\varphi + \langle F(x), D_x\varphi \rangle, \psi \rangle_{L^2(\mu)}, \ \forall \ \varphi, \psi \in D(A).$$

The form a is obviously continuous on $V \times V$. It is also coercive since, for all $\varphi \in W_Q^{1,2}(H,\mu)$, we have

$$a(\varphi,\varphi) \ \geq \ |Q^{1/2}D_x\varphi|^2_{L^2(\mu)}$$

$$- \ \|Q^{-1/2}F\|_0 \int_H |Q^{1/2}D_x\varphi(x)| |\psi(x)| \mu(dx)$$

$$\geq \ |Q^{1/2}D_x\varphi|^2_{L^2(\mu)}$$

$$- \ \varepsilon \|Q^{-1/2}F\|_0 |Q^{1/2}D_x\varphi|^2_{L^2(\mu)} - \frac{1}{\varepsilon} \|Q^{-1/2}F\|_0 |\varphi|^2_{L^2(\mu)}.$$

It follows that $D(A) = D(\mathcal{L})$. Since $\mathcal{L}^*$ is the variational operator (with respect to $V = W_Q^{1,2}(H,\mu)$) corresponding to the bilinear form

$$a^*(\varphi,\psi) = a(\psi,\varphi), \ \varphi, \psi \in V,$$

we have $D(\mathcal{L}^*) \subset W_Q^{1,2}(H,\mu)$. The proof is complete.   ∎

We consider finally a situation where even better regularity of the density can be obtained. For this we give a complete description of $D(\mathcal{L}^*)$ under the following hypothesis.

Hypothesis 8.7 *There exists a constant $C > 0$ such that*

$$|\langle x, AF(x) \rangle| \leq C, \ \sum_{k=1}^{\infty} \left| \frac{\partial F_k}{\partial x_k} \right| \leq C, \ for \ \mu\text{--}a.e. \ x \in H. \qquad (8.8.3)$$

Theorem 8.8.2 *Assume Hypotheses 8.1, 8.2 and 8.7. Then* $D(\mathcal{L}^*) = D(\mathcal{L})$ *and there exists an invariant measure* μ_F *for equation (8.1.1) such that*

$$\frac{d\mu_F}{d\mu} \in W_Q^{2,2}(H,\mu).$$

It is sufficient to show that $D(\mathcal{L}^*) = D(\mathcal{L})$ and the proof of this identity is a consequence of the following two lemmas.

Lemma 8.8.3 *Assume that an infinitesimal generator* $\mathcal{L}$ *on a Hilbert space* $\mathcal{H}$ *is of the form*

$$\mathcal{L} = \mathcal{A} + \mathcal{B},$$

where $\mathcal{A}$ *is a negative definite operator,* $D(\mathcal{A}) = D(\mathcal{L})$ *and* $\mathcal{B}$ *is a closed operator with domain* $D(\mathcal{B}) \supset D(-(\mathcal{A})^{1/2})$. *If* $D(\mathcal{B}^*) \supset D((-\mathcal{A})^{1/2})$ *then*

$$\mathcal{L}^* = \mathcal{A} + \mathcal{B}^*, \qquad D(\mathcal{L}^*) = D(\mathcal{A}).$$

Proof — It follows from the closed graph theorem that

$$\mathcal{B}, \mathcal{B}^* \in L(D((-\mathcal{A})^{1/2}); \mathcal{H}).$$

Consequently the operator $\mathcal{B}R(\lambda, \mathcal{A})$ is well defined for all $\lambda \geq 0$ and for a constant $C > 0$

$$\|\mathcal{B}R(\lambda, \mathcal{A})\| \leq C\| - \mathcal{A}^{1/2}R(\lambda, \mathcal{A})\|$$

$$\leq \frac{C}{\sqrt{\lambda}}, \ \lambda > 0.$$

Moreover

$$R(\lambda, \mathcal{A} + \mathcal{B}) = R(\lambda, \mathcal{A})(I - \mathcal{B}R(\lambda, \mathcal{A}))^{-1}.$$

So if $\lambda > C^2$ then

$$R(\lambda, \mathcal{A} + \mathcal{B}) = \sum_{n=0}^{\infty} R(\lambda, \mathcal{A})(\mathcal{B}(R(\lambda, \mathcal{A} + \mathcal{B})))^n.$$

In a similar way, considering $\mathcal{A} + \mathcal{B}^*$ with the domain $D(\mathcal{A})$,

$$R(\lambda, \mathcal{A} + \mathcal{B}^*) = \sum_{n=0}^{\infty} R(\lambda, \mathcal{A})(\mathcal{B}^*(R(\lambda, \mathcal{A})))^n.$$

But

$$R(\lambda, (\mathcal{A} + \mathcal{B})^*) \ = \ R^*(\lambda, \mathcal{A} + \mathcal{B})$$

$$= \ \sum_{n=0}^{\infty} [(\mathcal{B}R(\lambda, \mathcal{A}))^*]^n R(\lambda, \mathcal{A}).$$

Therefore, to show that $\mathcal{A} + \mathcal{B}^* = (\mathcal{A} + \mathcal{B})^*$ it is enough to prove that for arbitrary $n \in \mathbb{N} \cup \{0\}$

$$R(\lambda, \mathcal{A})(\mathcal{B}^*(R(\lambda, \mathcal{A})))^n = [(\mathcal{B}R(\lambda, \mathcal{A}))^*]^n R(\lambda, \mathcal{A}). \qquad (8.8.4)$$

We check (8.8.4) for $n = 1$, the general case follows by induction. Let $x, y \in \mathcal{H}$. Then

$$\langle R(\lambda, \mathcal{A})\mathcal{B}^* R(\lambda, \mathcal{A})x, y \rangle = \langle \mathcal{B}^* R(\lambda, \mathcal{A})x, R(\lambda, \mathcal{A})y \rangle$$

$$= \langle R(\lambda, \mathcal{A}), \mathcal{B}R(\lambda, \mathcal{A}) \rangle$$

because $D(\mathcal{B}^{**}) = D(\mathcal{B}) \supset D(\mathcal{A})$. Since

$$\langle (\mathcal{B}R(\lambda, \mathcal{A}))^* R(\lambda, \mathcal{A})x, y \rangle = \langle R(\lambda, \mathcal{A})x, \mathcal{B}R(\lambda, \mathcal{A})y \rangle.$$

Identity (8.8.4) for $n = 1$ follows.   ∎

Lemma 8.8.4 *If $\mathcal{B}$ is the operator given by*

$$\mathcal{B}\varphi = \langle F, D_x \varphi \rangle, \ \varphi \in D(\mathcal{B}) = W^{1,2}(H; \mu),$$

and Hypothesis 8.7 holds then

$$D(\mathcal{B}^*) \supset W^{1,2}(H; \mu)$$

and

$$\mathcal{B}^* \psi = -\langle F(\cdot), D_x \psi \rangle + [2\langle Ax, F(\cdot) \rangle - \text{Div } F(\cdot)]\psi, \qquad (8.8.5)$$

for all $\psi \in W^{1,2}(H; \mu)$, where

$$\text{Div } F(x) = \sum_{k=1}^{\infty} \frac{\partial F}{\partial x_k}(x), \quad x \in H.$$

Proof — Take φ smooth and depending only on $x_1, ..., x_N$, then

$$\mathcal{B}\varphi = \sum_{k=1}^{N} F_k(\cdot)\frac{\partial\varphi}{\partial x_k}.$$

If $\psi \in W^{1,2}(H,\mu)$ we have,

$$\int_H \mathcal{B}\varphi(x)\psi(x)\mu(dx) = \sum_{k=1}^{N} \int_H F_k\frac{\partial\varphi}{\partial x_k}\psi\mu_1(dx_1)\cdots\mu_N(dx_N)$$

$$= \sum_{k=1}^{N}\sqrt{\frac{1}{2\pi q_k}} \int_H \varphi(x)\frac{\partial}{\partial x_k}\left[F_k\psi(x)e^{-\frac{x_k^2}{2q_k}}\right]\mu_1(dx_1)\cdots dx_k\cdots\mu_N(dx_N)$$

$$= -\sum_{k=1}^{N} \int_H \varphi(x)\left[\frac{\partial F_k}{\partial x_k}\psi + F_k\frac{\partial\psi}{\partial x_k} - \frac{x_k}{q_k}F_k\right]\mu(dx),$$

and the conclusion follows. ∎

Proof of Theorem 8.8.2. — By Lemma 8.8.3 the conditions of Lemma 8.8.4 are satisfied and therefore $D(\mathcal{L})=D(\mathcal{L}^*)$. ∎

Part III

Invariant measures for specific models

Chapter 9

Ornstein–Uhlenbeck processes

This chapter is devoted to the existence and uniqueness of invariant measures for linear equations under various specific conditions on the coefficients and on the Wiener process, motivated by applications. In particular the stochastic wave equation, a linear equation of mathematical finance, and Ornstein–Uhlenbeck processes in random environments are considered.

9.1   Introduction

The Ornstein–Uhlenbeck processes introduced in §5.2 and §6.2 are of great importance in applications. Let H be a separable Hilbert space and $(\Omega, \mathcal{F}, \mathbb{P})$ a probability space with a filtration $\mathcal{F}_t$, $t \geq 0$. *Ornstein–Uhlenbeck* (O–U) processes are solutions $X(\cdot, x) = X$ to the equation

$$dX = AX + BdW, \quad X(0) = x \in H, \qquad (9.1.1)$$

where W is a cylindrical Wiener process, usually on another separable Hilbert space U, and $B \in L(U; H)$. We denote by P_t, $t \geq 0$, the transition semigroup

$$P_t\varphi(x) = \mathbb{E}\left[\varphi(X(t, x))\right], \ t \geq 0, \cdot x \in H, \ \varphi \in B_b(H).$$

We first summarize results on invariant measures for O–U processes obtained so far in Theorem 6.2.1, Theorem 7.2.1 and Proposition 7.2.3.

Theorem 9.1.1 *Assume Hypothesis* 6.1.

(i) There exists an invariant measure for (9.1.1) if and only if the process X is bounded in probability and if and only if X is bounded in second moment.

(ii) The invariant measure for (9.1.1) is unique if and only if the only invariant measure for the deterministic equation

$$z' = Az$$

is $\delta_{\{0\}}$.

(iii) Assume, in addition, that for some $s > 0$ P_s is a strong Feller transition semigroup. Then an invariant measure for (9.1.1) exists if and only if there exists $M > 0$, $\omega > 0$ such that

$$\|S(t)\| \le M e^{-\omega t}, \ t \ge 0. \tag{9.1.2}$$

Moreover if μ is an invariant measure then it is unique and

$$\lim_{t \to +\infty} P_t \varphi(x) = \langle \varphi, \mu \rangle, \tag{9.1.3}$$

for every $x \in H$, $\varphi \in B_b(H)$.

Remark 9.1.2 By Theorem 7.2.1, P_s is a strong Feller semigroup if and only if Hypothesis 7.3 with $t = s$ holds.

9.2 Ornstein–Uhlenbeck processes of wave type

9.2.1 General properties

We start from a general result on stochastic linear systems

$$\left. \begin{array}{l} dX(t) = AX(t) + BdW(t), \\[2mm] X(0) = x \in H, \end{array} \right\} \tag{9.2.1}$$

with the operator A generating a C_0– group of operators $S(t)$, $t \in \mathbb{R}$. Linear stochastic wave equations provide a typical application of this study. As in the preceding section, $W(\cdot)$ stands for the cylindrical white noise on a Hilbert space U, and $B \in L(U; H)$.

Theorem 9.2.1 *Assume that the operator A generates a C_0– group of linear operators $S(t)$, $t \in \mathbb{R}$.*

(i) Equation (9.2.1) has an H–valued solution if and only if B is a Hilbert–Schmidt operator.

(ii) If B is a Hilbert–Schmidt operator and $\dim H = +\infty$, the transition semigroup corresponding to (9.2.1) is never a strong Feller semigroup.

Proof — (i) If there exists a solution to (9.2.1) then, for any $T > 0$, the stochastic integral

$$\int_0^T S(T - s)B dW(s), \ t > 0,$$

should be a well defined Gaussian random variable. In particular its second moment

$$\mathbb{E} \left| \int_0^T S(T - s)B dW(s) \right|^2 = \int_0^T \|S(r)B\|_{HS}^2 \, ds,$$

should be finite. From the group assumption it follows that there exist constants $c_1 > 0$ and $c_2 > 0$ such that

$$c_2|x| \leq |S(r)x| \leq c_1|x|, \ r \in [0, T], \ x \in H.$$

Let $\{e_n\}$ be an orthonormal and complete basis in U. Then

$$\|S(r)B\|_{HS}^2 = \sum_{n=1}^{\infty} |S(r)Be_n|^2,$$

and consequently

$$c_2^2 \sum_{n=1}^{\infty} |Be_n|^2 \leq \|S(r)B\|_{HS}^2 \leq c_1^2 \sum_{n=1}^{\infty} |Be_n|^2, r \in [0, T].$$

Since $\int_0^t \|S(r)B\|_{HS}^2 ds < +\infty$, part (i) follows.

(ii) It is enough to show that the control system

$$y' = Ay + Bu, \ y(0) = x, \qquad (9.2.2)$$

with B a Hilbert–Schmidt operator is not null controllable in any time $T > 0$. For this purpose we recall a well known result, that we prove for the reader's convenience.

Proposition 9.2.2 *Assume that U and H are separable Hilbert spaces and B is a linear compact operator from U into H. Then the operator*

$$\mathcal{L} : L^2(0,T;U) \to H, \ \mathcal{L}u = \int_0^T S(r)Bu(r)\,dr,$$

is also compact.

Proof — Since the operator B is a limit, in the operator norm of linear operators with finite dimensional range, one can assume that $U = \mathbb{R}$ and

$$Bu = bu, \ u \in \mathbb{R},$$

where $b \in H$. Then the adjoint operator $\mathcal{L}^*$ to $\mathcal{L}$ has the form

$$\mathcal{L}^*x(r) = \langle S(r)b, x \rangle, \ x \in H, \ r \in [0,T].$$

It is enough to show that $\mathcal{L}^*$ is compact. Let $\{x_n\}$ be any bounded sequence in H and let $\{x_{n_k}\}$ be a sub–sequence weakly convergent to $\hat{x}$. Then for arbitrary $r \in [0,T]$, $\langle S(r)b, x_{n_k} \rangle \to \langle S(r)b, x \rangle$, and therefore, by the Lebesgue dominated convergence theorem,

$$\mathcal{L}^*x_{n_k} \to \mathcal{L}^*\hat{x} \ \text{in} \ L^2(0,T;\mathbb{R}).$$

This finishes the proof of the proposition. ∎

To complete the proof of (ii) notice that the solution of (9.2.2) at moment $T > 0$ is zero if and only if

$$-S(T)x = \mathcal{L}u.$$

Since the range of $S(T)$ is the whole of H, the operator $\mathcal{L}$ would be surjective. This is impossible if dim $H = +\infty$ since $\mathcal{L}$ is compact. ∎

Assume now that B is a Hilbert–Schmidt operator. Then a necessary condition for existence of an invariant measure for (9.2.1) is that

$$\int_0^{+\infty} \|S(r)B\|_{HS}^2 \, ds < +\infty.$$

In particular

$$\liminf_{r \to +\infty} \|S(r)B\|_{HS} = 0.$$

This means that if $B \neq 0$, then the semigroup $S(r)$, $r \geq 0$, should have a *dissipation* property, at least for some $x \neq 0$:

$$\liminf_{r \to +\infty} |S(r)x| = 0. \tag{9.2.3}$$

Such systems are considered in the next sub–section.

9.2.2 Second order dissipative systems

Let H be a separable Hilbert space, Λ a positive definite symmetric linear operator on H, and $C : U \to H$ a linear bounded operator. We consider the stochastic equation

$$\left.\begin{array}{l} dX(t) = Y(t)dt, \\[2mm] dY(t) = (-\Lambda X(t) - \alpha Y(t))dt + C\,dW(t), \\[2mm] X(0) = x, \ Y(0) = y. \end{array}\right\} \tag{9.2.4}$$

The system (9.2.4) is of the form (9.2.1) where the Hilbert space H should be replaced by $\mathcal{H} = D(\Lambda^{1/2}) \times H$, with the norm

$$\left\| \begin{pmatrix} x \\ y \end{pmatrix} \right\|_{\mathcal{H}}^2 = |\Lambda^{1/2}x|^2 + |y|^2.$$

Moreover the infinitesimal generator A can be defined by

$$D(A) = D(\Lambda) \times D(\Lambda^{1/2}),$$

$$A \begin{pmatrix} x \\ y \end{pmatrix} = \begin{pmatrix} 0 & I \\ -\Lambda & -\alpha \end{pmatrix} \begin{pmatrix} x \\ y \end{pmatrix}, \quad \begin{pmatrix} x \\ y \end{pmatrix} \in D(A),$$

and B is given by

$$B \begin{pmatrix} x \\ y \end{pmatrix} = \begin{pmatrix} 0 \\ Cx \end{pmatrix}.$$

If $\alpha = 0$ one can write an explicit formula for the corresponding group of transformations

$$S^0(t) \begin{pmatrix} x \\ y \end{pmatrix} = \begin{pmatrix} \cos \sqrt{\Lambda}\, t & \Lambda^{-1/2} \sin \sqrt{\Lambda}\, t \\ -\Lambda^{1/2} \sin \sqrt{\Lambda}\, t & \cos \sqrt{\Lambda}\, t \end{pmatrix} \begin{pmatrix} x \\ y \end{pmatrix}, \ t \geq 0.$$

For the general α write

$$\begin{pmatrix} x(t) \\ y(t) \end{pmatrix} = S(t) \begin{pmatrix} x \\ y \end{pmatrix}, \quad \begin{pmatrix} x \\ y \end{pmatrix} \in D(A).$$

Then by direct calculations we get

$$\frac{1}{2} \frac{d}{dt} \left\| \begin{pmatrix} x(t) \\ y(t) \end{pmatrix} \right\|_{\mathcal{H}}^2 = -\alpha |y(t)|^2, \ t \geq 0. \tag{9.2.5}$$

So, if $\alpha \leq 0$, then

$$\left\| S(t) \begin{pmatrix} x \\ y \end{pmatrix} \right\|_{\mathcal{H}} \geq \left\| \begin{pmatrix} x \\ y \end{pmatrix} \right\|_{\mathcal{H}},$$

for arbitrary $\begin{pmatrix} x \\ y \end{pmatrix} \in \mathcal{H}$, and (9.2.3) cannot hold. Therefore if $C \neq 0$ and there exists an invariant measure for (9.2.4), then $\alpha > 0$. But then, see e.g. A. Pritchard and J. Zabczyk [126, Proposition 3.5],

$$\|S(t)\|_{\mathcal{H}} \leq M e^{-\omega t}, \ t \geq 0,$$

for some positive M and ω. Taking into account Theorem 9.1.1 we get the following result.

Theorem 9.2.3 *Assume that $C : U \to H$ is a non–trivial Hilbert–Schmidt operator. Then there exists an invariant measure for (9.2.4) if and only if $\alpha > 0$. If $\alpha > 0$ then the invariant measure is unique and Gaussian.*

Example 9.2.4 (Damped wave equation) Formally this can be written in the form

$$
\left.
\begin{aligned}
\frac{\partial^2}{\partial t^2} X(t,\xi) &= \Delta X(t,x) - \alpha \frac{\partial}{\partial t} X(t,\xi) + \frac{\partial}{\partial t} W_Q(t,\xi), \\[2mm]
X(0,\xi) &= x(\xi), \ \frac{\partial}{\partial t} X(0,\xi) = y(\xi), \ \xi \in \mathcal{O}, \ t \ge 0, \\[2mm]
X(t,\xi) &= 0, \ \xi \in \mathcal{O}, \ t > 0,
\end{aligned}
\right\}
$$

where $\mathcal{O}$ is a bounded closed subset in $\mathbb{R}^d$, and W_Q is the Wiener process with covariance $Q = BB^*$. To reduce this equation to the form (9.2.4), we set $H = U = L^2(\mathcal{O})$ and define

$$
\Lambda Y = -\Delta y, \ y \in D(\Lambda) = H^2(\mathcal{O}) \cap H^1_0(\mathcal{O}).
$$

Then $D(\Lambda^{1/2}) = H^1_0(\mathcal{O})$ and the norm $\| \cdot \|_{\mathcal{H}}$ is merely the *energy* norm of the classical theory.

The operator Q is usually given in the integral form

$$
Qx(\xi) = \int_{\mathcal{O}} q(\xi, \eta) x(\eta) d\eta, \ \xi \in H, \ \xi \in \mathcal{O},
$$

with Q being the spatial covariance of the noise process.

9.2.3 Comments on nonlinear equations

Of great interest are nonlinear equations

$$
\left.
\begin{aligned}
dX(t) &= (AX(t) + F(X(t)))dt + BdW(t), \\[2mm]
X(0) &= 0,
\end{aligned}
\right\} \tag{9.2.6}
$$

with A generating a C_0–group $S(t)$, $t \in \mathbb{R}$, on a Hilbert space H.

If F is Lipschitz continuous there exists a unique solution $X(\cdot, x)$ of (9.2.6) by Theorem 5.3.1.

Since the operators $S(t)$ are not compact one cannot use Theorem 6.1.2 for establishing existence of invariant measures. By Theorem 9.2.1 we cannot expect that the solution to (9.2.6) will be a strong Feller one and uniqueness, if it occurs, should be investigated with different methods from those developed in Chapter 4. In some situations however dissipativity arguments from §6.3 can be adapted and used. Note for instance that by formula (9.2.5) the semigroup $S(t)$, $t \geq 0$, is dissipative. However, it can happen that an estimate of some moment of $X(\cdot, x)$ in a norm $\| \cdot \|_Y$ with Y compactly embedded in H is available. In this case the existence of an invariant measure follows from a tightness argument. See e.g. H. Crauel, A Debussche and F. Flandoli [30] where the existence of an invariant measure, and also of an attractor, for the wave equation is proved.

9.3   Ornstein–Uhlenbeck processes in finance

A slight modification of Theorem 9.1.1 is needed to discuss ergodic properties of a process describing the evolution of a term structure of interest rates as proposed by M. Musiela [121], see also A. Brace and M. Musiela [14].

Let $X(t, \xi)$, $t \geq 0$, $\xi \geq 0$, be the rate at which one could enter a contract at date $t > 0$ to borrow or to lend for a period of time at date $t + \xi$. The risk related to borrowing and lending is represented by a real, nonnegative volatility function $r(\xi)$, $\xi \geq 0$. The process X satisfies, see [121], the following equation:

$$
\left.
\begin{aligned}
dX(t, \xi) &= \left(\frac{\partial}{\partial \xi} X(t, \xi) + r(\xi) \int_0^\xi r(\eta) d\eta \right) dt + r(\xi) dW(t), \\[2ex]
X(0, \xi) &= x(\xi), \ \xi \geq 0, \ t \geq 0,
\end{aligned}
\right\}
$$

$$(9.3.1)$$

where $W(t)$, $t \geq 0$, is a one dimensional Wiener process. To cover a large class of functions, we choose as the state space H a weighted Hilbert space $L_\gamma^2 = L_\gamma^2(0, +\infty)$, $\gamma \geq 0$, with weight $e^{-\gamma \xi}$, $\xi \geq 0$. Thus $x \in L_\gamma^2(0, +\infty)$ if and only if

$$
\int_0^{+\infty} |x(\xi)|^2 e^{-\gamma \xi} d\xi = |x|_\gamma^2 < +\infty. \tag{9.3.2}
$$

Let A denote the infinitesimal generator of the (left-shift) semigroup

$$
S(t)x(\xi) = x(t + \xi), \ t \geq 0, \ \xi \geq 0, \ x \in L_\gamma^2.
$$

Then
$$\|S(t)\|_{\mathcal{L}(L_\gamma^2)} \leq e^{\frac{\gamma}{2}t}, \ t \geq 0.$$

Let moreover B be the one dimensional operator from $U = \mathbb{R}$ in L_γ^2 given by
$$Bu(\xi) = ur(\xi), \ u \in \mathbb{R}, \ \xi \geq 0.$$

Then equation (9.3.1) can be written in an abstract way:

$$\left. \begin{array}{l} dX = (AX + a)dt + BdW(t), \\[2mm] X(0) = x \in H, \end{array} \right\} \qquad (9.3.3)$$

where $a(\xi) = r(\xi)$.

One of the main concerns of the paper [121] was existence and uniqueness of an invariant measure for (9.3.1). The following result is a direct extension of Theorem 6.2.1.

Theorem 9.3.1 *(i) There exists an invariant measure for (9.3.3) if and only if*

(i1) $\sup_{t \geq 0} \mathrm{Tr} Q_t < +\infty.$

(i2) There exists an invariant measure for the deterministic system

$$z' = Az + a. \qquad (9.3.4)$$

(ii) If conditions (i1),(i2) are satisfied then any invariant measure for (9.3.3) is of the form

$$\mu = \nu * \mathcal{N}(0, Q_\infty)$$

where $Q_\infty = \int_0^{+\infty} S(r)QS^*(r)ds$, and ν is invariant for (9.3.4).

(iii) If (i1) holds and $a \in$ Image A then any invariant measure for (9.3.3) is of the form

$$\mu = \nu * \delta_{\{b\}} * \mathcal{N}(0, Q_\infty)$$

where ν is invariant for $z' = Az$ and $b \in D(A)$ is a vector such that $a = Ab$.

Proof — The first two parts can be proved in a similar way as Theorem 6.2.1 and therefore the proof will be omitted. To show (iii) notice that a function z is a solution to $z' = Az + a$ if and only if $y(t) = z(t) + b$, $t \geq 0$, is a solution to $y' = Ay$ and therefore the formula for the invariant measure for (9.3.4) follows.   ∎

To apply the theorem to equation (9.3.1) notice that the condition (i) is equivalent to

$$\int_0^{+\infty} |S(t)r|_\gamma^2 dt < +\infty. \qquad (9.3.5)$$

Moreover the condition that $a \in$ Image A means that $a \in L_\gamma^2$ and $b(\xi) = \int_0^\xi a(\eta)d\eta$, $\xi \geq 0$, belongs to L_γ^2 as well. Starting from these observations T. Vargiolu [152], proved the following result.

Proposition 9.3.2 *(i) Assume that $\gamma = 0$ and τ is a bounded function. Then a necessary and sufficient condition for the existence of an invariant measure for (9.3.3) is*

$$\int_0^{+\infty} u|\tau(u)|du < +\infty.$$

If this condition is satisfied then there exists exactly one invariant measure for (9.3.3).
    (ii) If $\gamma > 0$ then a necessary and sufficient condition for the existence of an invariant measure for (9.3.3) is

$$\int_0^{+\infty} \tau^2(u)du < +\infty.$$

If this condition is satisfied then there exists an uncountable number of invariant measures for (9.3.3).

9.4 Ornstein–Uhlenbeck processes in chaotic environment

We assume here that $H \supset U = L^2(\mathbb{R}^d)$ and that A is the operator

$$A = \Delta - mI,$$

where Δ is the Laplacian, and $m \geq 0$. In particular, if $H = U$ the corresponding heat semigroup is

$$S(t)x(\xi) = e^{-mt} \int_{\mathbb{R}^d} p_t(\xi - \eta)x(\eta)d\eta,$$

$$= e^{-mt}p_t * x(\xi), \quad x \in U, \; \xi \in \mathbb{R}^d,$$

(9.4.1)

where

$$p_t(\xi) = (4\pi t)^{-d/2}e^{-\frac{|\xi|^2}{4t}}, \; t \geq 0, \; \xi \in \mathbb{R}^d.$$

We will consider the equation

$$dZ = (\Delta - m)Z dt + J dW, \quad Z(0) = \xi, \qquad (9.4.2)$$

in two important cases: when $Q = I$ and when Q is a convolution operator,

$$Qu = q * u, \quad u \in U,$$

see Example 5.2.1. Moreover J is the embedding operator of U into H.

9.4.1 Cylindrical noise

If $Q = I$ then equation (9.4.2) has no solution in $H = L^2(\mathbb{R}^d)$ because

$$\mathrm{Tr}\, Q_T = \int_0^T e^{-2mt}\, \mathrm{Tr}\, S(2t)dt = +\infty,$$

for arbitrary $T > 0$. It is therefore of interest to look for solutions of (9.4.2) in larger spaces. As a natural candidate we consider weighted spaces $H = L_\rho^2(\mathbb{R}^d)$, where the weight ρ is a continuous, positive and integrable function. However, other choices are possible and have been considered by several authors. For instance the space $S'(\mathbb{R}^d)$ of tempered distributions is taken as the state space for a generalized Ornstein–Uhlenbeck process by R. A. Holley and D. Stroock [88]. This space is natural in some applications to branching diffusions and to statistical mechanics [89], see also D. A. Dawson [50], [51], D. A. Dawson and H. Salehi [53], D. A. Dawson and G. C. Papanicolau [52].

Proposition 9.4.1 *Assume that ρ is a positive, continuous and integrable function on $\mathbb{R}^d$ such that, for some $\omega \geq 0$,*

$$p_t * \rho \leq e^{\omega t}\rho, \ \forall \, t \geq 0. \tag{9.4.3}$$

Then, for arbitrary $p \geq 1$ and $m \in \mathbb{R}$, the formula

$$S(t)x = e^{-mt}(p_t * x), \ t \geq 0,$$

defines a C_0-semigroup on $L_\rho^p(\mathbb{R}^d)$ spaces, and

$$\|S(t)\|_{L(L_\rho^p(\mathbb{R}^d))} \leq e^{(\frac{\omega}{p}-m)t}, \ t \geq 0. \tag{9.4.4}$$

Proof — We can assume $m = 0$. Consider first $p > 1$. Setting $q = \frac{p}{p-1}$ and taking $x \in L_\rho^p(\mathbb{R}^d)$ we have

$$\left| \int_{\mathbb{R}^d} p_t(\xi - \eta)x(\eta)d\eta \right|^p$$

$$\leq \left(\int_{\mathbb{R}^d} p_t^{1/p}(\xi - \eta)|x(\eta)|p_t^{1/q}(\xi - \eta)d\eta \right)^p$$

$$\leq \left(\int_{\mathbb{R}^d} p_t(\xi - \eta)|x(\eta)|^p d\eta \right) \left(\int_{\mathbb{R}^d} p_t(\xi - \eta)d\eta \right)^{p/q}.$$

Consequently

$$\begin{aligned}
|S(t)x|_{L_\rho^p(\mathbb{R}^d)}^p &\leq \int_{\mathbb{R}^d} \int_{\mathbb{R}^d} p_t(\xi - \eta)|x(\eta)|^p \rho(\xi) \, d\eta \, d\xi \\
&\leq \int_{\mathbb{R}^d} |x(\eta)|^p \left(\int_{\mathbb{R}^d} p_t(\xi - \eta)\rho(\xi) \, d\xi \right) d\eta \\
&\leq e^{\omega t} \int_{\mathbb{R}^d} |x(\eta)|^p \rho(\eta) d\eta \\
&= e^{\omega t} |x|_{L_\rho^p(\mathbb{R}^d)}^p.
\end{aligned}$$

Also we see that

$$\|S(t)\|_{L(L_\rho^p(\mathbb{R}^d))} \leq e^{\omega t/p}, \ t \geq 0,$$

and therefore the estimate (9.4.4) holds. Continuity of $S(t)x$, $t \geq 0$, is straightforward for functions x which are continuous and with have compact support. The case of general x follows by a standard density argument. Finally, the case $p = 1$ can be handled in the same way.
∎

Definition 9.4.2 *A positive, continuous, integrable weight ρ is said to be* admissible *for the operator $\Delta - mI$ if and only if the operators $S(t)$, $t \geq 0$, form a C_0-semigroup in $L^2_\rho(\mathbb{R}^d)$.*

Remark 9.4.3 Nonnegative, continuous functions ρ satisfying (9.4.3) are called ω-*excessive* in probabilistic potential theory, R. M. Blumenthal and R. K. Getoor [10]. They are solutions of the inequality

$$\Delta \rho \leq \omega \rho, \qquad (9.4.5)$$

understood in the sense of distributions, see L. Schwartz [135].

We will need the following lemma.

Lemma 9.4.4 *Assume that $\rho(x) = \psi(|x|)$, $x \in \mathbb{R}^d$, where ψ is a C^1 function on $[0, +\infty[$, with ψ'' continuous and integrable on any bounded closed subinterval of $]0, +\infty[$. Then the estimate (9.4.3) holds if and only if*

$$\psi''(y) + \frac{d-1}{r}\psi'(y) \leq \omega\psi(y), \; y > 0,$$

and in addition if $d = 1$ then $\psi'(0) \leq 0$.

Proof — If $d = 1$ then, see L. Schwartz [135],

$$\Delta\rho(x) = 2\psi'(0)\delta_{\{0\}} + \psi''(|x|), \; x \in \mathbb{R}$$

and if $d > 1$ then

$$\Delta\rho(x) = \psi''(|x|) + \frac{d-1}{|x|}\psi'(|x|), \; x \in \mathbb{R}^d,$$

and the result follows.   ∎

We will consider two specific families of weights

$$\rho^\kappa(x) = e^{-\kappa|x|}, \ x \in \mathbb{R}^d,$$

where $\kappa > 0$ and

$$\rho_{\kappa,r}(x) = \frac{1}{1 + \kappa|x|^r}, \ x \in \mathbb{R}^d,$$

where $\kappa > 0$ and $r > d$. When the number r is fixed we will write for short ρ_κ.

We write $H^\kappa = L^2_{\rho^\kappa}(\mathbb{R}^d)$ and $H_r = L^2_{\rho_{\kappa,r}}(\mathbb{R}^d)$. The latter notation is justified since for fixed $r > d$ the weights $\rho_{\kappa,r}$, for all $\kappa > 0$, give rise to isomorphic spaces. If it is not stated otherwise H_r is equipped with the norm of $L^2_{\rho_{\kappa,r}}(\mathbb{R}^d)$.

For the weights introduced we have the following result.

Proposition 9.4.5 *For the heat semigroup $S(t), t \geq 0$, given by (9.4.1) we have the following estimates (i) If $\rho = \rho^\kappa$ and $p \geq 1$, then*

$$\|S(t)\|_{L^p_\rho(\mathbb{R}^d)} \leq e^{(\frac{\kappa^2}{p} - m)t}, \ t \geq 0.$$

(ii) If $\rho = \rho_{\kappa,r}$ and $p \geq 1$, then

$$\|S(t)\|_{L^p_\rho(\mathbb{R}^d)} \leq e^{(\frac{\kappa^{\frac{2}{r}}\delta}{p} - m)t}, \ t \geq 0,$$

where

$$\delta = \delta(r,d) = \frac{r - d + 2}{r}(r-1)^{2(1-\frac{1}{r})}.$$

Proof — The proof follows from the previous lemma. We give only basic steps of the calculations.

(i) Here $\psi(y) = e^{-\kappa y}, \ y \geq 0$. Consequently $\psi'(0) = -\kappa \leq 0$. Moreover

$$\psi''(y) + \frac{d-1}{y}\psi'(y) = \psi(y)\left[\kappa^2 - \frac{d-1}{y}\kappa\right]$$

and

$$\psi(y)\left(\kappa^2 - \frac{d-1}{y}\kappa\right) \leq \omega\psi(y), \ \forall \ y \geq 0,$$

if and only if $\kappa^2 \leq \omega$.

(ii) In this case

$$\psi(y) = \frac{1}{1 + \kappa y^r}, \ y \geq 0,$$

$$\psi'(y) = -\frac{\kappa r y^{r-1}}{(1 + \kappa y^r)^2}, \ y \geq 0,$$

$$\psi''(y) = 2\frac{\kappa^2 r^2 y^{2r-2}}{(1 + \kappa y^r)^3} - \frac{\kappa r (r-1)y^{r-2}}{(1 + \kappa y^r)^2}, \ y \geq 0.$$

Note that $\psi'(0) = 0$ and for all $y > 0$

$$\omega\psi(y) - \psi''(y) - \frac{d-1}{y}\psi'(y)$$

$$= \frac{1}{(1 + \kappa y^r)^3}\Big[\omega(1 + \kappa y^r)^2 - 2\kappa^2 r^2 y^{2r-2})$$

$$+ (1 + \kappa y^r)y^{r-2}\kappa r(r + d - 2)\Big].$$

Therefore

$$\omega\psi(y) - \psi''(y) - \frac{d-1}{y}\psi'(y) > 0$$

for all $y > 0$ if, for instance,

$$\omega(1 + \kappa y^r)^2 \geq 2\kappa^2 r^2 y^{2r-2} - \kappa^2 r(r + d - 2)y^{2r-2}, \ y > 0,$$

or equivalently, if

$$\omega \geq \kappa^2 r(r + d - 2)\frac{y^{2r-2}}{(1 + \kappa y^r)^2}, y \geq 0. \tag{9.4.6}$$

The right hand side of (9.4.6) attains its maximum for y such that $y^r = \frac{r-1}{\kappa}$. So (9.4.6) holds for all $y > 0$ if and only if

$$\omega \geq \kappa^{\frac{2}{r}} \frac{r - d + 2}{r}(r - 1)^{2(1-\frac{1}{r})}$$

as required.   ∎

Proposition 9.4.6 *(i) Assume that ρ is an admissible weight for $\Delta - mI$. Then the solution to equation (9.4.2) exists in $L^2_\rho(\mathbb{R}^d)$ if and only if $d = 1$.*

(ii) If $d = 1$, then there exists an invariant measure for equation (9.4.2) in H^κ if and only if $m > 0$.

(iii) If $d = 1$, and $m > \frac{\kappa^2}{2}$, then the invariant measure for equation (9.4.2) is unique in H^κ.

(iv) If $d = 1$, and $0 < m < \frac{\kappa^2}{4}$, then there are many invariant measures for equation (9.4.2) in H^κ.

(v) If $d = 1$ then there exists exactly one invariant measure for equation (9.4.2) in H_r.

Proof — (i) Let $\{e_j\}$ be an orthonormal and complete basis in $L^2(\mathbb{R}^d)$ and $\{\beta_j\}$ independent real–valued Wiener processes. Define, see §5.2, the cylindrical Wiener process $W(t)$ by the formula

$$W(t) = \sum_{j=1}^{\infty} e_j \beta_j(t).$$

Assume that $W_\Delta(t)$ is $L^2_\rho(\mathbb{R}^d)$–valued. Then

$$\mathbb{E}|W_\Delta(t)|_2^2 = \mathbb{E}\left(\int_{\mathbb{R}^d} \rho(\xi)|W_\Delta(t,\xi)|^2 d\xi\right)$$

$$= \int_{\mathbb{R}^d} \rho(\xi)\mathbb{E}\left|\int_0^t \sum_{j=1}^{\infty} S(t-s)e_j(\xi)d\beta_j(s)\right|^2 d\xi$$

$$= \int_{\mathbb{R}^d} \rho(\xi)\left(\int_0^t \sum_{j=1}^{\infty} |S(\sigma)e_j(\xi)|^2 d\sigma\right) d\xi$$

$$= \int_{\mathbb{R}^d} \rho(\xi) \int_0^t \sum_{j=1}^{\infty} |\langle e_j, p_\sigma(\xi - \cdot)\rangle|^2 d\sigma d\xi,$$

where $\langle e_j, p_\sigma(\xi - \cdot) \rangle$ is the scalar product in $L^2(\mathbb{R}^d)$ of e_j and $p_\sigma(\xi - \cdot)$. Consequently, by Plancherel's identity,

$$\mathbb{E}|W_\Delta(t)|_2^2 = \int_{\mathbb{R}^d} \rho(\xi) \int_0^t |p_\sigma(\xi - \cdot)|_{L^2(\mathbb{R}^d)}^2 \, d\sigma \, d\xi$$

$$= \int_{\mathbb{R}^d} \rho(\xi) \, d\xi \int_0^t |p_\sigma|_{L^2(\mathbb{R}^d)}^2 \, d\sigma.$$

But

$$|p_\sigma|_{L^2(\mathbb{R}^d)}^2 = \frac{1}{(4\pi\sigma)^d} \int_{\mathbb{R}^d} e^{-\frac{|\xi|^2}{2\sigma}} \, d\xi = \frac{1}{2^d} \frac{1}{\sqrt{(2\pi s)^d}}.$$

Consequently

$$\mathbb{E}|W_\Delta(t)|_2^2 < +\infty,$$

if and only if $d = 1$.

(ii) Assume $d = 1$. In the same way one has

$$\mathbb{E}|W_{\Delta - m}(t)|_2^2 = \int_{\mathbb{R}^d} \rho(\xi) \, d\xi \, \frac{1}{2\sqrt{2\pi}} \int_0^t e^{-2m\sigma} \frac{1}{\sqrt{\sigma}} \, d\sigma.$$

So if and only if $m > 0$ we have

$$\sup_{t \geq 0} \mathbb{E}|W_{\Delta - m}(t)|_2^2 < +\infty,$$

and the result follows.

(iii) It is enough to notice that for the semigroup $S(t)$, $t \geq 0$, corresponding to $\Delta - m$ one has

$$\|S(t)\|_{L(H^\kappa)} \leq e^{-(m - \frac{\kappa^2}{2})t}, \ t \geq 0,$$

with $m - \frac{\kappa^2}{2} > 0$.

(iv) Let $S(t)$, $t \geq 0$, be the semigroup corresponding to $\Delta - m$ with $m > 0$. Note that the function $\overline{x}(\xi) = e^{\sqrt{m}\xi}$, $\xi \in \mathbb{R}$, satisfies the equation $\Delta \overline{x} = m\overline{x}$. Moreover

$$\int_{\mathbb{R}} e^{-\kappa|\xi|} |\overline{x}(\xi)|^2 d\xi = \int_{\mathbb{R}} e^{-\kappa|\xi|} e^{2\sqrt{m}\xi} d\xi < +\infty$$

if and only if $\kappa - 2\sqrt{m} > 0$. If $\kappa - 2\sqrt{m} > 0$ then $S(t)\overline{x} = \overline{x}$ for all $t \geq 0$ and $\delta_{\{\overline{x}\}} \neq \delta_{\{0\}}$ is an invariant measure for $z' = (\Delta - m)z$. By Theorem 9.1.1(ii) this finishes the proof of (iv).

(v) Let us choose $\kappa > 0$ such that $\frac{\kappa^{\frac{2}{r}\delta}}{p} < m$. Then by Proposition 9.4.5(ii) the semigroup $S(t)$, $t \geq 0$, is exponentially stable in $L^2_{\rho_{\kappa,r}}$ and Theorem 9.1.1 implies the result.   ∎

9.4.2 Chaotic noise

We assume now that the covariance operator Q, of the Wiener process W, is of convolution type:

$$Q u = q * u, \ u \in U = H = L^2(\mathbb{R}^d), \tag{9.4.7}$$

where q is a continuous, integrable, positive definite function on $\mathbb{R}^d$, see Example 5.2.1. In fact we will assume the following:

Hypothesis 9.1 *The operator Q is of the form (9.4.7) where the function q is integrable and of the form*

$$q(\xi) = \int_{\mathbb{R}^d} e^{i\langle \xi, \eta \rangle} g(\eta) d\eta, \ \xi \in \mathbb{R}^d, \tag{9.4.8}$$

with g a nonnegative integrable function.

It follows from Hypothesis 9.1 that q and g are continuous and bounded. The representation (9.4.8) is natural in the light of Bochner's theorem.

Theorem 9.4.7 *(i) Assume that Hypothesis 9.1 holds and that ρ is an admissible weight for $\Delta - m$. Then equation (9.4.1) has a weak solution on $L^p_\rho(\mathbb{R}^d)$.*

(ii) If in addition $m > 0$ then there exists an invariant measure for equation (9.4.1). Moreover if $m = 0$ then there exists an invariant measure for equation (9.4.1) if and only if $\int_{\mathbb{R}^d} g(\eta) d\eta < +\infty$. In the latter case the invariant measure is never unique.

(iii) If $\rho = \rho^\kappa$ and $m > \frac{\kappa^2}{2}$ then invariant measures are unique and if $m < \frac{\kappa^2}{4d^2}$ then they are not unique.

(iv) If $r > d$ and $m > 0$ then there exists exactly one invariant measure for equation (9.4.2) in H_r.

Proof — Using the same notation as in the proof of Proposition 9.4.6 we have that

$$W(t,\xi) = \sum_{j=1}^{\infty} Q^{1/2} e_j(\xi)\beta_j(t), \ t \geq 0, \ \xi \in \mathbb{R}^d.$$

By Hypothesis 9.1, $Q^{1/2}u = q_1 * u, \ u \in L^2(\mathbb{R}^d)$, where

$$q_1(\xi) = \int_{\mathbb{R}^d} e^{i\langle \xi, \eta \rangle} \sqrt{g(\eta)} d\eta, \ \xi \in \mathbb{R}^d.$$

So we have, compare the proof of Proposition 9.4.6(i), that

$$\mathbb{E} |W_{\Delta-m}(t)|_2^2 = \mathbb{E} \left(\int_{\mathbb{R}^d} \rho(\xi) |W_{\Delta-m}(t,\xi)|^2 d\xi \right),$$

$$\mathbb{E} |W_{\Delta-m}(t,\xi)|_2^2 = \sum_{j=1}^{\infty} \mathbb{E} \left| \int_0^t e^{-m(t-s)} p_{t-s} * q_1 * e_j(\xi) d\beta_j(s) \right|^2$$

$$= \int_0^t e^{-2ms} \sum_{j=1}^{\infty} \left(\langle p_s * q_1(\xi - \cdot), e_j(\cdot) \rangle_{L^2}^2 \right) ds$$

$$= \int_0^t e^{-2ms} |p_s * q_1|_{L^2}^2 ds, \ \xi \in \mathbb{R}^d.$$

Consequently

$$\mathbb{E} |W_{\Delta-m}(t,\xi)|_{L_\rho^2}^2 = \int_{\mathbb{R}^d} \rho(\eta) d\eta \int_0^t e^{-2ms} |p_s * q_1|_{L^2}^2 ds$$

By Plancherel's theorem

$$|p_s * q_1|_{L^2}^2 = (2\pi)^d \int_{\mathbb{R}^d} e^{-2s|\xi|^2} g(\xi) d\xi.$$

Therefore

$$\mathbb{E} |W_{\Delta-m}(t,\xi)|_{L_\rho^2}^2 = (2\pi)^d \int_{\mathbb{R}^d} \rho(\eta) d\eta$$

$$\times \int_{\mathbb{R}^d} \left(\int_0^t e^{-2(m+|\xi|^2)s} ds \right) g(\eta) d\eta.$$

Since g is an integrable function,

$$\mathbb{E}\,|W_{\Delta-m}(t,\xi)|^2_{L^2_\rho} < +\infty \text{ for arbitrary } t \geq 0.$$

This shows (i).
To prove (ii) it is enough to remark that

$$\sup_{t\geq 0} \mathbb{E}\,|W_{\Delta-m}(t,\xi)|^2_{L^2_\rho} = (2\pi)^d \int_{\mathbb{R}^d} \rho(\eta)d\eta$$

$$= \frac{1}{2}\int_{\mathbb{R}^d} \frac{g(\eta)}{m+|\eta|^2}d\eta < +\infty,$$

provided that $m > 0$, and for $m = 0$ one has to require that

$$\int_{\mathbb{R}^d} \frac{g(\xi)}{|\xi|^2}d\xi < +\infty.$$

The proofs of (iii) and (iv) are completely analogous to those of
(iii),(iv) and (v) of Proposition 9.4.6 and will be omitted. ∎

We will now construct functions q and g such that (9.4.5) holds.
This way we will show that even if $m = 0$ invariant measures for
(9.4.1) may really exist.

Example 9.4.8 For arbitrary $\gamma \geq 1$ the even function

$$h(\xi) = \begin{cases} (1 - |\xi|^\gamma) & \text{if } |\xi| \leq 1, \\ 0 & \text{if } |\xi| \geq 1, \end{cases}$$

is concave on $[0,+\infty[$ and by a result of Polya, see W. Feller [64,
Vol. II], and Plancherel's theorem, it is a characteristic function of
an even, integrable, nonnegative continuous function ψ :

$$h(\xi) = \int_{\mathbb{R}} e^{i\xi\eta}\psi(\eta)d\eta, \; \xi \in \mathbb{R}.$$

We define
$$g(\xi) = \frac{1}{2}\left(h(\xi-1) + h(\xi+1)\right), \; \xi \in \mathbb{R}.$$

Then

$$q(\eta) = \int_{\mathbb{R}} e^{i\xi\eta} g(\xi) d\xi + \frac{1}{2} \int_{\mathbb{R}} e^{i\xi\eta} \left(h(\xi - 1) + h(\xi + 1) \right) d\xi$$

$$= \frac{1}{2} e^{i\eta} \int_{\mathbb{R}} e^{i(\xi-1)\eta} h(\xi - 1) d\xi + \frac{1}{2} e^{-i\eta} \int_{\mathbb{R}} e^{i(\xi+1)\eta} h(\xi + 1) d\xi$$

$$= 2\pi \cos \eta \psi(\eta), \quad \eta \in \mathbb{R}.$$

It is therefore clear that Hypothesis 9.1 is satisfied. If, in addition, $\gamma > 1$ then

$$\int_{\mathbb{R}} \frac{g(\xi)}{|\xi|^2} d\xi \leq \int_{|\xi| \leq 1} \frac{1}{|\xi|^{2-\gamma}} d\xi + \int_{|\xi| \geq 1} \frac{1}{|\xi|^2} d\xi < +\infty,$$

so the condition (9.4.5) holds as well. ∎

Chapter 10

Stochastic delay systems

In this chapter we study invariant measures for stochastic delay equations by the dissipativity method.

10.1   Introduction

Let $d, m \in \mathbb{N}, r > 0$, $a(\cdot)$ be a finite $d \times d$ matrix valued measure on $[-r, 0]$, b a $d \times m$ matrix, f a mapping from $\mathbb{R}^d$ into $\mathbb{R}^d$ and $W(\cdot)$ a standard m dimensional Wiener process. In this section we are concerned with a stochastic delay equation of the form

$$
\left.
\begin{aligned}
dy(t) &= \left(\int_{-r}^{0} a(d\theta)y(t + \theta) + f(y(t)) \right) dt + b dW(t), \\
y(0) &= x_0 \in \mathbb{R}^d, \\
y(\theta) &= x_1(\theta),\ \theta \in [-r, 0].
\end{aligned}
\right\}
\qquad (10.1.1)
$$

For simplicity of the presentation we assume that f is Lipschitz continuous. It is known, see A. Chojnowska–Michalik [24], that if $H = \mathbb{R}^d \times L^2(-r, 0, \mathbb{R}^d)$ then the H–valued process X,

$$
X(t) = \begin{pmatrix} y(t) \\ y_t(\cdot) \end{pmatrix},\ t \geq 0,
$$

where

$$
y_t(\theta) = y(t + \theta),\ t \geq 0,\ \theta \in [-r, 0],
$$

199

is a mild solution of the equation

$$dX(t) = (AX(t) + F(X(t)))dt + BdW(t) \qquad (10.1.2)$$

where the operators A, F and B are defined as follows

$$
\left.
\begin{aligned}
A \left(\begin{array}{c} \varphi(0) \\ \varphi \end{array} \right) &= \left(\begin{array}{c} \int_{-r}^{0} a(d\theta)\varphi(\theta) \\ \frac{d\varphi}{d\theta} \end{array} \right), \\[2ex]
D(A) &= \left\{ \left(\begin{array}{c} \varphi(0) \\ \varphi \end{array} \right) : \varphi \in W^{1,2}(-r,0,\mathbb{R}^d) \right\},
\end{aligned}
\right\}
\qquad (10.1.3)
$$

$$F \left(\begin{array}{c} x_0 \\ x_1 \end{array} \right) = \left(\begin{array}{c} f(x_0) \\ 0 \end{array} \right), \quad \left(\begin{array}{c} x_0 \\ x_1 \end{array} \right) \in H, \qquad (10.1.4)$$

$$Bu = \left(\begin{array}{c} bu \\ 0 \end{array} \right), \quad u \in \mathbb{R}^m. \qquad (10.1.5)$$

The process X will be called an *extended* or a *complete* solution to (10.1.1) and denoted by

$$X(t,x) = \left(\begin{array}{c} y(t; x_0, x_1) \\ y_t(\cdot; x_0, x_1) \end{array} \right), \quad x = \left(\begin{array}{c} x_0 \\ x_1 \end{array} \right) \in H.$$

Stochastic delay equations can be studied and have been studied as equations on $\mathbb{R}^d$, without any reference to the theory of stochastic evolution equations. Basic work in this direction is due to K. Ito and M. Nisio [92]. From more recent publications let us mention M. Scheutzov [131], [132], [133] and S. A. Mohammed and M. Scheutzov [119].

In the present chapter we apply general results developed in the first two parts of the book. We are concerned with linear systems and with applicability of dissipativity method to nonlinear ones.

10.2 Linear case

We start from linear delay equations for which the abstract version is

$$
\left.
\begin{aligned}
dX(t) &= AX(t)dt + BdW(t), \\[2ex]
X(0) &= x \in H.
\end{aligned}
\right\}
\qquad (10.2.1)
$$

Note that the system (10.2.1) is genuinly infinite dimensional while the noise process W is finite dimensional.

To simplify the following presentation we restrict our considerations to the discrete delay case,

$$\left. \begin{aligned} dy(t) &= \left(a_0 y(t) + \sum_{i=1}^{N} a_i y(t + \theta_i) \right) dt + b dW(t), \\ y(0) &= x_0, \; y(\theta) = x_1(\theta), \; \theta \in [-r, 0], \end{aligned} \right\} \quad (10.2.2)$$

when $-r = \theta_1 < \theta_2 < \cdots < \theta_N < 0 = \theta_{N+1}$. Then the measure $a(\cdot)$ is of the form

$$a(\Gamma) = a_0 \delta_0(\Gamma) + \sum_{i=1}^{N} a_i \delta_{\theta_i}(\Gamma), \; \Gamma \in \mathcal{B}([-r, 0]). \quad (10.2.3)$$

We have the following result on the generator A and the corresponding semigroup $S(t)$, $t \geq 0$, relevant to the asymptotic behaviour of (10.2.1).

Proposition 10.2.1 *(i) The operator A given by (10.1.3) generates a C_0–semigroup $S(t)$, $t \geq 0$, such that the operators $S(t)$, $t > r$, are compact.*

(ii) The spectrum $\sigma(A)$ of A is of the form

$$\sigma(A) = \left\{ \lambda \in \mathbb{C} : \quad \det \left[\lambda I - a_0 - \sum_{j=1}^{N} e^{\lambda \theta_j} a_j \right] = 0 \right\}, \quad (10.2.4)$$

and the semigroup S is exponentially stable if and only if

$$\sup\{\operatorname{Re} \lambda : \; \lambda \in \sigma(A)\} < 0. \quad (10.2.5)$$

For the proof we refer to J. Hale [82].

Example 10.2.2 Assume that $d = 1$, and $N = 1$, then the stability condition (10.2.5) holds (for the corresponding generator A) if and only if

$$a_0 < 1, \; a_0 < -a_1 < \sqrt{\gamma^2 + a_0^2}, \quad (10.2.6)$$

where $0 < \gamma < \pi$, and $\gamma \coth \gamma = a_0$, see N. D. Hayes [84].

Theorem 10.2.3 *Let P_t, $t \geq 0$, be the transition semigroup on $B_b(H)$ corresponding to (10.2.2).*

(i) P_t, $t \geq 0$, is a strong Feller semigroup for all $t > r$ if and only if

$$\text{rank } [\lambda I - \sum_{i=1}^{N} e^{\lambda \theta_i} a_i, b] = d, \qquad (10.2.7)$$

for all $\lambda \in \mathbb{C}$.

(ii) P_t, $t \geq 0$, is irreducible for all $t > r$ if and only if

$$\text{rank } [\lambda I - a_0 - \sum_{i=1}^{N} e^{\lambda \theta_i} a_i, b] = d, \quad \text{rank } [a_1, b] = n, \qquad (10.2.8)$$

for all $\lambda \in \mathbb{C}$.

Proof — It follows from A. W. Olbrot and L. Pandolfi [122] that (10.2.7) is equivalent to the null controllability of the control system corresponding to (10.2.2). Then by Theorem 7.3.1 (i) is true. In a similar way the proof of (ii) follows from a characterization of the approximate controllability due to A. Manitius [110] (for an alternative proof see R. F. Curtain and H. J. Zwart [32]), and the fact that the corresponding control system is approximately controllable in time t if and only if the image of $Q_t = \int_0^t S(t)BB^*S^*(t)dt$ is dense in H. ■

Proposition 10.2.4 *A sufficient condition for the strong Feller property of P_t, $t \geq 0$, for all $t > r$ is that*

$$\text{rank } [b, a_0 b, ..., a_0^{d-1} b] = d, \text{ and Image } b \supset \text{ Image } a_i, \ i = 1, ..., N.$$
$$(10.2.9)$$

Proof — Consider the control system

$$z'(t) = a_0 z(t) + \sum_{i=1}^{N} a_i z(t + \theta_i) + b u(t), \left.\begin{array}{c}\\\\\\\end{array}\right\}$$

$$z(0) = x_0, \ z(\theta) = x_1(\theta), \ \theta \in [-r, 0], \qquad (10.2.10)$$

and introduce new control functions u_0 and v such that $u(t) = u_0(t) + v(t)$, $t \geq 0$, and $\sum_{i=1}^{N} a_i z(t + \theta_i) + b u(t) = b v(t), t \geq 0$. Then (10.2.10) becomes $z'(t) = a_0 z(t) + bv(t)$, and, by the rank condition, the system can be steered to zero. ∎

The main result of the present section is formulated in the following theorem.

Theorem 10.2.5 *(i) Assume that*

$$\sup \left\{ \operatorname{Re} \lambda : \det \left[\lambda I - \sum_{i=1}^{N} e^{\lambda \theta_i} a_i \right] = 0 \right\} < 0, \qquad (10.2.11)$$

then there exists exactly one invariant measure for (10.2.2).
(ii) If there exists an invariant measure for (10.2.2), and

$$\operatorname{rank} \left[\lambda I - \sum_{i=1}^{N} e^{\lambda \theta_i} a_i, b \right] = d$$

for all $\lambda \in \mathbb{C}$ then (10.2.11) holds.

Proof — The result is a direct consequence of Theorems 9.1.1, 10.2.3 and Proposition 10.2.4. ∎

Remark 10.2.6 (i) If $d = 1$ and $b \neq 0$ then there exists an invariant measure for (10.2.2) if and only if (10.2.6) holds.
(ii) If condition (10.2.9) holds then there exists an invariant measure for (10.2.2) if and only if (10.2.11) holds.

10.3 Nonlinear equations

We conjecture that if there exists a solution to (10.1.1) which is bounded in probability then there exists an invariant measure for (10.1.1). However, this result does not follow from Theorem 6.1.2 since the operators $S(t)$ are compact only for sufficiently large $t > 0$. Some results in this direction can be found in K. Ito and M. Nisio [92]. It is also possible that if the transition semigroup for the linear equation is a strong Feller one, see §4.2, then the semigroup corresponding to the nonlinear equation (10.1.1) is likewise but the proof

of this result is also laking. Consequently, for the proof of the uniqueness of the invariant measure, Doob's theorem cannot be applied at
the moment. Results on this topic, based on Girsanov's theorem, are
contained in M. Scheutzov [132].

In this section we discuss applicability of the dissipativity method
for the equation

$$
\left.\begin{array}{l}
dy(t) = \left(a_0 y(t) + \displaystyle\sum_{j=1}^{N} a_j y(t + \theta_j) + f(y(t)) \right) dt + b dW(t), \\[4mm]
y(0) = x_0 \in \mathbb{R}^d, \\[2mm]
y(\theta) = x_1(\theta), \ \theta \in [-r, 0].
\end{array}\right\}
$$

$$(10.3.1)$$

We start from the crucial question: under what conditions does exist
a positive number ω such that the operator $A + \omega I$ is dissipative,
compare §6.3.2, Unfortunately, with respect to the original norm in
$H = \mathbb{R}^d \times L^2(-r, 0; \mathbb{R}^d)$, the operator A does not necessarily have this
property. If, however, we replace the original norm by an equivalent
one of the form

$$
\| \begin{pmatrix} x_0 \\ x_1 \end{pmatrix} \|^2 = |x_0|^2 + \int_{-r}^{0} \tau(\theta) |x_1(\theta)|^2 d\theta, \ \begin{pmatrix} x_0 \\ x_1 \end{pmatrix} \in H,
$$

with the function τ satisfying

$$
\gamma_1 \leq \tau(\theta) \leq \gamma_2, \ \theta \in [-r, 0],
$$

γ_1 and γ_2 being positive numbers, the situation can change dramat-
ically. It has been shown in G. F. Webb [161] that if

$$
\tau(\theta) = \sum_{j=1}^{N} (|a_1| + \cdots + |a_j|) \chi_{]\theta_j, \theta_{j+1}]}, \ \theta \in [-r, 0],
$$

and

$$
\langle a_0 z, z \rangle \leq -\alpha_0 |z|^2, \ z \in \mathbb{R}^d,
$$

then

$$
\| S(t) \| \leq e^{\gamma t}, \ t \geq 0, \tag{10.3.2}
$$

for

$$\gamma = \max\left(0, \frac{|a_1|}{2} - \alpha_0\right).$$

Thus in the estimate (10.3.2), the speed γ is nonnegative, a result not satisfactory for our purposes. We will therefore introduce a different function τ,

$$\tau(\theta) = \gamma e^{2\beta\theta} + \sum_{j=1}^{N} \tau_j \chi_{]\theta_j, \theta_{j+1}]}, \quad \theta \in [-r, 0], \tag{10.3.3}$$

where $\beta > \alpha$ and

$$\tau_j = \gamma\left(\frac{\beta}{\alpha} - 1\right) e^{2\beta\theta_j}, \quad j = 1, 2, ..., N. \tag{10.3.4}$$

Theorem 10.3.1 *Assume that the matrix* $a_0 + \alpha_0 I_d$ ([1]) *is dissipative on* $\mathbb{R}^d$ *with* $\alpha_0 > 0$ *and that for positive numbers* $\beta > \alpha > 0$,

$$\alpha_0 \geq \alpha + \left(\frac{\beta - \alpha}{\alpha} e^{2\beta\theta_N} + 1\right)^{1/2}$$

$$\times \left(\frac{\alpha}{\beta} \frac{|a_1|^2}{e^{2\beta\theta_1}} + \frac{\alpha}{\beta - \alpha} \sum_{j=2}^{N} \frac{|a_j|^2}{e^{2\beta\theta_j} - e^{2\beta\theta_{j-1}}}\right)^{1/2}. \tag{10.3.5}$$

Then there exists $\gamma > 0$ *such that in the corresponding norm* $\|\cdot\|$ *one has*

$$\|S(t)\| \leq e^{-\alpha t}, \quad t \geq 0 \tag{10.3.6}$$

Proof — Denote by $\prec \cdot, \cdot \succ$ the scalar product

$$\prec \begin{pmatrix} x_0 \\ x_1 \end{pmatrix}, \begin{pmatrix} z_0 \\ z_1 \end{pmatrix} \succ = \langle x_0, z_0 \rangle + \int_{-r}^{0} \tau(\theta)\langle x_1(\theta), z_1(\theta)\rangle d\theta,$$

where $\begin{pmatrix} x_0 \\ x_1 \end{pmatrix}, \begin{pmatrix} z_0 \\ z_1 \end{pmatrix} \in H$. Then the operator $A + \alpha I_d$ is dissipative with respect to the scalar product $\prec \cdot, \cdot \succ$ if and only if

$$\prec A \begin{pmatrix} x(0) \\ x \end{pmatrix}, \begin{pmatrix} x(0) \\ x \end{pmatrix} \succ \leq -\alpha\left(|x(0)|^2 + \int_{-r}^{0} \tau(\theta)|x(\theta)|^2 d\theta\right), \tag{10.3.7}$$

---

[1] I_d represents the identity on $\mathbb{R}^d$.

for arbitrary $x \in W^{1,2}(-r, 0; \mathbb{R}^d)$. Note that

$$
\prec A \left(\begin{array}{c} x(0) \\ x \end{array} \right), \left(\begin{array}{c} x(0) \\ x \end{array} \right) \succ
$$

$$
= \langle a_0 x(0), x(0) \rangle + \sum_{j=1}^{N} \langle a_j x(\theta_j), x(0) \rangle
$$

$$
+ \int_{\theta_1}^{0} \tau(\theta) \left\langle \frac{dx}{d\theta}(\theta), x(\theta) \right\rangle d\theta
$$

$$
= J_1 + J_2 + J_3.
$$

We have

$$
\begin{aligned}
J_3 &= \int_{\theta_1}^{0} \tau(\theta) \left\langle \frac{dx}{d\theta}(\theta), x(\theta) \right\rangle d\theta \\
&= \gamma \int_{\theta_1}^{0} e^{2\beta\theta} \left\langle \frac{dx}{d\theta}(\theta), x(\theta) \right\rangle d\theta \\
&+ \sum_{j=1}^{N} \tau_j \int_{\theta_j}^{\theta_{j+1}} \left\langle \frac{dx}{d\theta}(\theta), x(\theta) \right\rangle d\theta.
\end{aligned}
$$

But

$$
\int_{\theta_1}^{0} e^{2\beta\theta} \left\langle \frac{dx}{d\theta}(\theta), x(\theta) \right\rangle d\theta = \frac{1}{2} \int_{\theta_1}^{0} e^{2\beta\theta} \frac{d}{d\theta} |x(\theta)|^2 d\theta
$$

$$
= \frac{1}{2} \left(|x(0)|^2 - e^{2\beta\theta_1} |x(\theta_1)|^2 \right) - \beta \int_{\theta_1}^{0} e^{2\beta\theta} |x(\theta)|^2 d\theta,
$$

and

$$
\int_{\theta_j}^{\theta_{j+1}} \left\langle \frac{dx}{d\theta}(\theta), x(\theta) \right\rangle d\theta = \frac{1}{2} \left(|x(\theta_{j+1})|^2 - |x(\theta_j)|^2 \right), \ j = 1, \cdots, N.
$$

Moreover

$$
J_1 \leq -\alpha_0 |x(0)|^2, \ J_2 \leq \sum_{j=1}^{N} |a_j| \, |x(\theta_j)| \, |x(0)|.
$$

Therefore

$$\prec A \begin{pmatrix} x(0) \\ x \end{pmatrix}, \begin{pmatrix} x(0) \\ x \end{pmatrix} \succ$$

$$\leq -\alpha_0 |x(0)|^2 + \sum_{j=1}^{N} |a_j| \, |x(\theta_j)| \, |x(0)|$$

$$+ \frac{1}{2} \sum_{j=1}^{N} \tau_j \left(|x(\theta_{j+1})|^2 - |x(\theta_j)|^2 \right) + \frac{\gamma}{2} \left(|x(0)|^2 - e^{2\beta\theta_1} |x(\theta_1)|^2 \right)$$

$$- \gamma\beta \int_{\theta_1}^{0} e^{2\beta\theta} |x(\theta)|^2 d\theta$$

$$\leq \left(-\alpha_0 + \frac{\tau_N + \gamma}{2} \right) |x(0)|^2$$

$$+ \frac{1}{2} \sum_{j=2}^{N} (\tau_{j-1} - \tau_j) |x(\theta_j)|^2 - \frac{1}{2} \left(\tau_1 + \gamma e^{2\beta\theta_1} \right) |x(\theta_1)|^2$$

$$+ \sum_{j=1}^{N} |a_j| \, |x(\theta_j)| \, |x(0)| - \gamma\beta \int_{\theta_1}^{0} e^{2\beta\theta} |x(\theta)|^2 d\theta.$$

Consequently, for (10.3.7) to hold it is sufficient that

$$\left(\alpha_0 - \alpha - \frac{\tau_N + \gamma}{2} \right) |x(0)|^2$$

$$+ \frac{1}{2} \sum_{j=2}^{N} (\tau_j - \tau_{j-1}) |x(\theta_j)|^2 + \frac{1}{2} \left(\tau_1 + \gamma e^{2\beta\theta_1} \right) |x(\theta_1)|^2 \qquad (10.3.8)$$

$$- \sum_{j=1}^{N} |a_j| \, |x(\theta_j)| \, |x(0)| \geq 0,$$

and

$$\int_{\theta_1}^{0} \left(\gamma\beta e^{2\beta\theta} - \alpha\tau(\theta) \right) |x(\theta)|^2 d\theta \geq 0, \qquad (10.3.9)$$

for all $x \in W^{1,2}(-r, 0; \mathbb{R}^d)$. The inequality (10.3.9) holds if and only if

$$\gamma \frac{\beta - \alpha}{\alpha} e^{2\beta\theta} \geq \sum_{j=1}^{N} \tau_j \chi_{]\theta_j, \theta_{j+1}]}, \quad \theta \in \,]\theta_1, \theta_{N+1}].$$

Taking into account (10.3.5) one easily sees that (10.3.9) is satisfied. Moreover if $|x(0)| = 0$ then the inequality (10.3.8) holds because $\tau_j \geq \tau_{j-1}$ for $j = 2, ..., N$ and $\tau_1 + \gamma e^{2\beta\theta_1} \geq 0$. If $|x(0)| \neq 0$ then dividing both sides of (10.3.8) by $|x(0)|^2$ we get

$$\frac{1}{2} \sum_{j=1}^{N} (\tau_j - \tau_{j-1}) \left(\frac{|x(\theta_j)|}{|x(0)|} \right)^2 + \frac{1}{2}(\tau_1 + \gamma e^{2\beta\theta_1}) \left(\frac{|x(\theta_1)|}{|x(0)|} \right)^2$$

$$- \sum_{j=1}^{N} a_j \left(\frac{|x(\theta_j)|}{|x(0)|} \right) \geq \alpha + \frac{\tau_N + \gamma}{2} - \alpha_0.$$

$$(10.3.10)$$

By the definition (10.3.5), the minimal value of the left hand side of (10.3.10) is

$$-\frac{1}{4} \left(\frac{2|a_1|^2}{\tau_1 + \gamma e^{2\beta\theta_1}} + \sum_{j=2}^{N} \frac{2|a_j|^2}{\tau_j - \tau_{j-1}} \right)$$

$$= -\frac{1}{2\gamma} \left(\frac{\alpha}{\beta} \frac{|a_1|^2}{e^{2\beta\theta_1}} + \frac{\alpha}{\beta - \alpha} \sum_{j=2}^{N} \frac{|a_j|^2}{e^{2\beta\theta_j} - e^{2\beta\theta_{j-1}}} \right).$$

Therefore (10.3.10) holds provided there exists $\gamma > 0$ such that

$$\alpha_0 - \alpha \geq \frac{\gamma}{2} \left(\frac{\beta - \alpha}{\alpha} e^{2\beta\theta_N} + 1 \right)$$

$$(10.3.11)$$

$$+ \frac{1}{2\gamma} \left(\frac{\alpha}{\beta} \frac{|a_1|^2}{e^{2\beta\theta_1}} + \frac{\alpha}{\beta - \alpha} \sum_{j=2}^{N} \frac{|a_j|^2}{e^{2\beta\theta_j} - e^{2\beta\theta_{j-1}}} \right).$$

Minimizing the right hand side of (10.3.11) with respect to $\gamma > 0$ one gets the desired result.   ∎

Remark 10.3.2 One gets even more explicit conditions implying (10.3.6) if one takes for instance $\beta = 2\alpha$ (or $\beta = 3\alpha$, $\beta = 4\alpha$,...). Different choices for the weight τ are possible. For instance if $N = 1$ and

$$\tau(\theta) = \gamma e^{2\alpha\theta}, \ \theta \in [-r, 0],$$

one finds that the inequality

$$\alpha_0 \geq \alpha + e^{\alpha r}|a_1|$$

implies (10.3.6).

The main result of this section is a direct consequence of Theorem 6.3.3, see Remark 6.3.4, and of Theorem 10.3.1.

Theorem 10.3.3 *Assume that condition (10.3.5), holds and that the Lipschitz constant of f is smaller than α. Then there exists exactly one invariant measure μ for the extended solution of (10.3.1). Moreover there exists $\omega > 0$ such that for all* $x = \begin{pmatrix} x_0 \\ x_1 \end{pmatrix} \in H$, *$\varphi \in B_b(\mathbb{R}^d)$ and $t > 0$*

$$\left| \mathbb{E}\left(\varphi(y(t; x_0, x_1)) - \int_{\mathbb{R}^d} \varphi(z)\mu_d(dz) \right) \right| \leq (C + 2|x|)e^{-\omega t}\|\varphi\|_{\mathrm{Lip}},$$

where μ_d is the projection of μ on the $\mathbb{R}^d$ component of H.

Chapter 11

Reaction–Diffusion equations

In this chapter we study invariant measures for stochastic reaction–diffusion equations. First equations on a finite space interval are studied and then equations on the whole of $\mathbb{R}^d$ and with the space homogeneous Wiener process are investigated.

11.1 Introduction

Deterministic reaction–diffusion equations are usually written as systems of equations

$$
\left.
\begin{aligned}
&\frac{\partial y_k}{\partial t}(t,\xi) = \sigma_k \Delta y_k(t,\xi) + f_k(\xi, y_1(t,\xi), \cdots, y_K(t,\xi)), \ t > 0, \\
&y_k(0,\xi) = x_k(\xi), \ \xi \in \mathcal{O} \subset \mathbb{R}^d, \ k = 1, ..., K,
\end{aligned}
\right\}
$$

$$(11.1.1)$$

where f_k, $k = 1, ..., K$, are given real functions from $\mathcal{O} \times \mathbb{R}^K$ to $\mathbb{R}$, σ_k, $k = 1, ..., K$, are positive constants, and $\mathcal{O}$ is an open subset of $\mathbb{R}^d$. If $\mathcal{O} \neq \mathbb{R}^d$ one also adds boundary conditions of Dirichlet, Neumann or mixed types.

The stochastic version of (11.1.1) is of the form

$$
\begin{aligned}
dX_k(t,\xi) &= [\sigma_k \Delta X_k(t,\xi) + f_k(\xi, X_1(t,\xi), \cdots, X_K(t,\xi))]\, dt \\[2mm]
&\quad + b_k(\xi, X_1(t,\xi), \cdots, X_K(t,\xi))dW_k(t,\xi), \\[2mm]
X_k(0,\xi) &= x_k(\xi),\ \xi \in \mathcal{O} \subset \mathbb{R}^d,\ k = 1, ..., K,\ t > 0,
\end{aligned}
\tag{11.1.2}
$$

plus boundary conditions if $\mathcal{O} \neq \mathbb{R}^d$. Moreover $W_1, \cdots, W_K$ are independent Wiener processes and b_k, $k = 1, ..., K$, are given real functions.

Setting

$$
X(t,\xi) = \begin{pmatrix} X_1(t,\xi) \\ \cdots \\ X_K(t,\xi) \end{pmatrix}, \quad
A = \begin{pmatrix} \sigma_1\Delta \\ \cdots \\ \sigma_K\Delta \end{pmatrix}, \quad
W(t) = \begin{pmatrix} W_1(t) \\ \cdots \\ W_K(t) \end{pmatrix}
$$

$$
F(x)(\xi) = \begin{pmatrix} f_1(\xi, x_1(\xi), \cdots, x_d(\xi)) \\ \cdots \\ f_d(\xi, x_1(\xi), \cdots, x_d(\xi)) \end{pmatrix}, \quad
x(\xi) = \begin{pmatrix} x_1(\xi) \\ \cdots \\ x_d(\xi) \end{pmatrix}
$$

$$
(B(x)u)(\xi) = \begin{pmatrix} b_1(\xi, x_1(\xi), \cdots x_d(\xi))u_1 \\ \cdots \\ b_d(\xi, x_1(\xi), \cdots x_d(\xi))u_d \end{pmatrix}, \quad
u(\cdot) = \begin{pmatrix} u_1(\cdot) \\ \cdots \\ u_d(\cdot) \end{pmatrix},
$$

one can write problem (11.1.2) in the standard form

$$
\begin{aligned}
dX(t) &= (AX + F(X))dt + B(X)dW(t), \\[2mm]
X(0) &= x.
\end{aligned}
\tag{11.1.3}
$$

To formalize this procedure one should give precise definitions of the operators A, F and B, of the state space E (or H in the Hilbert space case) and of the space U on which the cylindrical Wiener processes are defined.

We will consider only one equation, $K = 1$, and distinguish two main cases:

(i) $d = 1$, $\mathcal{O} = \,]a, b[$, $-\infty < a < b < +\infty$,

(ii) $d \geq 1$, $\mathcal{O} = \mathbb{R}^d$.

In both cases the operator A will be the infinitesimal generator of a strongly continuous semigroup, $S(t)$, $t \geq 0$. For each case various conditions on F and B can be considered. The diffusion mapping B can be independent of the state variable (then we have the so-called *additive noise* case) or can depend on the state variable in an essential manner. In the latter case the dependence might be of Lipschitz type or be less regular.

A number of different conditions can be imposed on the mapping F. F can act as a Lipschitz mapping and therefore have linear growth, or it can have nonlinear growth. F can be locally Lipschitz or merely continuous or even less regular.

All the above mentioned types of assumptions have an important impact on the existence and uniqueness of the solutions to (11.1.2). Additional requirements will be needed for existence and uniqueness of invariant measures. In some cases it will be possible to throw additional light on the limit behaviour of the transition semigroup.

As our aim here is to illustrate applicability of the various techniques discussed in Part II, we will restrict our considerations to typical situations and examples. Possible extensions will be indicated in remarks and comments.

11.2 Finite interval. Lipschitz coefficients

In this section we assume that $K = d = 1$. We start from the analysis of the equation

$$
\left.
\begin{aligned}
&dX(t, \xi) = \left[\frac{\partial^2 X}{\partial \xi^2}(t, \xi) + f(\xi, X(t, \xi)) \right] dt + b(\xi, X(t, \xi))dW(t, \xi), \\[2mm]
&X(t, 0) = X(t, 1) = 0, \\[2mm]
&X(0, \xi) = x(\xi),
\end{aligned}
\right\}
$$

$$(11.2.1)$$

for $t > 0, \xi \in]0,1[$. We will require that the Wiener process W is cylindrical on $U = L^2(0,1)$. This space will also be our state space H.

In this section it is natural to assume that

$$Ax = \frac{d^2 x}{d\xi^2}, \quad x \in D(A) = H_0^{1,2}(0,1) \cap H^{2,2}(0,1),$$

$$F(x)(\xi) = f(x(\xi)), \quad (B(x)u)(\xi) = b(x(\xi))u(\xi).$$

Thus problem (11.2.1) is equivalent to problem (11.1.3).

11.2.1   Existence and uniqueness of solutions

We assume that f and b are Lipschitz continuous mappings. Consequently, F is likewise, but B is not from H into $L(H)$. We will check, however, that conditions of Theorem 5.3.1 are satisfied.

As the functions

$$e_n(\xi) = \sqrt{2}\sin\pi n\xi, \; n \in \mathbb{N}, \; \xi \in [0,1],$$

form an orthonormal sequence of eigenfunctions of the operator A, corresponding to the eigenvalues $-\pi^2 n^2$, $n \in \mathbb{N}$, we have

$$\|S(t)(B(x) - B(y))\|_{HS}^2 = \sum_{n,m=1}^{\infty} |\langle (B(x) - B(y))e_n, S(t)e_m \rangle|^2$$

$$= \sum_{m=1}^{\infty} e^{-2t\pi^2 m^2} |(B(x) - B(y))e_m|^2.$$

But

$$|(B(x) - B(y))e_m|^2 = 2\int_0^1 (b(x(\xi)) - b(y(\xi)))^2 (\sin m\pi\xi)^2 d\xi$$

$$\leq 2\|b\|_{\text{Lip}}^2 |x - y|^2.$$

Consequently

$$\|S(t)(B(x) - B(y))\|_{HS} \leq \sqrt{2}\,\|b\|_{\text{Lip}}\,|x - y|\,\|S(t)\|_{HS},$$

so we see that Hypothesis 5.1(iii) holds.

Moreover, for arbitrary $T > 0$

$$\int_0^T \|S(t)\|_{HS}^2 dt = \sum_{n=1}^\infty \int_0^T e^{-2t\pi^2 n^2} dt \leq \frac{1}{2\pi^2} \sum_{n=1}^\infty \frac{1}{n^2} < +\infty.$$

Thus Theorem 5.2.5 is applicable.

In this way we have proved the following result, see J.B. Walsh [158], M. I. Freidlin [69], G. Da Prato and J. Zabczyk [44], R. Sowers [144].

Theorem 11.2.1 *If f and b are Lipschitz continuous, then equation (11.2.1) has a unique mild solution and the corresponding transition semigroup is a strong Feller one.*

11.2.2 Existence and uniqueness of invariant measures

We will prove the following result, see R. Sowers [144], as in G. Da Prato, Gątarek and J. Zabczyk [39].

Theorem 11.2.2 *Assume that f and b are Lipschitz continuous functions such that*

(i) $\|f\|_{\text{Lip}} < \pi^2$

(ii) b is bounded.

Then there exists an invariant measure for (11.1.3) on H.

Proof — The operators $S(t)$, $t > 0$, are compact. Then to apply Theorem 6.1.2 it is enough to check condition (6.1.2).

For this we use Proposition 6.1.6. Assumption (ii) of this proposition is automatically satisfied because

$$\int_0^T \|S(t)\|_{HS}^2 dt = \sum_{n=1}^\infty \int_0^\infty e^{-2t\pi^2 n^2} dt = \frac{1}{2\pi^2} \sum_{n=1}^\infty \frac{1}{n^2} < +\infty,$$

and $\sup_x \|B(x)\| = \sup_\xi |b(\xi)| < +\infty$.

To check condition (i) notice that

$$\langle Ax, x \rangle \leq -\pi^2 |x|^2, \quad x \in D(A),$$

and

$$\langle F(x+y), x \rangle = \langle F(x+y) - F(y), x \rangle + \langle F(y), x \rangle$$

$$\leq \|f\|_{\text{Lip}}|x|^2 + |F(y)||x|$$

$$\leq \|f\|_{\text{Lip}}|x|^2 + \left(\varepsilon|x|^2 + \frac{1}{\varepsilon}|F(y)|^2 \right), \ x, y \in H, \ \varepsilon > 0.$$

It is therefore enough to use the fact that F has linear growth and choose $\varepsilon > 0$ sufficiently small.   ∎

Remark 11.2.3 Using the factorization procedure one can prove that solutions of (11.1.3) take values, for $t > 0$, in $E = C^\alpha([0, 1])$ with $\alpha \in]0, 1/2[$. Consequently if μ is an invariant measure for (11.1.3) then

$$\mu(E) = \int_H P_t(x, E)\mu(dx) = 1, \ t > 0,$$

and the measure μ is concentrated on E, see R. Sowers [144].

For uniqueness we have the following theorem, see R. Sowers [144], C. Mueller [120], S. Peszat and J. Zabczyk [124]. The present proof and the limit formula for P_t, $t > 0$, are taken from [124].

Theorem 11.2.4 *Assume that f and b are Lipschitz continuous functions and for some constants $\kappa, k > 0$*

$$\kappa \leq |b(\xi)| \leq k, \ \xi \in \mathbb{R}.$$

Then there is at most one invariant measure for (11.1.3). If μ is an invariant measure for (11.1.3) then, for arbitrary $x \in H$, $\Gamma \in \mathcal{B}(H)$,

$$\lim_{t \to \infty} P_t(x, \Gamma) = \mu(\Gamma).$$

Proof — By Theorem 4.2.1 and Proposition 4.1.1 it is enough to show that the transition semigroup corresponding to the solution of (11.1.3) is a strong Feller one and irreducible.

The strong Feller property is not a direct consequence of Theorem 7.1.1 because B is not a Lipschitz transformation from H into

$L(H)$. However, as we saw in subsection 11.2.1, B satisfies Hypothesis 5.1(iii), and the proof of Theorem 7.1.1 workes with only one exception. That is we do not know if the mappings $B_n(x)$, $x \in H$, are invertible. But in the present particular case their invertibility is a simple consequence of the fact that either $b(\xi) \geq m$ for all ξ or $b(\xi) \leq -m$ for all ξ. Irreducibility follows from Theorem 7.3.1. ∎

11.3 Equations with non–Lipschitz coefficients

As in §11.2 we set $K = d = 1$; however, we no longer assume that the function f is Lipschitz continuous, but to simplify presentation we set the diffusion coefficient to be equal to 1. Thus we will consider now the problem

$$
\left.
\begin{aligned}
dX(t,\xi) &= \left[\frac{\partial^2 X}{\partial \xi^2}(t,\xi) + f(\xi, X(t,\xi)) \right] dt + dW(t,\xi), \\[2mm]
X(t,0) &= X(t,1) = 0, \\[2mm]
X(0,\xi) &= x(\xi),
\end{aligned}
\right\}
\qquad (11.3.1)
$$

for $t > 0, \xi \in \,]0,1[$. We will need the following conditions on f.

Hypothesis 11.1 *(i) The function f is locally Lipschitz continuous and*

$$
(f(\xi + \eta) + \xi + \eta)\, \mathrm{sgn}\, \xi \leq a(|\eta|), \quad \textit{for all } \xi, \eta \in \mathbb{R},
$$

where

$$
a(r) \leq c_0 + c_1 r^p,
$$

for some $c_0 > 0, c_1 > 0$ and $p \geq 1$.
(ii) For some $\omega \in \mathbb{R}$, the function $f - \omega$ is decreasing.

By K we denote the space $C_0([0,1])$ of all continuous functions on $[0,1]$ vanishing at 0 and 1 with the usual supremum norm. As in §11.2 we take $H = L^2(0,1)$. We define

$$
Ax = \frac{d^2 x}{d\xi^2} - x, \; x \in D(A) = H^2(0,1) \cap H_0^1(0,1),
$$

and

$$F(x)(\xi) = f(x(\xi)) + x(\xi), \; \xi \in [0,1].$$

Theorem 11.3.1 *Under Hypothesis 11.1(i) the solution of problem (11.3.1) on K has an invariant measure concentrated on K. If in addition Hypothesis 11.1(ii) holds then the solution of equation (11.3.1) on H has an invariant measure concentrated on K.*

Proof — We will check the hypotheses of Theorem 6.3.5 and Theorem 6.3.6. The semigroup $S(t)$, $t \geq 0$, generated by A is of the form

$$S(t)x = \sum_{n=1}^{\infty} e^{-\alpha_n t} e_n \langle e_n, x \rangle, \; x \in H, \qquad (11.3.2)$$

where $\alpha_n = -\pi^2 n^2 - 1$, $n \in \mathbb{N}$, and $e_n(\xi) = \sqrt{2} \sin n\pi\xi$, $n \in \mathbb{N}$. The same formula defines the restriction $S_K(t)$, $t \geq 0$, of $S(t)$, $t \geq 0$, to K. Both semigroups are compact repectively in H and K and

$$|S(t)| \leq e^{-2t}, \; \|S_K(t)\|_K \leq e^{-t}, \; t \geq 0.$$

It is also clear that the transformation F has the desired properties. Therefore it is enough to check that the process

$$W_A(t) = \int_0^t S(t - s)dW(s), \; t \geq 0,$$

has an E-continuous version such that for arbitrary $p \geq 1$

$$\sup_{t \geq 0} \mathbb{E}\|W_A(t)\|_K^p < +\infty.$$

This follows however from Theorem 5.2.9.

Remark 11.3.2 In I. Gyöngy and E. Pardoux [81], C. Donati Martin and E. Pardoux [55], and I. Gyöngy [80], equation (11.3.1) is studied under weaker assumptions by using a comparison theorem for SPDEs. For a result about the existence of an invariant measure in this setting see G. Da Prato and E. Pardoux [41].

11.4 Reaction–diffusion equations on d dimensional spaces

In this section we use the notation of §9.2. We will consider equations of the form

$$\left.\begin{array}{l} dX(t,\xi) = [(\Delta - m)X(t,\xi) + f(X(t,\xi))]\,dt + dW(t,\xi),\ t > 0, \\[2mm] x(0,\xi) = x(\xi),\ \xi \in \mathbb{R}^d, t > 0, \end{array}\right\}$$

$$(11.4.1)$$

where Δ is the Laplace operator, m a nonnegative constant and the covariance operator Q of the Wiener process W is either I or a convolution operator

$$Qu(\xi) = q * u(\xi),\ \xi \in \mathbb{R}^d,\ u \in U,$$

satisfying Hypothesis 9.1.

As far as the function $f : \mathbb{R} \to \mathbb{R}$ is concerned we will require that the following is satisfied.

Hypothesis 11.2 *The function f is of the form*

$$f = f_0 + f_1$$

where $f_0(\xi)+\eta\xi$, $\xi \in \mathbb{R}$, is continuous and decreasing for some $\eta \in \mathbb{R}$, and for some $s \geq 1$ and $c_0 > 0$

$$|f_0(\xi)| \leq c_0\left(1 + |\xi|^s\right),\ \xi \in \mathbb{R}.$$

Moreover f_1 is Lipschitz continuous.

For instance if f_1 is Lipschitz continuous and

$$f_0(\xi) = -\xi^{2n+1} + \sum_{k=0}^{2n} b_k \xi^{2n-k},\ \xi \in \mathbb{R},$$

then Hypothesis 11.2 is satisfied with $s = 2n + 1$.

We start from an existence result.

Theorem 11.4.1 *Assume that f satisfies Hypothesis 11.2 and either*

(i) $d = 1$, and either $Q = I$ or Hypothesis 9.1 holds

or

(ii) $d > 1$ and Hypothesis 9.1 holds.

Then equation (11.4.1) has a unique generalized solution in H^κ, $\kappa > 0$, and H_r, $r > 0$. If in addition $x \in L_\rho^{2s}(\mathbb{R}^d)$ where $\rho = \rho^\kappa$ or $\rho = \rho_r$, then the generalized solution is mild.

Proof — We will apply Theorem 5.5.8 with $H = L_{\rho^\kappa}^2(\mathbb{R}^d)$, $H = L_{\rho_r}^2(\mathbb{R}^d)$, and $K = L_{\rho^\kappa}^{2s}(\mathbb{R}^d)$, $K = L_{\rho_r}^{2s}(\mathbb{R}^d)$.

By Proposition 9.4.5 the heat semigroup can be extended to semigroups on all those spaces. If $A_{\kappa,2}$, $A_{r,2}$, $A_{\kappa,2s}$, $A_{r,2s}$ are the generators of the semigroups then, again by Proposition 9.4.5, the operators $A_{\kappa,2} + \eta$, $A_{r,2} + \eta$, $A_{\kappa,2s} + \eta$, $A_{r,2s} + \eta$ are m–dissipative for sufficiently small η.

To see that the operator $F + \eta$ where

$$\left. \begin{array}{rcl} F(x)(\xi) & = & f(x(\xi)), \quad \xi \in \mathbb{R}^d, \\[2mm] D(F) & = & \{x \in L_\rho^2(\mathbb{R}^d) : f(x(\cdot)) \in L_\rho^2(\mathbb{R}^d)\}, \end{array} \right\}$$

sastisfies Hypothesis 5.4, note that $D(F) \supset K$ and the domain $D(F_K)$ of the part F_K of F in K contains $L_\rho^{2s}(\mathbb{R}^d)$. It is also easy to see that $F + \eta$ and $F_K + \eta$ are m–dissipative for small η and that F maps bounded sets in K into bounded sets in H. Then Hypothesis 5.4 holds. The following proposition shows that Hypothesis 5.5 is satisfied as well and consequently finishes the proof of Theorem 11.4.1.

Proposition 11.4.2 *Let $A_{\rho,p}$, where $\rho = \rho^\kappa$, $\kappa > 0$, or $\rho = \rho_r$, $r > 0$, and $p \geq 1$, be the generator of the heat semigroup in $L_\rho^p(\mathbb{R}^d)$. Then the Ornstein–Uhlenbeck process $W_{A_{\rho,p}}(t)$, $t \geq 0$, has continuous trajectories in $L_\rho^p(\mathbb{R}^d)$, provided that condition (i) or condition (ii) of Theorem 11.4.1 holds.*

Proof — We will consider only the more difficult case (ii). Without any loss of generality, we can assume that $p \geq 2$ and that

$$W(t,\xi) = \sum_{j=1}^{\infty} Q^{1/2} e_j(\xi) \beta_j(t), \quad t \geq 0, \ \xi \in \mathbb{R}^d.$$

We will use the factorization formula, Theorem 5.2.5, and investigate first the process

$$Y_\delta(t) = \int_0^t S(t-s)(t-s)^{-\delta} dW(s), \quad t \geq 0,$$

where $\delta \in]\frac{1}{p}, \frac{1}{2}[$. By Gaussianity of $Y_\delta(t,\xi)$, $t \geq 0$, $\xi \in \mathbb{R}^d$, for a suitable constant $c_1 > 0$,

$$\mathbb{E}|Y_\delta(t)|^p_{L^p_\rho(\mathbb{R}^d)} = \mathbb{E}\int_{\mathbb{R}^d} \rho(\xi)|Y_\delta(t,\xi)|^p d\xi$$

$$= \int_{\mathbb{R}^d} \rho(\xi)\mathbb{E}\left|\int_0^t (t-s)^{-\delta}\sum_{j=1}^\infty (S(t-s)Q^{1/2}h_j)(\xi)d\beta_j(s)\right|^p d\xi$$

$$\leq c_1 \int_{\mathbb{R}^d} \rho(\xi)\left(\mathbb{E}\left|\int_0^t (t-s)^{-\delta}\sum_{j=1}^\infty (S(t-s)Q^{1/2}h_j)(\xi)d\beta_j(s)\right|^2\right)^{p/2} d\xi$$

$$\leq c_1 \int_{\mathbb{R}^d} \rho(\xi)\left(\int_0^t \sigma^{-2\delta}\sum_{j=1}^\infty |S(s)Q^{1/2}h_j(\xi)|^2 ds\right)^{p/2} d\xi.$$

But

$$S(s)Q^{1/2}h_j(\xi) = e^{-\alpha s}p_s * q_1 * h_j(\xi), \quad \xi \in \mathbb{R}^d,$$

where

$$Q^{1/2}u = q_1 * u, \quad u \in U.$$

Consequently

$$\left|S(s)Q^{1/2}h_j(\xi)\right|^2 = e^{-2\alpha s}\langle p_s * q_1(\xi - \cdot), h_j \cdot\rangle_U$$

and by Parseval's and Plancherel's identities

$$\sum_{j\in\mathbb{R}^d}\left|S(s)Q^{1/2}h_j(\xi)\right|^2 = e^{-2\alpha s}\|p_s * q_1\|_U$$

$$= e^{-2\alpha s}\int_{\mathbb{R}^d} e^{-2s|\eta|^2}g(\eta)d(\eta).$$

Therefore, for a constant c_2,

$$\mathbb{E}|Y_\delta(t)|^p_{L^p_\rho(\mathbb{R}^d)} \le c_2 \left[\int_{\mathbb{R}^d} g(\eta) \left(e^{-2s(m+|\eta|^2)} s^{-2\alpha} ds \right) d(\eta) \right]^{p/2}.$$

Since $m > 0$, for another constant c_3,

$$\mathbb{E}|Y_\delta(t)|^p_{L^p_\rho(\mathbb{R}^d)} \le \left[c_3 \int_{\mathbb{R}^d} \frac{g(\eta)}{(m+|\eta|^2)^{2\delta+1}} d\eta \right]^{p/2} < +\infty$$

for arbitrary $t \ge 0$.

Taking into account, see G. Da Prato and J. Zabczyk [44, p.128], that

$$W_{A_{\rho,p}}(t) = \frac{\sin \pi \delta}{\pi} G_\delta Y_\delta(t), \ t \ge 0,$$

where G_δ is the operator given by

$$G_\delta \varphi(t) = \int_0^t S(t-s)(t-s)^{\delta-1} \varphi(s) ds, \ t \in [0,T], \ \varphi \in L^p_\rho(0,T; L^p(\mathbb{R}^d)),$$

we see that

$$\|W_{A_{\rho,p}}(t)\|_{L^p_\rho(\mathbb{R}^d)} \le \frac{\sin \pi \delta}{\pi} \left(\int_0^t s^{q(\delta-1)} \|S(s)\|^q_{L(L^p_\rho(\mathbb{R}^d))} ds \right)^{1/q}$$

$$\times \left(\int_0^t \|Y_\delta(s)\|^p_{L^p_\rho(\mathbb{R}^d)} ds \right)^{1/p},$$

where $q = \frac{p}{p-1}$.

The function $\|S(s)\|^q_{L(L^p(\mathbb{R}^d))}$, $s \ge 0$, is locally bounded and $q(\delta - 1) > -1$. Therefore for arbitrary $T > 0$ there exists a constant c_4 such that

$$\mathbb{E} \left(\sup_{t\in[0,T]} \|W_{A_{\rho,p}}(t)\|^p_{L^p_\rho(\mathbb{R}^d)} \right) \le c_4 \int_0^T \mathbb{E}\|Y_\delta(s)\|_{L^p_\rho(\mathbb{R}^d)} ds < +\infty,$$

and the required continuity follows.   ■

The next theorem is devoted to the existence and uniqueness of the invariant measures.

Theorem 11.4.3 *In addition to the conditions of Theorem 11.4.1 assume that the function f_0 is decreasing and*

$$m - \|f_1\|_{\text{Lip}} > \omega > 0.$$

Then there exists $\gamma_0 > 0$ such that the transition semigroup P_t, $t \geq 0$, corresponding to the solution of (11.4.1) has a unique invariant measure in both $H = L^2_{\rho^\kappa}(\mathbb{R}^d)$ and $H = L^2_{\rho_r}(\mathbb{R}^d)$, for any $\kappa, r \in]0, \kappa_0[$. Moreover there exists $c > 0$ such that for any bounded Lipschitz function φ on H, all $t > 0$ and all $x \in H$

$$\left| P_t\varphi(x) - \int_H \varphi(x)\mu(dx) \right| \leq (c + 2|x|)e^{-\omega t} \|\varphi\|_{\text{Lip}}.$$

Proof — We apply Theorem 6.3.3. Arguing as in the proof of Theorem 11.4.1 one easily shows that condition (i) of Hypothesis 6.2 holds. It remains to check that (ii) holds as well. We will prove that for arbitrary $p \geq 1$

$$\sup_{t \geq 0} \mathbb{E}\left(\|W_{A_{\rho,p}}(t)\|_{L^p_\rho(\mathbb{R}^d)} \right) < +\infty.$$

Note, compare the proof of Theorem 11.4.1, that for all $t \geq 0$

$$\mathbb{E}\|W_{A_{\rho,p}}(t)\|^p_{L^p_\rho(\mathbb{R}^d)} = \int_{\mathbb{R}^d} \rho(\xi) \left(\left| \int_0^t \sum_{j=1}^\infty S(t-s)Q^{1/2}h_j(\xi)d\beta_j(s) \right|^p \right) d\xi$$

$$\leq c_1 \int_{\mathbb{R}^d} \rho(\xi) \left(\int_0^t \sum_{j=1}^\infty |S(t-s)Q^{1/2}h_j(\xi)|^2 ds \right)^{p/2} d\xi$$

$$\leq c_1 \int_{\mathbb{R}^d} \rho(\xi) \left(\frac{g(\eta)}{2(m+|\eta|^2)} d\eta \right)^{p/2} d\xi \qquad \blacksquare$$

Chapter 12

Spin systems

Invariant measures for classical and quantum spin systems on d dimensional lattices $\mathbb{Z}^d$ are considered. Using the dissipativity method, conditions for the existence of invariant measures are given. The exponential convergence of the transition probabilities to the invariant measure is established as well.

12.1 Introduction

Let $\mathbb{Z}^d$ be the d dimensional lattice whose elements will be interpreted as atoms. A *configuration* is basically any real function x defined on $\mathbb{Z}^d$. The value $x(\gamma)$ of the configuration x at the point γ can be viewed as the state of the atom γ and it is sometime called the *spin* of γ. Physical properties of a lattice system are determined by a matrix $(a_{\gamma,j})_{\gamma,j \in \mathbb{Z}^d}$ of *global interactions* and a function $g : \mathbb{R} \to \mathbb{R}$, of *local interactions*. For an arbitrary finite set $\Gamma \subset \mathbb{Z}^d$ the matrix $(a_{\gamma,j})_{\gamma,j \in \mathbb{Z}^d}$ is assumed to be negative definite. The Hamiltonian H_Γ of the *restricted* spin system is given by the formula

$$H_\Gamma(x) = - \sum_{\gamma,j \in \Gamma^d} a_{\gamma,j} x_\gamma x_j - \sum_{\gamma \in \Gamma} g(x_\gamma), \ x \in \mathbb{R}^{\mathbb{Z}^d}, \qquad (12.1.1)$$

and the corresponding *restricted Gibbs measure* μ_Γ has a density, with respect to the Lebesgue measure on $\mathbb{R}^\Gamma$, of the form

$$C_\Gamma e^{\{-H_\Gamma(x)\}} = C_\Gamma e^{\{\sum_{\gamma,j \in \mathbb{Z}^d} a_{\gamma,j} x_\gamma x_j + \sum_{\gamma \in \Gamma} g(x_\gamma)\}}, \ x \in \mathbb{R}^\Gamma, \quad (12.1.2)$$

225

where C_Γ is a normalizing constant such that

$$\mu_\Gamma(\mathbb{R}^\Gamma) = 1. \qquad (12.1.3)$$

For this to be true some conditions on the behaviour of g at infinity have to be imposed. Identifying the space $\mathbb{R}^\Gamma$ with the set of those configurations $x \in \mathbb{R}^{\mathbb{Z}^d}$ which vanish outside of Γ, every measure μ_Γ can be regarded as a probability measure on $\mathbb{R}^{\mathbb{Z}^d}$. Assume now that $\{\Gamma_n\}$ is any increasing sequence of finite subsets of $\mathbb{Z}^d$. All probability measures on the space of configurations that are weak limits of $\{\mu_{\Gamma_n}\}$ are called *Gibbs measures*. Of special interest are those Gibbs measures which are concentrated on configurations with polynomial growth.

Denote by ν_Γ the Gaussian measure on $\mathbb{R}^\Gamma$ with mean vector zero and covariance matrix

$$S_\Gamma = \frac{1}{2}\left((-a_{\gamma,j})_{\gamma,j \in \Gamma}\right)^{-1}.$$

Then for a possibly different constant C'_Γ

$$\mu_\Gamma(dx) = C'_\Gamma e^{\sum_{\gamma \in \Gamma} g(x_\gamma)}\nu_\Gamma(dx). \qquad (12.1.4)$$

In general the sequence of densities

$$\{C'_\Gamma e^{\sum_{\gamma \in \Gamma} g(x_\gamma)}\}_{n \in \mathbb{N}}$$

does not converge to a density with respect to the limit of Gaussian measures μ_{Γ_n}. This makes the problem of finding a Gibbs measure rather difficult. Let f be a primitive function of $\frac{1}{2}g$:

$$\frac{df}{dz}(z) = \frac{1}{2}\,g(z), \ z \in \mathbb{R}. \qquad (12.1.5)$$

Under general conditions on g the measure μ_Γ is the unique invariant measure for the stochastic differential equation

$$\left.\begin{array}{l} dX_\gamma(t) = \left(\displaystyle\sum_{j \in \Gamma} a_{\gamma j}X_j(t) + f(X_\gamma(t))\right) dt + dW_\gamma(t), \\[2em] \gamma \in \Gamma, \ t \geq 0, \end{array}\right\} \qquad (12.1.6)$$

where W_γ, $\gamma \in \Gamma$, are independent Wiener processes. It is therefore reasonable to expect that Gibbs measures corresponding to $((a_{\gamma,j}), g)$ can be invariant measures for the infinite system of equations

$$\left. \begin{aligned} dX_\gamma(t) &= \left(\sum_j a_{\gamma j} X_j(t) + f(X_\gamma(t)) \right) dt + dW_\gamma(t), \\[2mm] X_\gamma(0) &= x_\gamma, \ \gamma \in \mathbb{Z}^d, \ t \geq 0. \end{aligned} \right\} \qquad (12.1.7)$$

In the following section we will study system (12.1.7) regarded as a stochastic evolution equation

$$\left. \begin{aligned} dX &= (AX + F(X))dt + BdW(t), \\[2mm] X(0) &= x \in H \end{aligned} \right\} \qquad (12.1.8)$$

on a properly chosen Hilbert space H of functions on $\mathbb{Z}^d$. We not only prove existence of solutions to (12.1.7) but give conditions under which the corresponding transition semigroup has a unique invariant measure. Generalization to quantum spin systems will be the object of §12.3. We follow G. Da Prato and J. Zabczyk [48].

12.2 Classical spin systems

Let $W(t) = \{W_\gamma(t)\}_{\gamma \in \mathbb{Z}^d}$, $t \geq 0$, be a Wiener process on $(\Omega, \mathcal{F}, \mathbb{P})$ with values in $U = \ell^2(\mathbb{R}^d)$ and with covariance operator $Q \in L(U)$. In particular if Q is the identity operator then the processes W_γ, $\gamma \in \mathbb{Z}^d$, are independent standard real valued Wiener processes. However, Q can be any nonnegative definite operator on U and for any $t > 0$ the family $\{W_\gamma(t)\}_{\gamma \in \mathbb{Z}^d}$ can form a general Gaussian random field including stationary ones if Q is shift invariant.

To study system (12.1.7) we apply results on dissipative systems from subsections 5.5.2 and 6.3.2 to (12.1.8). The operators A and F will be given by the formulae

$$\left. \begin{aligned} A(x_\gamma) &= \left(\sum_{j \in \mathbb{Z}^d} a_{j\gamma} x_j \right), \ x = (x_\gamma) \in H, \\[2mm] F(x_\gamma) &= (f(x_\gamma)), \ x = (x_\gamma) \in H, \end{aligned} \right\} \qquad (12.2.1)$$

and $H = \ell^2_\rho(\mathbf{Z}^d)$ will be a weighted Hilbert space of sequences (x_γ) with positive summable weight $\rho : \mathbf{Z}^d \to \mathbf{R}^+$. B will denote the embedding operator J from U into H. $C_b(H)$ and $B_b(H)$ are respectively the space of all bounded continuous functions in H and the space of all bounded Borel functions in H.

The following proposition, which goes back to Schur, see P. R. Halmos [83, Pag. 36], gives sufficient conditions under which the matrix $(a_{\gamma j})$ defines a bounded operator on $\ell^p_\rho(\mathbf{Z}^d)$ for arbitrary $p \in [1, +\infty]$.

Proposition 12.2.1 *Assume that*

(I) $\displaystyle \sup_{\gamma \in \mathbf{Z}^d} \sum_{j \in \mathbf{Z}^d} |a_{\gamma j}| = \alpha < +\infty,$

(II) There exists $\beta > 0$ such that

$$\sum_{\gamma \in \mathbf{Z}^d} |a_{\gamma j}| \rho(\gamma) \le \beta \rho(j), \quad j \in \mathbf{Z}^d.$$

Then the formula

$$A_p x = \left(\sum_{j \in \mathbf{Z}^d} a_{\gamma j} x_j \right), \quad x \in \ell^p_\rho(\mathbf{Z}^d), \qquad (12.2.2)$$

defines a linear bounded operator on $\ell^p_\rho(\mathbf{Z}^d)$, for all $p \in [1, +\infty]$, with norm not greater thAn

$$\alpha^{1/q} \beta^{1/p}, \quad \frac{1}{p} + \frac{1}{q} = 1.$$

Proof— Assume, for instance, that $p \in \,]1, +\infty[$ and that $x = (x_j)$ has only a finite number of coordinates different from 0. By Hölder's inequality and (i), for arbitrary $\gamma \in \mathbf{Z}^d$

$$\left| \sum_{j \in \mathbf{Z}^d} a_{\gamma j} x_j \right|^p \le \left(\sum_{j \in \mathbf{Z}^d} |a_{\gamma j}| \, |x_j| \right)^p \le \left(\sum_{j \in \mathbf{Z}^d} (|a_{\gamma j}|^{1/p} |x_j|) |a_{\gamma j}|^{1/q} \right)^p$$

$$\le \left(\sum_{j \in \mathbf{Z}^d} |a_{\gamma j}| \, |x_j|^p \right) \left(\sum_{j \in \mathbf{Z}^d} |a_{\gamma j}| \right)^{p/q} \le \alpha^{p/q} \sum_{j \in \mathbf{Z}^d} |a_{\gamma j}| \, |x_j|^p.$$

Consequently, by (ii)

$$\sum_{\gamma \in \mathbb{Z}^d} \rho(\gamma) \left| \sum_{j \in \mathbb{Z}^d} a_{\gamma j} x_j \right|^p \leq \alpha^{p/q} \sum_{\gamma, j \in \mathbb{Z}^d} |a_{\gamma j}| \rho(\gamma) |x_j|^p$$

$$\leq \alpha^{p/q} \beta \sum_{j \in \mathbb{Z}^d} \rho(j) |x_j|^p,$$

and the result follows.   ∎

We will impose several conditions on the matrix $(a_{\gamma,j})_{\gamma, j \in \mathbb{Z}^d}$ and on the positive weight ρ, compare S. Albeverio, Y. G. Kondratiev and T. V. Tsycalenko [4], and B. Zegarlinski [170]. We will assume that there exist constants $R, M > 0$ such that

$$\left.\begin{array}{l} a_{\gamma, j} = 0 \ \text{if} \ |\gamma - j| > R, \\[2mm] |a_{\gamma, j}| \leq M \ \text{for all} \ \gamma, j \in \mathbb{Z}^d \end{array}\right\} \tag{12.2.3}$$

and

$$\left.\begin{array}{l} \left| \dfrac{\rho(\gamma)}{\rho(j)} \right| \leq M \ \text{if} \ |\gamma - j| \leq R, \\[4mm] \displaystyle\sum_{\gamma \in \mathbb{Z}^d} \rho(\gamma) < +\infty. \end{array}\right\} \tag{12.2.4}$$

The next result follows directly from Proposition 12.2.1.

Proposition 12.2.2 *If conditions (12.2.3) and (12.2.4) hold then the operators A_p given by (12.2.1) are bounded.*

Condition (12.2.4) is satisfied for two specific families of weights,

$$\rho^\kappa(\gamma) = e^{-\kappa |\gamma|}, \ \gamma \in \mathbb{Z}^d, \ \kappa > 0, \tag{12.2.5}$$

and

$$\rho_r(\gamma) = \rho_{\kappa r}(\gamma) = \frac{1}{1 + \kappa |\gamma|^r}, \ \gamma \in \mathbb{Z}^d, \ \kappa > 0, \ r > d. \tag{12.2.6}$$

If $r > d$ is fixed then all spaces $\ell^p_{\rho_{\kappa r}}(\mathbb{R}^d)$, $\kappa > 0$, are isomorphic. Therefore the index κ will often be dropped.

Remark 12.2.3 It is useful to notice that for arbitrary $\kappa > 0$ and $r > d$

$$\ell^2_{\rho^\kappa}(\mathbf{Z}^d) \supset \mathcal{H}_\infty \supset \ell^2_{\rho_\kappa}(\mathbf{Z}^d),$$

where $\mathcal{H}_\infty$ is the space of configurations with at most polynomial growth. Moreover the two spaces $\ell^2_{\rho^\kappa}(\mathbf{Z}^d)$ and $\ell^2_{\rho_\kappa}(\mathbf{Z}^d)$ are shift invariant: if $x \in \ell^2_{\rho^\kappa}(\mathbf{Z}^d)$ (resp. $\ell^2_{\rho_\kappa}(\mathbf{Z}^d)$, then for arbitrary $\gamma \in \mathbf{Z}^d$, $x(\cdot + \gamma) \in \ell^2_{\rho^\kappa}(\mathbf{Z}^d)$, (resp. $\ell^2_{\rho_\kappa}(\mathbf{Z}^d)$, and therefore they are appropriate for physical considerations.)

As far as the function $f : \mathbb{R} \to \mathbb{R}$ is concerned we will require that it satisfies Hypothesis 11.2.

Our first theorem is the following existence result.

Theorem 12.2.4 *Assume that conditions (12.2.3), (12.2.4), and Hypothesis 11.2 are satisfied. Let $H = \ell^2_\rho(\mathbf{Z}^d)$ and let the operators A and F be given by (12.2.1).*

(i) For arbitrary $x \in \ell^{2s}_\rho(\mathbf{Z}^d)$, there exists a unique strong solution $X(t,x)$, $t \geq 0$ of (12.1.8) and (12.1.7).

(ii) For arbitrary $x \in H = \ell^2_\rho(\mathbf{Z}^d)$, there exists a unique generalized solution $X(t,x)$, $t \geq 0$, of (12.1.8) and (12.1.7), and the transition semigroup

$$P_t\varphi(x) = \mathbb{E}(\varphi(X(t,x))), \ t \geq 0, x \in H, \ \varphi \in C_b(H),$$

has the Feller property.

Proof —
We will apply Theorem 5.5.8 with

$$H = \ell^2_\rho(\mathbf{Z}^d) \ \text{and} \ K = l^{2s}_\rho(\mathbf{Z}^d).$$

Note that, by Proposition 12.2.1, the linear operators A and A_{2s} are bounded on H and K respectively and therefore, without any loss of generality, we can assume that $A = 0$. The operator $F + \eta_1$, with the domain

$$D(F) = \left\{ x \in \ell^2_\rho : \sum_{\gamma \in \mathbf{Z}^d} \rho(\gamma)|f(x_\gamma)|^2 < +\infty \right\},$$

is m–dissipative in H and its part in K is m–dissipative in K provided η_1 is small enough. It is also clear that F maps bounded sets in K into bounded sets in H. So Hypothesis 5.4 is satisfied in our situation.

We show now that Hypothesis 5.2 holds as well. Let $Q^{1/2} = (q_{\gamma j})$ be the nonnegative square root of Q. Then

$$W_\gamma(t) = \sum_{j \in \mathbf{Z}^d} q_{\gamma j} \widehat{W}_\gamma(t), \ \gamma \in \mathbf{Z}^d,$$

where $\widehat{W}_\gamma$, $\gamma \in \mathbf{Z}^d$, are independent real valued Wiener processes. By Gaussianity of $\widehat{W}$, for any $p \geq 1$,

$$\mathbb{E}\,|W_\gamma(t)|^p = c_p \left(\mathbb{E}\,|W_\gamma(t)|^2 \right)^{p/2}$$

$$= c_p t^{p/2} \left(\sum_{j \in \mathbf{Z}^d} q_{\gamma j}^2 \right)^{p/2}, \ t \geq 0.$$

Consequently

$$\mathbb{E}\|W(t)\|^p_{\ell^p_\rho(\mathbf{Z}^d)} = c_p t^{p/2} \sum_{\gamma \in \mathbf{Z}^d} \rho(\gamma) \left(\sum_{j \in \mathbf{Z}^d} q_{\gamma j}^2 \right)^{p/2}, \ t \geq 0.$$

Since $Q^{1/2}$ is a bounded operator on $\ell^2(\mathbf{Z}^d)$, we have

$$\sup_{\gamma \in \mathbf{Z}^d} \sum_{j \in \mathbf{Z}^d} q_{\gamma j}^2 < +\infty,$$

and $W(t)$, $t \geq 0$, is an $\ell^p_\rho(\mathbf{Z}^d)$-valued Wiener process. By Kolmogorov's theorem W has a continuous version in any $\ell^p_\rho(\mathbf{Z}_d), p \geq 1$, and Hypothesis 5.5 is satisfied. ∎

Remark 12.2.5 Similar results, but with different concepts of solutions, are contained e.g. in H. Doss and G. Royer [57] and in G. Royer [128]. Existence of strong solutions is treated in G. Da Prato and J. Zabczyk [48]. The conditions we impose on f are slightly weaker than those in B. Zegarlinski [170] as we do not require that f_0 and f_1 are differentiable. By (ii) the semigroup P_t, $t \geq 0$ has the Feller property. Note however, that by Theorem 9.2.1 it is not a strong Feller one even in the linear case, $f = 0$. Moreover if $f = 0$ the transition probabilities $P_t(x, \cdot)$, $x \in H$, $t > 0$, are in general singular and therefore Doob's theorem, see §4.2, is not applicable in the present context.

The following theorem gives precise information on the rate of convergence to equilibrium.

Theorem 12.2.6 *Assume, in addition to the conditions of Theorem 12.2.4, that for an $\eta > 0$, the operator $A + \eta$, restricted to $\ell^2(\mathbb{Z}^d)$, is dissipative, that f_0 is decreasing and $\eta - \|f_1\|_{\text{Lip}} > \omega > 0$. Then there exists $\kappa_0 > 0$ such that in the spaces $\ell_\rho^2(\mathbb{Z}^d) = H$, with $\rho = \rho^\kappa$ given by (12.2.5), or with $\rho = \rho_r$ given by (12.2.6), and $\kappa, r \in \,]0, \kappa_0[$, the equation (12.1.8) has unique generalized solutions. The semigroup P_t, $t \geq 0$, has a unique invariant measure μ on H and there exists $c > 0$ such that, for any bounded and Lipschitz function φ on H, all $t > 0$ and all $x \in H$,*

$$\left| P_t\varphi(x) - \int_H \varphi(x)\mu(dx) \right| \leq (c + 2|x|_H)e^{-\omega t}\|\varphi\|_{\text{Lip}}.$$

To prove the theorem we need the following two elementary lemmas.

Lemma 12.2.7 *Assume that conditions (12.2.3) and (12.2.4) are satisfied and in addition*

$$\left| \sqrt{\frac{\rho(\gamma)}{\rho(j)}} - 1 \right| \leq \delta, \ \ for \ |\gamma - j| \leq R.$$

If

$$\langle Ax, x \rangle_{\ell^2(\mathbb{Z}^d)} \leq \alpha\|x\|_{\ell^2(\mathbb{Z}^d)}^2, \ x \in \ell^2(\mathbb{Z}^d),$$

then

$$\langle Ax, x \rangle_{\ell_\rho^2(\mathbb{Z}^d)} \leq (\alpha + \frac{\delta}{2M}(2R + 1)^d)\|x\|_{\ell_\rho^2(\mathbb{Z}^d)}^2, \ x \in \ell_\rho^2(\mathbb{Z}^d).$$

Proof — Take $x \in \ell_\rho^2(\mathbb{Z}^d)$, then

$$
\langle Ax, x \rangle_{\ell_\rho^2(\mathbb{Z}^d)} = \sum_\gamma \rho(\gamma) \left(\sum_j a_{\gamma,j} x_j \right) x_\gamma
$$

$$
= \sum_{|\gamma - j| \leq R} \sqrt{\frac{\rho(\gamma)}{\rho(j)}} a_{\gamma,j} \sqrt{\rho(j)} x_j \sqrt{\rho(\gamma)} x_\gamma
$$

$$
\leq \sum_{|\gamma - j| \leq R} \left(\sqrt{\frac{\rho(\gamma)}{\rho(j)}} a_{\gamma,j} - a_{\gamma,j} \right) \sqrt{\rho(j)} x_j \sqrt{\rho(\gamma)} x_\gamma
$$

$$
+ \sum_{|\gamma - j| \leq R} a_{\gamma,j} \sqrt{\rho(j)} x_j \sqrt{\rho(\gamma)} x_\gamma
$$

$$
\leq \frac{\delta}{2M} \left[\sum_{|\gamma - j| \leq R} \left(\rho(j) x_j^2 + \rho(\gamma) x_\gamma^2 \right) \right] + \alpha \|x\|_{\ell_\rho^2(\mathbb{Z}^d)}^2
$$

$$
= \frac{\delta}{2M} (2R + 1)^d |x|_{\ell_\rho^2(\mathbb{Z}^d)}^2 + \alpha \|x\|_{\ell_\rho^2(\mathbb{Z}^d)}^2.
$$

■

Lemma 12.2.8 *For weights ρ^κ and $\rho_{\kappa,r}$ given by (12.2.5) and by (12.2.6) we have*

$$
\lim_{\kappa \to 0} \left[\sup_{|\gamma - j| \leq R} \left| \sqrt{\frac{\rho^\kappa(\gamma)}{\rho^\kappa(j)}} - 1 \right| \right] = 0, \quad \lim_{\kappa \to 0} \left[\sup_{|\gamma - j| \leq R} \left| \sqrt{\frac{\rho_{\kappa,r}(\gamma)}{\rho_{\kappa,r}(j)}} - 1 \right| \right] = 0.
$$

The proof follows by direct calculations.

Proof of Theorem 12.2.6 — We will consider only the weights ρ^κ as the case of $\rho = \rho_{\kappa,r}$ can be treated in the same way.

It is clear that

$$
\langle F(x) - F(y), x - y \rangle \leq \|f_1\|_{\text{Lip}} \|x - y\|^2, \quad x, y \in D(F),
$$

and therefore $F - \|f_1\|_{\text{Lip}}$ is m–dissipative. By Lemma 12.2.7 and Lemma 12.2.8 for arbitrary $\varepsilon > 0$ there exists $\kappa_\varepsilon > 0$ such that for

$\kappa \in \,]0, \kappa_\varepsilon[$

$$\langle A(x-y)+F(x)-F(y)+(\eta-\varepsilon)(x-y), x-y\rangle_{\ell^2_{\rho^\kappa}(\mathbb{Z}^d)} \leq 0, \ x, y \in \ell^2_{\rho^\kappa}(\mathbb{Z}^d).$$

Consequently if $\kappa \in \,]0, \kappa_\varepsilon[$ the operator $A + \eta - \varepsilon$ is m–dissipative. Since $-\|f_1\|_{\text{Lip}} + \eta - \varepsilon > \omega$ for sufficiently small $\varepsilon > 0$ Hypothesis 6.2 is fulfilled.

We will show now that the process W_A has bounded p–moments in $\ell^p_\rho(\mathbb{Z}^d)$ spaces for all $p \geq 1$:

$$\sup_{t>0} \mathbb{E}|W_A(t)|^p_{\ell^p_\rho(\mathbb{Z}^d)} < +\infty. \tag{12.2.7}$$

Write $W_A(t) = (Z_\gamma(t))^d_{\gamma \in \mathbb{Z}}$ and $S(t) = (s_\gamma(t))^d_{j\gamma \in \mathbb{Z}}$. Then

$$Z_\gamma(t) = \sum_{j \in \mathbb{Z}^d} \int_0^t s_{\gamma j}(u) dW_j(u),$$

and

$$\mathbb{E}\|W_A(t)\|^p_{\ell^p_\rho(\mathbb{Z}^d)} = \sum_{\gamma \in \mathbb{Z}^d} \mathbb{E}\left(|Z_\gamma(t)|^p\right).$$

By Gaussianity and assuming, to simplify notation, that $Q = I$,

$$\mathbb{E}|Z_\gamma(t)|^p = c_p \left(\mathbb{E}|Z_\gamma(t)|^2\right)^{p/2}$$

$$\leq c_p \left(\int_0^t \sum_{j \in \mathbb{Z}^d} s^2_{\gamma j}(u) du\right)^{p/2}.$$

However,

$$\sum_{j \in \mathbb{Z}^d} s^2_{\gamma j}(u) \leq \|S(u)\|^2_{L(\ell^2(\mathbb{Z}^d))}$$

and therefore

$$\mathbb{E}\|W_A(t)\|^p_{\ell^p_\rho(\mathbb{Z}^d)} \leq c_p \sum_{\gamma \in \mathbb{Z}^d} \rho(\gamma) \left(\int_0^t \|S(u)\|^2_{L(\ell^2(\mathbb{Z}^d))} du\right)^{p/2}.$$

Since

$$\|S(u)\|_{L(\ell^2(\mathbb{Z}^d))} \leq e^{-\eta u}, \ u \geq 0,$$

the estimate (12.2.7) follows. From (12.2.7)

$$\sup_{t>0} \left(\mathbb{E}\|W_A(t)\| + \|F(W_A(t))\| \right) < +\infty.$$

This completes the proof. ∎

Remark 12.2.9 A theorem similar to Theorem 12.2.6 was proved earlier by B. Zegarlinski [170, Theorem 4.2]. His method, based on logarithmic Sobolev inequalities, does not require that $\|f_1\|_{\text{Lip}}$ is small. On the other hand we can cover less regular local interactions f like $f(\xi) = -\ \text{sgn}\ \xi |\xi|^{1/2}$, $\xi \in \mathbb{R}$, and our basic estimate holds for a more general class of functions φ. Since for all $r > 0$ the spaces $\ell_\rho^2(\mathbb{Z}^d)$ are identical as sets, see Remark 12.2.3, the invariant measure μ is supported by sequences (x_γ) such that

$$\sum_{\gamma \in \mathbb{Z}^d} \frac{|x_\gamma|^2}{1 + |\gamma|^r} < +\infty$$

for arbitrary $r > d$. This set is smaller than $\mathcal{H}_\infty$. Thus working with weighted Hilbert spaces not only is technically convenient but gives additional information about the spin system. More general nonlinearities could be treated with the help of appropriate Orlicz spaces (replacing $\ell_\rho^p(\mathbb{Z}^d)$).

We remark finally, that the noise process W can be absent in our approach, but not in the one based on logarithmic Sobolev inequalities.

12.3   Quantum lattice systems

Quantum lattice systems are a mixture of systems studied in §12.1 and §12.2 They were introduced in S. Albeverio, Y. G. Kondratiev and T. V. Tsycalenko [4] and are described by systems of equations of the form

$$
\left.
\begin{aligned}
dX_\gamma(t) &= \left(\mathcal{A}X_\gamma(t) + \sum_{j \in \mathbb{Z}^d} a_{\gamma j} X_j(t) + \mathcal{F}(X_\gamma(t)) \right) dt, \\[2mm]
&+ \ dW_\gamma(t) \\[2mm]
X_\gamma(0) &= x_\gamma \in \mathcal{H}, \ \gamma \in \mathbb{Z}^d, \ t \ge 0.
\end{aligned}
\right\}
$$
$$(12.3.1)$$

In the equation (12.3.1), $\mathcal{A}$ and $\mathcal{F}$ are respectively linear and nonlinear, in general unbounded, operators on a Hilbert space $\mathcal{H}$, $(a_{\gamma j})_{\gamma, j \in \mathbb{Z}^d}$ is a given matrix with real elements and $(W_\gamma)_{\gamma \in \mathbb{Z}^d}$ is a family of independent, cylindrical Wiener processes on $\mathcal{H}$. Following [4] we will require that the space $\mathcal{H}$ and the operators $\mathcal{A}$ and $\mathcal{F}$ are of special character although it will be clear that our general schema works in the more abstract setting. We will thus assume that

$$\mathcal{H} = L^2(0,1), \qquad\qquad (12.3.2)$$

$$
\left.
\begin{aligned}
&\mathcal{A} = \frac{d^2}{d\xi^2} - \alpha, \\[2mm]
&D(\mathcal{A}) = \{ x \in H^2(0,1) : \ x(0) = x(1), \ x'(0) = x'(1) \},
\end{aligned}
\right\}
\quad (12.3.3)
$$

$$
\left.
\begin{aligned}
&\mathcal{F}(x)(\xi) = f(x(\xi)), \ \xi \in [0,1], \\[2mm]
&D(\mathcal{F}) = \{ x \in L^2(0,1) : \ f(x) \in L^2(0,1) \},
\end{aligned}
\right\}
\quad (12.3.4)
$$

and on f and on the matrix $(a_{\gamma j})_{\gamma, j \in \mathbb{Z}^d}$ we will impose Hypothesis 11.2 and conditions (12.2.3)–(12.2.4).

To see that (12.3.1) is of the general form (12.1.8) define

$$H = \ell_\rho^2(L^2(0,1)) = \{ (x_\gamma) \in \mathcal{H}^{(\mathbb{Z}^d)} : \sum_{\gamma \in \mathbb{Z}^d} \rho(\gamma) \|x_\gamma\|_{\mathcal{H}}^2 < +\infty \},$$

$$
\begin{aligned}
K &= \ell_\rho^{2s}(L^{2s}(0,1)) \\[2mm]
&= \{ (x_\gamma) \in (L^{2s}(0,1))^{(\mathbb{Z}^d)} : \sum_{\gamma \in \mathbb{Z}^d} \rho(\gamma) \|x_\gamma\|_{L^{2s}(0,1)}^{2s} < +\infty \},
\end{aligned}
$$

where $\rho = \rho^\kappa$ or $\rho = \rho_r$, $\kappa, r > 0$, see the previous section.

Let $A = A_0 + A_1$ where A_1 is a bounded linear operator on H given by

$$A_1(x_\gamma) = \left(\sum_{j \in \mathbb{Z}^d} a_{\gamma j} x_j \right), \ x \in D(A_1) = H,$$

and

$$\left. \begin{array}{l} A_0(x_\gamma) = (Ax_\gamma), \ x = (x_\gamma) \in D(A_0). \\[2mm] D(A_0) = \{(x_\gamma) \in H : \sum_{\gamma \in \mathbb{Z}^d} \rho(\gamma)\|Ax_\gamma\|_{\mathcal{H}}^2 < +\infty\}. \end{array} \right\}$$

Boundedness of A_1 follows from an obvious generalization of Proposition 12.2.1.
Let, in addition,

$$\left. \begin{array}{l} F(x_\gamma) = (\mathcal{F}x_\gamma), \ x = (x_\gamma) \in D(F), \\[2mm] D(F) = \{(x_\gamma) \in H : \sum_{\gamma \in \mathbb{Z}^d} \rho(\gamma)\|\mathcal{F}x_\gamma\|_{\mathcal{H}}^2 < +\infty\}. \end{array} \right\}$$

It is easy to check that the operator $A_0 + \eta$ on H and its restriction $A_{0p} + \eta$ to K are m–dissipative for sufficiently small η.
Let $S(t)$, $S_0(t)$, $t \geq 0$, be C_0–semigroups on H generated by A and A_0 and write

$$W_A(t) = \int_0^t S(t-s)dW(s), \ t \geq 0,$$

$$W_{A_0}(t) = \int_0^t S_0(t-s)dW(s), \ t \geq 0,$$

where $W(t)$, $t \geq 0$, is the Wiener process $(W_\gamma(\cdot))_{\gamma \in \mathbb{Z}^d}$ embedded into H.

Proposition 12.3.1 *The processes $W_{A_0}(t)$, $t \geq 0$, $W_A(t)$, $t \geq 0$, have continuous versions with values in $\ell_\rho^p(L^p(0,1))$, $p \geq 1$.*

Proof — Existence of a continuous version of $W_{A_0}(t)$, $t \geq 0$, in $\ell_\rho^p(L^p(0,1))$ can be obtained by factorization, as in the proof of Proposition 11.4.2, and using the diagonal character of the semigroup $S_0(\cdot)$.

Note that for $Z(t) = W_A(t)$, $t \geq 0$,

$$Z(t) = W_{A_0}(t) + \int_0^t S_0(t - s)A_1 Z(s)ds, \quad t \geq 0. \qquad (12.3.5)$$

Since the semigroup $S_0(\cdot)$ has a C_0 –continuous restriction to $\ell_\rho^p(L^p(0,1))$, the equation (12.3.1) has a unique solution in

$$C([0, T]; \ell_\rho^p(L^p(0, 1))),$$

for arbitrary $T > 0$ by an elementary fixed point argument. This proves the result.   ∎

As a corollary from our discussion and Proposition 12.3.1 we set the following result.

Theorem 12.3.2 *Assume that Hypothesis 11.2 and conditions (12.3.2)–(12.3.4) hold. Then for arbitrary $x \in H$ the equation (12.3.1) has a unique generalized solution $X(\cdot, x)$. If $x \in K$ the solution is mild.*

As in §11.4 and in §12.2 one can derive a result on exponential decay for the transition semigroup P_t, $t \geq 0$, corresponding to the solution $X(\cdot, x)$ of (12.3.1).

Theorem 12.3.3 *In addition to the conditions of Theorem 12.3.2, assume that f_0 is decreasing and that $\alpha - \|f_1\|_{\text{Lip}} > \omega > 0$. Then there exists $\kappa_0 > 0$ such that the semigroup P_t, $t \geq 0$, corresponding to the solution of (12.3.1) has a unique invariant measure μ both on $H = L_{\rho^\kappa}^2$ and on $H = L_{\rho_\kappa}^2$. Moreover there exists $c > 0$ such that for any bounded and Lipschitz function φ on H, all $t > 0$ and all $x \in H$,*

$$\left| P_t\varphi(x) - \int_H \varphi(y)\mu(dy) \right| \leq (c + 2\|x\|)e^{-\omega t}\|\varphi\|_{\text{Lip}}.$$

Proof — As in the proof of Theorem 11.4.3 it is enough to check the assumptions of Theorem 6.3.3. We will show that

$$\sup_{t \geq 0} \mathbb{E}\left(|W_A(t)|_H\right) < +\infty, \qquad (12.3.6)$$

$$\sup_{t \geq 0} \mathbb{E}\left(|F(W_A(t))|_H\right) < +\infty, \qquad (12.3.7)$$

as the remaining conditions hold in an obvious way.

To prove (12.3.6) note that for $t \geq 1$,

$$\mathbb{E}\,|W_A(t)|_H = \mathbb{E}\left|\int_0^t S(t-s)dW(s)\right|_H = \mathbb{E}\left|\int_0^t S(s)dW(s)\right|_H$$

$$\leq \sum_{k=0}^{[t]-1} \mathbb{E}\left|\int_k^{k+1} S(s)dW(s)\right|_H + \mathbb{E}\left|\int_{[t]}^t S(s)dW(s)\right|_H$$

$$\leq \sum_{k=0}^{[t]-1} \|S(k)\|_H \mathbb{E}\left|\int_0^1 S(s)dW(s)\right|_H + \sup_{u \leq 1}\mathbb{E}\left|\int_0^u S(s)dW(s)\right|_H$$

$$\leq \sum_{k=0}^{[t]-1} e^{-\alpha k}\mathbb{E}\left|\int_0^1 S(s)dW(s)\right|_H + \sup_{u \leq 1}\mathbb{E}\left|\int_0^u S(s)dW(s)\right|_H$$

$$\leq \frac{2e^\alpha}{e^\alpha - 1}\sup_{u \leq 1}\mathbb{E}\left|\int_0^u S(s)dW(s)\right|_H, t \geq 1.$$

To show (12.3.7) note first that

$$\mathbb{E}\,|F(W_A(t))|_H \leq \left(\mathbb{E}\,|F(W_A(t))|_H^2\right)^{1/2} \leq (\mathbb{E}\,|W_A(t)|_K^s)^{1/2}.$$

Since the distribution of $W_A(t)$ on K is Gaussian, there exists a constant $\hat{c}$ such that

$$\mathbb{E}\,\|W_A(t)\|_K^s \leq \hat{c}(\mathbb{E}\,\|W_A(t)\|_K)^s.$$

Moreover there exists a constant $\hat{\alpha} > 0$ such that for the restriction $\hat{S}(t), t \geq 0$, of the semigroup $S(t), t \geq 0$, to K, one has

$$\|\hat{S}(t)\|_K \leq e^{-\hat{\alpha}t},\ t \geq 0.$$

It remains now to repeat the proof of (12.3.6). ∎

Remark 12.3.4 Theorem 12.3.3, under slightly stronger conditions, was proved earlier in [4] using specific properties of the one dimensional heat equation. It is, however, clear that the dissipativity method allows us to treat spaces $\mathcal{H}$ more general than $L^2(0,1)$. Existence of an invariant measure μ for (12.3.1) was also obtained in S. Albeverio, Y. G. Kondratiev and T. V. Tsycalenko [4] with more stringent conditions imposed on f, ρ and $(a_{\gamma j})$. Exponential estimates in Theorem 12.3.3 were obtained in G. Da Prato and J. Zabczyk [48].

Chapter 13

Systems perturbed through the boundary

In some applications the noise can effect the evolution of a system only through the boundary of a region. For such systems conditions are given for existence and uniqueness of invariant measures. Here once more the dissipativity method is used.

13.1   Introduction

In several applications external forces enter the system through the boundary of a region where the system evolves. The same is true if the forces have a stochastic character. Consider for instance the nonlinear heat equation in the region $G =]0, \pi[^d \subset \mathbb{R}^d$, with the noise affecting only a portion Γ_0

$$\Gamma_0 = \{\xi = (\xi_1, ..., \xi_d) \in \partial G : \xi_d = 0\},$$

of the boundary Γ of G. Thus the state equation is of the form

$$
\left.
\begin{array}{l}
\dfrac{\partial X}{\partial t}(t,\xi) = (\Delta - m)X(t,\xi) + f(X(t,\xi)), \; t > 0, \\[4mm]
X(0,\xi) = x(\xi), \; \xi \in G, \; \dfrac{\partial X}{\partial \nu}(t,\xi) = 0, \; \xi \in \Gamma \backslash \Gamma_0, \; t > 0, \\[4mm]
\dfrac{\partial X}{\partial \nu}(t,\xi) = \dfrac{\partial V}{\partial t}(t,\xi), \; \xi \in \Gamma_0, \; t > 0,
\end{array}
\right\}
$$
$$(13.1.1)$$

where ν is the inner normal vector to the boundary at those points $\xi \in \Gamma$ where it is well defined. The process $V(t)$, $t \geq 0$, is a Wiener process on $U = L^2(\Gamma_0)$ with the covariance operator Q not necessarily of trace class. The operator Q could be of integral type,

$$Qu(\xi) = \int_{\Gamma_0} g(\xi,\eta)u(\eta)d\eta, \; u \in U, \; \xi \in \Gamma_0, \qquad (13.1.2)$$

with the kernel g being the correlation function of the random field $V(1,\xi)$, $\xi \in \Gamma_0$. Of special interest is the case when the field $V(1,\xi)$, $\xi \in \Gamma_0$, is the restriction of a stationary random field on $\mathbb{R}^{d-1}$. Then there exists an integrable function $q_0 : \mathbb{R}^{d-1} \to \mathbb{R}$ such that

$$q(\xi,\eta) = q_0(\xi - \eta), \; \xi, \eta \in \Gamma_0. \qquad (13.1.3)$$

Moreover q_0 is the Fourier transform of the spectral density of the random field.

System (13.1.1) is our main example. We will establish existence and uniqueness of Markov solutions using the dissipativity method. We will show also that under additional conditions on f there exists a unique invariant measure for the solution. Asymptotic properties of the corresponding transition semigroup will be established as well.

We start from general information on parabolic systems with non–homogeneous boundary data. We formulate an integral version of a generalization of (13.1.1) and give conditions under which it has a solution. Then we apply the general theory for the specific model (13.1.1). We obtain in this way explicit conditions, in terms of the spectral density of the noise, implying the existence of solutions to (13.1.1) and the existence of a unique strongly mixing invariant measure.

13.2   Equations with non–homogeneous boundary conditions

We start from the linear case. Let $G \subset \mathbb{R}^d$ be a fixed region, $\Gamma = \partial G$ its boundary and let $\mathcal{A}$ be a linear second order differential operator and τ a linear first order differential operator. A non–homogeneous boundary problem is of the form

$$\left.\begin{array}{l} \dfrac{\partial y}{\partial t}(t,\xi) = \mathcal{A}y(t,\xi),\ (t,\xi) \in\,]0,+\infty[\, \times G, \\[2ex] \tau y(t,\xi) = u(t,\xi),\ (t,\xi) \in\,]0,+\infty[\, \times \Gamma, \\[2ex] y(0,\xi) = x(\xi),\ \xi \in \Gamma, \end{array}\right\} \qquad (13.2.1)$$

where u is a U–valued function and $U = L^2(\Gamma)$. Under some conditions on the data the solution to problem (13.2.1) is given by the formula

$$y(t) = S(t)x + (\lambda - A) \int_0^t S(t-s)\mathcal{D}u(s)ds,\ t > 0, \qquad (13.2.2)$$

where λ is a real number, A the infinitesimal generator of a strongly continuous semigroup $S(t)$, $t \geq 0$, on a Hilbert space $H = L^2(G)$ and $\mathcal{D}$ a properly defined linear bounded operator from U into H. Assume for instance that the linear operator

$$\left.\begin{array}{l} Ax = \mathcal{A}x, \\[2ex] D(A) = \{x \in H^{2,2}(G) : \tau = 0\}, \end{array}\right\} \qquad (13.2.3)$$

generates a strongly continuous semigroup $S(t)$, $t \geq 0$. on H. Let λ be a real number such that the problem

$$\lambda w - \mathcal{A}w = 0,\ \tau w = u \qquad (13.2.4)$$

has a unique solution $w = \mathcal{D}u \in H$ for arbitrary $u \in U$. If $u(\cdot)$ is a twice continuously differentiable function with values in U and $x - \mathcal{D}u(0) \in D(A)$ then there exists a strong solution of the equation

$$\left.\begin{array}{l} z'(t) = Az(t) + \lambda\mathcal{D}u(t) - \mathcal{D}u'(t),\ t > 0, \\[2ex] z(0) = x - \mathcal{D}u(0). \end{array}\right\} \qquad (13.2.5)$$

In particular $\tau(z(t)) = 0$ for all $t \geq 0$. Define

$$y(t) = z(t) + \mathcal{D}u(t), \ t \geq 0.$$

Then

$$\tau y(t) = \tau(\mathcal{D}u(t)) = u(t), \ t \geq 0,$$

and

$$\frac{d}{dt}(y(t) - \mathcal{D}u(t)) = (\mathcal{A} - \lambda)(y(t) - \mathcal{D}u(t)) + \lambda y(t) - \mathcal{D}\frac{d}{dt}u(t).$$

Taking into account that $(\mathcal{A} - \lambda)y(t) = 0$, for all $t \geq 0$, we see that

$$\left.\begin{array}{l} \dfrac{\partial y}{\partial t}(t) = \mathcal{A}y(t), \ t > 0, \\[2mm] \tau y(t) = u(t), \ t > 0, \\[2mm] y(0) = x. \end{array}\right\} \qquad (13.2.6)$$

Consequently the function $y(t)$, $t \geq 0$, solves the problem (13.2.1) and is given by the formula

$$y(t) = S(t)(x - \mathcal{D}u(0)) \int_0^t S(t-s)[\lambda \mathcal{D}u(s) - \mathcal{D}u'(s)]ds + \mathcal{D}u(t), \ t > 0.$$

Proposition 13.2.1 *If $u(\cdot)$ is a twice continuously differentiable function and $x - \mathcal{D}u(0) \in D(A)$ then a solution $y(t)$, $t \geq 0$, of the abstract boundary problem (13.2.6) is of the form*

$$y(t) = S(t)x + (\lambda - A) \int_0^t S(t - s)\mathcal{D}u(s)ds, \ t > 0.$$

The proof of the proposition is an immediate consequence of the following well known lemma.

Lemma 13.2.2 *Assume that the operator A generates a strongly continuous semigroup $S(t)$, $t \geq 0$, on a Banach space E and that $\varphi \in W^{1,1}(0, T; E)$ for some $T > 0$. If $x \in D(A)$ and*

$$z(t) = S(t)x + \int_0^t S(t - s)\varphi(u(s))ds, \ t \in [0, T], \qquad (13.2.7)$$

then $z \in C^1([0,T]; E) \cap C([0,T]; D(A))$. Moreover

$$Az(t) = S(t)x + \int_0^t S(t-s)\varphi'(u(s))ds + S(t)\varphi(0) - \varphi(t), \quad t \in [0,T].$$
(13.2.8)

Proof — For the completeness of the presentation we sketch a proof. One can assume that $x = 0$ and that

$$\varphi(t) = \varphi(0) + \int_0^t \psi(r)dr, \quad t \in [0,T],$$

where $\psi \in L^1(0,T;E)$. Consequently

$$
\begin{aligned}
z(t) &= \int_0^t S(t-s)\varphi(0)ds + \int_0^t S(t-s)\left(\int_0^s \psi(r)dr\right)ds \\
&= \int_0^t S(t-s)\varphi(0)ds + \int_0^t \left(\int_r^t S(t-s)\psi(r)ds\right)dr.
\end{aligned}
$$

For arbitrary $r \le t$, $\int_r^t S(t-s)\psi(r)ds \in D(A)$ and

$$A\int_r^t S(t-s)\psi(r)ds = (S(t-r) - I)\psi(r).$$

Therefore

$$Az(t) = (S(t) - I)\varphi(0) + \int_0^t (S(t-r) - I)\varphi'(r)dr,$$

which is the required formula (13.2.8). Moreover $z \in C([0,T]; D(A))$ and by direct differentiation

$$z'(t) = S(t)\varphi(0) + \int_0^t S(t-r)\psi(r)dr, \quad t \in [0,T].$$

The proof of the lemma is complete. ∎

The formula (13.2.2) can be extended by continuity to less regular functions $u(\cdot)$.

In particular assume that $W(t)$, $t \ge 0$, is a cylindrical Wiener process defined in a probability space $(\Omega, \mathcal{F}, \mathbb{P})$ adapted to a filtration

$\mathcal{F}_t$ and with values in U. If Q is a nonnegative bounded operator on U such that the process

$$Z(t) = (\lambda - A) \int_0^t S(t-s)\mathcal{D}Q^{1/2}dW(s), \ t > 0, \qquad (13.2.9)$$

is a well defined H–valued process, then it will be accepted as a solution to the linear problem

$$\left. \begin{array}{l} \dfrac{\partial Z}{\partial t}(t) = AZ(t), \ t > 0, \\[3mm] \tau Z(t) = Q^{1/2} \dfrac{\partial W}{\partial t}(t), \ t > 0, \\[3mm] Z(0) = 0. \end{array} \right\} \qquad (13.2.10)$$

It will be called an *Ornstein–Uhlenbeck process* in analogy to the classical noise case. A sufficient condition for (13.2.9) to be a well defined process is that

$$\int_0^t \|AS(r)\mathcal{D}Q^{1/2}\|^2_{L_2(U;H)}dr < +\infty, \ t > 0, \qquad (13.2.11)$$

where $L_2(U;H)$ stands for the space of Hilbert–Schmidt operators acting from U into H, see G. Da Prato and J. Zabczyk [44] and [45].

Let now F be a nonlinear, usually unbounded, operator on H. An $\mathcal{F}_t$–adapted H–valued and H–continuous process X is the mild solution of the problem

$$\left. \begin{array}{l} \dfrac{\partial X}{\partial t}(t) = AX(t) + F(X(t)), \ t > 0, \\[3mm] \tau X(t) = Q^{1/2} \dfrac{\partial W}{\partial t}(t), \ t > 0, \\[3mm] X(0) = x, \end{array} \right\} \qquad (13.2.12)$$

if it is a solution of the following integral equation:

$$X(t) = S(t)x + Z(t) + \int_0^t S(t-s)F(X(s))ds, \ t \geq 0. \qquad (13.2.13)$$

In particular we have the following result.

Proposition 13.2.3 *Assume that the operators A and F satisfy Hypotheses 5.4 and 5.5. Then equation (13.2.13) has a unique Markovian generalized solution.*

We finish this section by giving sufficient conditions for continuity of the Ornstein–Uhlenbeck process $Z(\cdot)$ in some important subspaces of H. They will be used in §13.4.

To simplify our general considerations we assume from now on that the operator A is self-adjoint and that there exists a complete orthonormal basis of eigenvectors $\{g_n\}$ of A corresponding to the sequence $\{-\lambda_n\} \downarrow -\infty$ of non positive eigenvalues. Let $H^{2\alpha} = D((-A)^\alpha)$. If a stronger version of (13.2.13) holds, namely

$$\exists\, \gamma > 0 \int_0^t r^{-\gamma} \|AS(r)\mathcal{D}Q^{1/2}\|^2_{L_2(U;H^{2\alpha})} dr < +\infty, \; t > 0, \quad (13.2.14)$$

then the process Z has a version which is $H^{2\alpha}$–continuous. In particular we have the following result, see G. Da Prato and J. Zabczyk [45].

Proposition 13.2.4 *If for some $\gamma > 0$*

$$\sum_{n=1}^{\infty} \lambda_n^{2\alpha+\gamma+1} \|Q^{1/2}\mathcal{D}^* g_n\|^2 < +\infty \qquad (13.2.15)$$

then the process Z has an $H^{2\alpha}$–continuous version.

Proof — Note that

$$\int_0^t r^{-\gamma} \|AS(r)\mathcal{D}Q^{1/2}\|^2_{L_2(U;H^{2\alpha})} dr$$

$$= \int_0^t r^{-\gamma} \|((-A)^{1+\alpha})S(r)\mathcal{D}Q^{1/2}\|^2_{L_2(U;H)} dr$$

$$= \int_0^t r^{-\gamma} \|Q^{1/2}\mathcal{D}^*[((-A)^{1+\alpha})S(r)]\|^2_{L_2(U;H^{2\alpha})} dr$$

$$= \sum_{n=1}^{\infty} \lambda_n^{\alpha+1} \|Q^{1/2}\mathcal{D}^* g_n\|^2 \int_0^t r^\gamma e^{-2\lambda_n r} dr.$$

So the conclusion follows. ∎

13.3   Equations with Neumann boundary conditions

We will apply the general approach of §13.2 to problem (13.1.1) with $V(t) = Q^{1/2}W(t)$, $t \geq 0$.

For simplicity we set $m = 0$ and define

$$H = L^2(G), \quad U = L^2(\Gamma_0).$$

Moreover the operator $\mathcal{A}$ is the Laplacian and τ is the trace in Γ_0 of the normal derivative operator $\frac{\partial}{\partial\nu}$ and

$$\left. \begin{array}{l} D(A) = \{x \in H^2(G) : \frac{\partial}{\partial\nu}(\xi) = 0, \ \xi \in \Gamma\}, \\[2mm] Ax = \Delta x. \end{array} \right\}$$

It is well known that A is a self–adjoint generator of a semigroup $S(t)$, $t \geq 0$. Moreover the eigenvectors of A are of the form

$$g_{n_1,\dots,n_d} = g_{n_1}(\xi_1) \cdots g_{n_d}(\xi_d), \ (\xi_1,\dots,\xi_d) \in G, \qquad (13.3.1)$$

where $n_1, \dots, n_d \in \mathbb{N} \cup \{0\}$ and

$$g_0(\xi) = \frac{1}{\sqrt{\pi}}, \ g_n(\xi) = \frac{2}{\sqrt{\pi}} \cos n\xi, \ \xi \in [0,\pi], \ n \in \mathbb{N} \cup \{0\}, \quad (13.3.2)$$

and the corresponding eigenvalues are

$$\lambda_{n_1,\dots,n_d} = -(n_1^2 + \dots + n_d^2), \ n_1, \dots, n_d \in \mathbb{N} \cup \{0\}.$$

We will now find an explicit formula for the operator $\mathcal{D}$. For this purpose it is convenient to consider elements $u \in U$ as functions in $L^2(\Gamma)$ equal to 0 outside of Γ_0. We will solve (13.2.4) with $\lambda = 1$. An arbitrary element $u \in U$ can be uniquely represented by the expansion with respect to the orthonormal and complete system

$$\{g_{n_1,\dots,n_{d-1}} : n_1, \dots, n_{d-1} \in \mathbb{N} \cup \{0\}\},$$

namely

$$u = \sum_{n_1,\dots,n_{d-1}=0}^{\infty} \langle u, g_{n_1,\dots,n_{d-1}} \rangle_U \, g_{n_1,\dots,n_{d-1}},$$

and therefore the solution $w = \mathcal{D}u$ of (13.2.4) is of the form

$$w = \sum_{n_1,\ldots,n_{d-1}=0}^{\infty} \langle u, g_{n_1,\ldots,n_{d-1}} \rangle_U \, w_{n_1,\ldots,n_{d-1}}$$

where

$$w_{n_1,\ldots,n_{d-1}} = \mathcal{D}g_{n_1,\ldots,n_{d-1}}, \quad n_1,\ldots,n_{d-1} \in \mathbb{N} \cup \{0\}\}.$$

By direct calculation we find that

$$w_{n_1,\ldots,n_{d-1}}(\xi_1,\ldots,\xi_d) = g_{n_1,\ldots,n_{d-1}}(\xi_1,\ldots,\xi_{d-1})h_{n_1,\ldots,n_{d-1}}(\xi_d),$$

where

$$h_{n_1,\ldots,n_{d-1}}(\xi_d) = -\frac{\cosh\left[\sqrt{1 + n_1^2 + \cdots + n_{d-1}^2}\,\right](\xi_d - \pi)}{\sqrt{1 + n_1^2 + \cdots + n_{d-1}^2}\,\sinh\left[\sqrt{1 + n_1^2 + \cdots + n_{d-1}^2}\,\right]},$$

$(\xi_1,\ldots,\xi_d) \in G$ and $n_1,\ldots,n_{d-1} \in \mathbb{N} \cup \{0\}$.

It is clear that D is a bounded operator from U into H. By the very definition

$$(\mathcal{D}^* g_{n_1,\ldots,n_d})(\xi_1,\ldots,\xi_{d-1})$$

$$= g_{n_1,\ldots,n_{d-1}}(\xi_1,\ldots,\xi_{d-1})\int_0^{\pi} g_{n_d}(\xi)h_{n_1,\ldots,n_{d-1}}(\xi)d\xi$$

$$= -\frac{\delta(n_d)}{1 + n_1^2 + \cdots + n_d^2}g_{n_1,\ldots,n_{d-1}}(\xi_1,\ldots,\xi_{d-1}), \quad (\xi_1,\ldots,\xi_d) \in G,$$

$$(13.3.3)$$

where

$$\delta(n) = \begin{cases} \frac{1}{\sqrt{\pi}} & \text{if } n = 0, \\[2mm] \frac{2}{\sqrt{\pi}} & \text{if } n \in \mathbb{N}. \end{cases}$$

Theorem 13.3.1 *The Ornstein–Uhlenbeck process $Z(t)$, $t \geq 0$, has a continuous version in the space $H^{2\alpha,2}(G)$ provided there exists $\gamma > 0$ such that*

$$\sum_{n_1,\ldots,n_{d-1}=0}^{\infty} \left\|Q^{1/2}g_{n_1,\ldots,n_{d-1}}\right\|^2 \left(1 + n_1^2 + \cdots + n_{d-1}^2\right)^{2\alpha+\gamma-\frac{1}{2}} < +\infty.$$

$$(13.3.4)$$

Proof— We can identify the spaces $H^{2\alpha,2}(G)$ with $H^{2\alpha} = D((-A)^\alpha)$, $\alpha \geq 0$. We will apply Proposition 13.2.4. Note that

$$\left\| Q^{1/2} \mathcal{D}^* g_{n_1,\ldots,n_d} \right\|^2 = \frac{\delta^2(n_d)}{(1 + n_1^2 + \cdots + n_d^2)^2} \left\| Q^{1/2} g_{n_1,\ldots,n_{d-1}} \right\|^2,$$

$n_1,\ldots,n_{d-1} \in \mathbb{N} \cup \{0\}$. Consequently the sum

$$\sum_{n_1,\ldots,n_{d-1}=0}^{\infty} \lambda_{n_1,\ldots,n_d}^{2\alpha+\gamma+1} \left\| Q^{1/2} \mathcal{D}^* g_{n_1,\ldots,n_d} \right\|^2$$

is finite if and only if

$$\sum_{n_1,\ldots,n_{d-1}=0}^{\infty} \left\| Q^{1/2} \mathcal{D}^* g_{n_1,\ldots,n_d} \right\|^2 \sum_{n=0}^{\infty} \frac{(n_1^2 + \cdots + n_{d-1}^2 + n^2)^{2\alpha+\gamma+1}}{(1 + n_1^2 + \cdots + n_{d-1}^2 + n^2)^2} < +\infty$$

However, there exist constants $C_1 > 0$ and $C_2 > 0$ such that for arbitrary $a \geq 0$ and $2\alpha + \gamma < \frac{1}{2}$

$$C_2(1 + a^2)^{2\alpha+\gamma-\frac{1}{2}} \leq \sum_{n=0}^{\infty} \frac{(a^2 + n^2)^{2\alpha+\gamma+1}}{(1 + a^2 + n^2)^2} \leq C_1(1 + a^2)^{2\alpha+\gamma-\frac{1}{2}},$$

and therefore the result follows. ∎

Remark 13.3.2 If $d = 1$ then $\Gamma_0 = \{0\}$ and the condition (13.3.4) is automatically satisfied.

We finish this section by giving conditions on the space homogeneous noise implying continuity of the Ornstein–Uhlenbeck process in all $L^p\left((0,\pi)^d\right)$ spaces. Thus we assume that the operator Q is of the form (13.1.2)–(13.1.3) with q_0 being the Fourier transform of the symmetric spectral density g, see §5.2:

$$q_0(\xi) = \int_{\mathbb{R}^{d-1}} e^{i\xi\zeta} g(\zeta) d\zeta, \ \xi \in \mathbb{R}.$$

We will call g the *spectral density* of the noise process V.

We need the following lemma.

Lemma 13.3.3 *The function*

$$h_{d,\alpha}(\zeta) = \left(\prod_{j=1}^{d} \zeta_j^2 \right) \sum_{n_1,\dots,n_d=0}^{\infty} \Big\{ (1 + n_1^2 + \cdots + n_d^2)^{2\alpha - \frac{1}{2}}$$

$$\times \prod_{j=1}^{d} \frac{1}{1 + (\zeta_j + n_j)^2} \frac{1}{1 + (\zeta_j - n_j)^2} \Big\}, \ \zeta \in \mathbb{R}^d,$$

is finite and continuous if and only if $\alpha < 1$.

Proof — There exists a constant $C > 0$ such that

$$(1 + n_1^2 + \cdots + n_d^2)^{2\alpha - \frac{1}{2}} \leq C + C \prod_{j=1}^{d} n_j^{2(2\alpha - \frac{1}{2})}.$$

Since

$$\sum_{n_1,\dots,n_d=0}^{\infty} \prod_{j=1}^{d} \frac{n_j^{2(2\alpha - \frac{1}{2})}}{[1 + (\zeta_j + n_j)^2][1 + (\zeta_j - n_j)^2]}$$

$$= \prod_{j=1}^{d} \sum_{n_j=0}^{\infty} \frac{n_j^{2(2\alpha - \frac{1}{2})}}{[1 + (\zeta_j + n_j)^2][1 + (\zeta_j - n_j)^2]}$$

and $4 - 2(2\alpha - 1/2) > 1$ we get that if $\alpha < 1$, the continuity of $h_{d,\alpha}$ follows. If $\alpha \geq 1$ and $\zeta_j \neq 0$, $j = 1, \dots, d$, then $h_{d,\alpha}(\zeta_1, \dots, \zeta_d) = +\infty$. ∎

We have the following continuity result.

Theorem 13.3.4 *Assume that $\alpha < 1$ and for $\gamma \in]0, 1 - \alpha[$:*

$$\int_{\mathbb{R}^{d-1}} g(\zeta) h_{d-1,\alpha+\gamma}(\zeta) d\zeta < +\infty. \qquad (13.3.5)$$

Then the Ornstein–Uhlenbeck process $Z(t)$, $t \geq 0$, has a continuous version in $H^{2\alpha+2,2}(G)$.

Proof — We use Proposition 13.2.4 and check that the condition

$$\sum_{n_1,\dots,n_{d-1}=0}^{\infty} \left\| Q^{1/2} g_{n_1,\dots,n_{d-1}} \right\|^2 \left(1 + n_1^2 + \cdots + n_{d-1}^2 \right)^{2\alpha+\gamma-\frac{1}{2}} < +\infty$$

holds.

In the present situation

$$\left\| Q^{1/2} g_{n_1,\ldots,n_{d-1}} \right\|^2 = \langle Q g_{n_1,\ldots,n_{d-1}}, g_{n_1,\ldots,n_{d-1}} \rangle, \ n_1, \ldots, n_{d-1} \in \mathbb{N} \cup \{0\}.$$

Let

$$\widetilde{g}_{n_1,\ldots,n_{d-1}}(\xi) = \begin{cases} g_{n_1,\ldots,n_{d-1}}(\xi) & \text{for } \xi \in \Gamma_0, \\[2mm] 0 & \text{for } \xi \notin \Gamma_0. \end{cases}$$

Then

$$\langle Q g_{n_1,\ldots,n_{d-1}}, g_{n_1,\ldots,n_{d-1}} \rangle_U = \langle q_0 * \widetilde{g}_{n_1,\ldots,n_{d-1}}, \widetilde{g}_{n_1,\ldots,n_{d-1}} \rangle_{L^2(\mathbb{R}^{d-1})}$$

and by Plancherel's theorem

$$\langle Q g_{n_1,\ldots,n_{d-1}}, g_{n_1,\ldots,n_{d-1}} \rangle_{\mathbb{R}^{d-1}} = \frac{1}{\sqrt{(2\pi)^{d-1}}} \int_{\mathbb{R}^{d-1}} g(\zeta) \left| \widehat{\widetilde{g}}_{n_1,\ldots,n_{d-1}}(\zeta) \right|^2 d\zeta$$

where

$$\widehat{\widetilde{g}}_{n_1,\ldots,n_{d-1}}(\zeta) = \frac{1}{\sqrt{(2\pi)^{d-1}}} \int_{\mathbb{R}^{d-1}} e^{i\zeta\xi} \widetilde{g}_{n_1,\ldots,n_{d-1}}(\xi) d\xi$$

$$= \frac{1}{\sqrt{(2\pi)^{d-1}}} \int_{[0,\pi]^{d-1}} e^{i\zeta\xi} g_{n_1,\ldots,n_{d-1}}(\xi) d\xi.$$

Integrating by parts we obtain that

$$\int_0^\pi e^{i\zeta\xi} \cos n\xi d\xi = \frac{i\zeta}{n^2 - \zeta^2} \left(e^{i(\zeta-n)\pi} - 1 \right) \tag{13.3.6}$$

for all $n \in \mathbb{N} \cup \{0\}$ and $\zeta \in \mathbb{R}$; if $\zeta^2 = n^2$ one should take in (13.3.6) the limiting value $\frac{\pi}{2}$ if $n \in \mathbb{N}$ and π if $n = 0$. Taking into account the definitions (13.3.1) and (13.3.2) one gets for a constant $C_1 > 0$ that

$$\sum_{n_1,\dots,n_{d-1}=0}^{\infty} \left\|Q^{1/2}g_{n_1,\dots,n_{d-1}}\right\|^2 \left(1 + n_1^2 + \cdots + n_{d-1}^2\right)^{2\alpha+\gamma-\frac{1}{2}}$$

$$= \frac{1}{(\sqrt{2})^{d-1}} \int_{\mathbb{R}^{d-1}} g(\zeta) \sum_{n_1,\dots,n_{d-1}=0}^{\infty} \left(1 + n_1^2 + \cdots + n_{d-1}^2\right)^{2\alpha+\gamma-\frac{1}{2}}$$

$$\times \left|\widehat{\widehat{g}}_{n_1,\dots,n_{d-1}}(\zeta)\right|^2 d\zeta$$

$$\leq C_1 \int_{\mathbb{R}^{d-1}} g(\zeta) \Big[\sum_{n_1,\dots,n_{d-1}=0}^{\infty} \left(1 + n_1^2 + \cdots + n_{d-1}^2\right)^{2\alpha+\gamma-\frac{1}{2}}$$

$$\times \prod_{j=1}^{d} \zeta_j^2 \frac{1-\cos(\zeta_j+n_j)\pi}{[(\zeta_j+n_j)\pi]^2} \times \frac{1-\cos(\zeta_j-n_j)\pi}{[(\zeta_j-n_j)\pi]^2}\Big] d\zeta.$$

$$\leq C_1 \int_{\mathbb{R}^{d-1}} g(\zeta)h_{d-1,\alpha+\frac{\gamma}{2}}(\zeta)d\zeta.$$

Therefore the result follows. ∎

Remark 13.3.5 Knowing the asymptotic property of $h_{d-1,\alpha+\gamma}(\zeta)$ as $|\zeta| \to +\infty$ one can replace (13.3.5) by more specific conditions. If for istance $d = 2$ then

$$h_{1,\alpha+\gamma} \leq \sum_{n=0}^{\infty}(\zeta^2 + n^2)^{2\alpha+2\gamma-1}\frac{\zeta^2}{(n+\zeta)^2} \leq C\zeta^{4(\alpha+\gamma)},$$

for large $|\zeta|$.

As a corollary we obtain sufficient conditions for continuity of Z in L^p spaces.

Theorem 13.3.6 *The Ornstein–Uhlenbeck process* $Z(t)$, $t \geq 0$, *has a continuous version in* $L^p(G)$ *provided that* (13.3.5) *and that one of the following conditions hold*

(i) $d = 1,2,3$, $p \geq 1$, $\frac{d}{4} < \alpha$,

(ii) $d \geq 4$, $1 \leq p \leq \frac{2d}{d-4\alpha}$, $\alpha < 1$.

Proof — By Sobolev's embedding theorems

$$H^{2\alpha+2,2}(G) \subset L^{\infty}(G) \text{ if } d < 4\alpha$$

and

$$H^{2\alpha+2,2}(G) \subset L^p(G) \text{ if either } d < 4\alpha, \text{ or } d > 4\alpha \text{ and } p \leq \frac{2d}{d-4\alpha}.$$

So the result follows.   ∎

Remark 13.3.7 If $d = 1, 2, 3$, and $\frac{3}{4} < \alpha < 1$ then condition (13.3.5) is sufficient for continuity of $Z(t)$, $t \geq 0$, in all $L^p(G)$ spaces, $p \geq 1$.

13.4 Ergodic solutions

In a similar way as in Chapter 11 we will apply the dissipativity method developed in §5.5 and §6.3.

Theorem 13.4.1 *Assume that the function f satisfies Hypothesis 11.2. If in addition for $p = 2s^2$ the conditions of Theorem 13.3.6 are satisfied, then problem (13.1.1) has a generalized Markovian solution in $L^2(G)$.*

Proof — The operator $A = \Delta - m$ with the domain $D(A) = \{x \in H^{2,2}(G) : \frac{dx}{d\nu} = 0\}$ generates a dissipative semigroup in $L^2(G)$ if $m > 0$. This semigroup has dissipative restriction for all spaces $K = L^{2s}(G)$ and $K = L^{2s^2}(G)$ Also the nonlinear operators $F - \eta$, where

$$F(x)(\xi) = f(x(\xi)), \ \xi \in G,$$

with natural domains, are dissipative in $L^2(G)$ and H if η is sufficiently small. Moreover by Theorem 13.3.6 the process $F(Z(t))$, $t \geq 0$, has continuous versions in $L^{2s}(G)$. Consequently all the assumptions of Proposition 13.2.3 are satisfied and the proof of the theorem is complete.   ∎

In a similar way we can prove the following theorem on invariant measure.

Theorem 13.4.2 *In addition to the conditions of Theorem 13.4.1 assume that the function f_0 is decreasing and*

$$m - \|f_1\|_{\text{Lip}} > \omega > 0,$$

Then the transition semigroup P_t, $t \geq 0$ corresponding to the solution of problem (13.1.1) has a unique invariant measure μ in $L^2(G)$. Moreover there exists $c > 0$ such that for any bounded Lipschitz function φ on H, all $t \geq 0$ and all $x \in H$ we have

$$\left| P_t\varphi(x) - \int_H \varphi(y)\mu(dy) \right| \leq (c + 2|x|)e^{-\omega t}\|\varphi\|_{\text{Lip}}.$$

Proof — We apply Theorem 6.3.3 and argue in a similar way as in the proof of Theorem 12.3.3. ∎

For other results on invariant measures for stochastic equations with boundary noise see B. Maslowski [118].

Chapter 14

Burgers equation

This chapter is devoted to the Burgers equation on a finite interval, perturbed by the cylindrical Wiener process. Existence of a unique solution is established. It is also shown that there exists a unique invariant measure by proving a kind of boundedness in probability of some solutions and by establishing strong Feller and irreducibility properties of the corresponding transition semigroup.

14.1 Introduction

An important role in fluid dynamics is played by the following *Burgers* equation, see J.M. Burgers [15].

$$\frac{\partial}{\partial t}X(t,\xi) = \nu\frac{\partial^2}{\partial\xi^2}X(t,\xi) + \frac{\lambda}{2}\frac{\partial}{\partial\xi}\left[X^2(t,\xi)\right], \qquad (14.1.1)$$

with $t > 0$, $\xi \in \mathbb{R}$. It is known, however, that the equation is not a good model for turbulence. It does not display any chaotic phenomena; even when a force is added to the right hand side all solutions converge to a unique stationary solution as time goes to infinity. The situation is different when the force is random. Several authors have indeed suggested using the stochastic Burgers equation as a simple model for turbulence, D. H. Chambers et al. [21], H. Choi et al. [22], Dah–Teng Jeng [33].

The equation has also been proposed in M. Kardar, M. Parisi and J. C. Zhang [94] for studying the growing of a onedimensional interface. Let us recall their model. Let $h(t,\xi)$ be the height of the interface at the point $\xi \in \mathbb{R}$, then we have

$$\frac{\partial}{\partial t}h(t,\xi) = \nu\frac{\partial^2}{\partial\xi^2}h(t,\xi) + \frac{\lambda}{2}\left[\frac{\partial}{\partial\xi}h(t,\xi)\right]^2 + \frac{\partial}{\partial t}W(t,\xi), \quad (14.1.2)$$

where $W(\cdot,\cdot)$ is the space–time white noise. Setting

$$X(t,\xi) = \frac{\partial}{\partial\xi}h(t,\xi),$$

we arrive, after differentiation with respect to ξ, at the equation

$$\frac{\partial}{\partial t}X(t,\xi) = \nu\frac{\partial^2}{\partial\xi^2}X(t,\xi) + \frac{\lambda}{2}\frac{\partial}{\partial\xi}\left[X^2(t,\xi)\right] + \frac{\partial^2}{\partial t\partial\xi}W(t,\xi), \quad (14.1.3)$$

In §14.2 we shall prove existence and uniqueness of a solution, following G. Da Prato, A. Debussche and R. Temam [36], of equation (14.1.3) supplemented by Dirichlet boundary conditions

$$X(t,0) = X(t,1) = 0, \ t > 0, \quad (14.1.4)$$

and with the initial condition

$$X(0,\xi) = x(\xi), \ \xi \in [0,1]. \quad (14.1.5)$$

Moreover in §14.3 we prove that the corresponding transition semi-group is a strong Feller one and in §14.4 the existence and uniqueness of an invariant measure. Here we use methods in G. Da Prato and D. Gątarek [38] (where a more general case with multiplicative noise is treated) and from Chapter 7. However, the proofs we present here are new.

For a different approach on existence and uniqueness, based on the Feynman–Kač formula, see L. Bertini, N. Cancrini and G. Jona-Lasinio [8]. Equation (14.1.2) has also been studied in several space dimensions using the Hopf–Cole transform

$$h(t,\xi) = \log\left(\frac{\lambda}{\nu}v(t,\xi)\right),$$

which reduces equation (14.1.2) to heat equations with a random force. In this connection we quote H. Holden, T. Lindstrøm , B.

Øksendal, J. Ubøe, & T. S. Zhang [87], A. Dermoune [54], Yu. Kifer [97]. We mention finally the papers by L. Bertini and N. Cancrini [7], and G. Tessitore and J. Zabczyk [151], where equation (14.1.2) with $\frac{\partial}{\partial t}W(t,\xi)$ replaced by $\frac{\partial^2}{\partial t \partial \xi}W(t,\xi)$ is studied.

In the sequel we will set $\nu = \lambda = 1$ for simplicity. Then we write the problem in abstract form. We define

$$Au = u'', \ \forall\, u \in D(A) = H^2(0,1) \cap H_0^1(0,1).$$

A is a negative self–adjoint operator on $L^2(0,1)$ and

$$D((-A)^\sigma) = H^{2\sigma}(0,1) \text{ if } \sigma \in\ ]0, 1/4[.$$

We denote by $S(t)$, $t \geq 0$, the semigroup on $L^2(0,1)$ generated by A, by $\{e_k\}$ the complete orthonormal system on $L^2(0,1)$ which diagonalizes A and by $\{\lambda_k\}$ the corresponding sequence of eigenvalues. We have

$$e_k(\xi) = \sqrt{\frac{2}{\pi}} \ \sin k\pi\xi, \ \xi \in [0,1], \ k \in \mathbb{N},$$

and

$$\lambda_k = -\pi^2 k^2, \ k \in \mathbb{N}.$$

Setting $X(t) = X(t,\cdot)$, $t \geq 0$, we can write (14.1.3)–(14.1.5) as

$$\left.\begin{array}{l} dX(t) = \left[AX(t) + \dfrac{1}{2}D_\xi(X^2(t))\right] dt + dW(t), \\[4mm] X(0) = x, \end{array}\right\} \qquad (14.1.6)$$

where W is defined by

$$W(t) = \sum_{h=1}^{\infty} \beta_h e_h, \qquad (14.1.7)$$

and $\{\beta_h\}$ is a sequence of mutually independent real Brownian motions in a fixed probability space $(\Omega, \mathcal{F}, \mathbb{P})$ adapted to a filtration $\mathcal{F}_t$, $t \geq 0$.

14.2   Existence of solutions

Here we are concerned with the problem

$$
\left.
\begin{aligned}
dX(t) &= \left[AX(t) + \frac{1}{2}D_\xi(X^2(t))\right]dt + dW(t), \\[2mm]
X(0) &= x,
\end{aligned}
\right\}
\qquad (14.2.1)
$$

which we write in the *mild* form

$$
X(t) = S(t)x + \frac{1}{2}\int_0^t S(t-s)D_\xi(X^2(s))ds + W_A(t),
$$

where $W_A(t)$ is given by

$$
W_A(t) = \int_0^t S(t-s)dW(s).
$$

Now, setting $Y(t) = X(t) - W_A(t)$, $t \geq 0$, we obtain a deterministic equation, a.s. in Ω,

$$
Y(t) = S(t)x + \frac{1}{2}\int_0^t S(t-s)D_\xi[(Y(s) + W_A(s))^2]\,ds, \qquad (14.2.2)
$$

which can be considered as the mild form of the evolution equation

$$
\left.
\begin{aligned}
Y'(t) &= AY(t) + \tfrac{1}{2}D_\xi[(Y(t) + W_A(t))^2], \quad t \geq 0, \\[2mm]
Y(0) &= x.
\end{aligned}
\right\}
\qquad (14.2.3)
$$

In order to give a precise meaning to equation (14.2.3), we introduce the mapping

$$
F_\sigma(u)(t) = \int_0^t (-A)^\sigma S(t-s)D_\xi u(s)\,ds, \ t \in [0,T], \qquad (14.2.4)
$$

for any $\sigma \in [0,1[$. If $u \in C([0,T]; H_0^1(0,1))$ then $F_\sigma(u)$ is certainly well defined and belongs to $C([0,T]; L^2(0,1))$.

Lemma 14.2.1 *The mapping F_σ defined by (14.2.4) can be extended to a bounded linear mapping from $C([0,T]; L^1(0,1))$ into $C([0,T]; H^{2\sigma}(0,1))$, for arbitrary $\sigma \in [0,1/4[$.*

Proof — An essential tool for the proof is the following well known estimate:

$$\|D_\xi S(t)\varphi\|_\infty \le C_1 t^{-\frac{3}{4}} \|\varphi\|_2, \quad \varphi \in L^2(0,1), \qquad (14.2.5)$$

for some constant C_1. Let us choose $\psi \in H^{2\sigma}(0,1)$ and consider the scalar product

$$I = \left\langle (-A)^\sigma \psi, \int_0^t S(t-s)D_\xi u(s)ds \right\rangle,$$

where $u \in C([0,T]; H_0^1(0,1))$. Then

$$I = \int_0^t \langle S(t-s)(-A)^\sigma \psi, D_\xi u(s)\rangle \, ds$$

$$= -\int_0^t \langle D_\xi(S(t-s)(-A)^\sigma \psi), u(s)\rangle \, ds.$$

Therefore

$$|I| \le \int_0^t \|D_\xi(S(t-s)(-A)^\sigma \psi)\|_\infty \, \|u(s)\|_1 ds.$$

But

$$\|D_\xi(S(r)(-A)^\sigma \psi)\|_\infty = \|D_\xi[S(r/2)(S(r/2)(-A)^\sigma \psi)]\|_\infty$$

$$\le C_1 \left(\frac{r}{2}\right)^{-3/4} \|S(r/2)(-A)^\sigma \psi)\|_2$$

$$\le C_1 \left(\frac{r}{2}\right)^{-3/4} C_2 \left(\frac{r}{2}\right)^{-\sigma} \|\psi\|_2$$

$$\le C_3 r^{-(3/4+\sigma)} \|\psi\|_2, r \in [0,T],$$

where C_2, C_3 are constants.

Consequently

$$|I| \le C_3 \int_0^t (t-s)^{-(3/4+\sigma)} \|u(s)\|_1 ds \, \|\psi\|_2 \qquad (14.2.6)$$

It follows from (14.2.6) that $\int_0^t S(t-s)D_\xi u(s)ds \in D((-A)^\sigma)$ and

$$\left\|(-A)^\sigma \int_0^t S(t-s)D_\xi u(s)ds\right\|_2 \leq C_3 \int_0^t (t-s)^{-(3/4+\sigma)}\|u(s)\|_1 ds$$

$$\leq C_3 \frac{4}{1-4\sigma}t^{\frac{1}{4}-\sigma}\sup_{0\leq s\leq t}\|u(s)\|_1.$$

This easily implies the conclusion.   ∎

Remark 14.2.2 In particular we see that $F = F_0$ is well defined as a transformation from $C([0,T];L^1(0,1))$ into $C([0,T];L^2(0,1))$.

Remark 14.2.3 In the same way we can show that if $\frac{1}{p} + \sigma < \frac{1}{4}$ then for a constant $C_4 > 0$ and all $t > 0$

$$\left\|(-A)^\sigma \int_0^t S(t-s)D_\xi u(s)ds\right\|_p \leq C_4 t^{4(\frac{1}{4}-\frac{1}{p}-\sigma)}\left(\int_0^t \|u(s)\|_1^p ds\right)^{1/p}.$$
$$(14.2.7)$$

Now we can rewrite equation (14.2.3), as

$$Y(t) = S(t)x + \frac{1}{2}F\left((Y(\cdot)+W_A(\cdot))^2\right)(t), \ t \in [0,T]. \qquad (14.2.8)$$

We say that $X \in C([0,T];L^2(0,1))$ is a *mild* solution of problem (14.2.1), if and only if $Y(\cdot) = X(\cdot) - W_A(\cdot)$ is a solution of equation (14.2.8).

Now we prove an existence and uniqueness result.

Theorem 14.2.4 *For any $x \in L^2(0,1)$, there exists a unique mild solution X of equation (14.2.1), and for all $T > 0$*

$$X \in C([0,T];L^2(0,1)) \cap L^2(0,T;C([0,1])), \ \mathbb{P}\text{--a.s.}$$

Proof —   We fix $T > 0$. The proof is divided into two steps.

Step 1—   Local existence

It is not difficult to prove existence of a mild solution of equation (14.2.1), in a small interval, depending on $\omega \in \Omega$, by using the contraction principle on the Banach space $C([0,T]; L^2(0,1))$. This follows from Step 1 and the fact that,

$$W_A(\cdot) \in C([0,T]; C([0,1])), \qquad (14.2.9)$$

see G. Da Prato and J. Zabczyk [44, page 141].

Step 2— A priori bound

We first consider equation (14.2.8) with $W_A(\cdot)$ replaced by a regular function $\varphi \in C([0,T]; H_0^1(0,1))$ and $x \in D(A)$:

$$\left. \begin{array}{l} Y'(t) = AY(t) + \dfrac{1}{2}D_\xi[(Y(t) + \varphi)^2], \ t \geq 0, \\[3mm] Y(0) = x. \end{array} \right\} \qquad (14.2.10)$$

As is easily seen equation (14.2.10) can be solved by a fixed point argument in $C([0,T]; H_0^1(0,1))$, so it has a classical solution.

We find now an a priori estimate on the solution Y which will be independent of regularity assumptions on φ and x.

Multiplying both sides of equation (14.2.10) by Y and integrating with respect to ξ in $[0,1]$ we get

$$\frac{1}{2}\frac{d}{dt}\int_0^1 |Y(t)|^2 d\xi = \int_0^1 Y_{\xi\xi}(t)Y(t)d\xi + \frac{1}{2}\int_0^1 D_\xi[(Y(t)+\varphi(t))^2]Y(t)d\xi.$$

It follows, integrating by parts, that

$$\frac{1}{2}\frac{d}{dt}\int_0^1 |Y(t)|^2 d\xi + \int_0^1 Y_\xi^2(t)d\xi = -\frac{1}{2}\int_0^1 (Y(t)+\varphi(t))^2 Y_\xi(t)d\xi$$

$$= -\frac{1}{2}\int_0^1 Y^2(t)Y_\xi(t)d\xi - \int_0^1 \varphi(t)Y(t)Y_\xi(t)d\xi - \frac{1}{2}\int_0^1 \varphi^2(t)Y_\xi(t)d\xi.$$

Noting that

$$\int_0^1 Y^2(t)Y_\xi(t)d\xi = \frac{1}{3}\left[Y^3(t)\right]_0^1 = 0,$$

we get

$$\frac{1}{2}\frac{d}{dt}\|Y(t)\|_2^2 + \int_0^1 Y_\xi^2(t)d\xi$$

$$\leq \|\varphi(t)\|_\infty \|Y_\xi(t)\|_2 \|Y(t)\|_2 + \frac{1}{2}\|\varphi(t)\|_\infty^2 \|Y_\xi(t)\|_2$$

$$\leq \frac{1}{2}\|Y_\xi(t)\|_2^2 + \|\varphi(t)\|_\infty^4 + 4\|\varphi(t)\|_\infty^2\|Y(t)\|_2^2.$$

In conclusion we find

$$\frac{d}{dt}\|Y(t)\|_2^2 + \|Y_\xi(t)\|_2^2 \leq 8\|\varphi(t)\|_\infty^2\|Y(t)\|_2^2 + 2\|\varphi(t)\|_\infty^4, \quad (14.2.11)$$

and thus

$$\begin{aligned}
\|Y(t)\|_2^2 \;\leq\; & e^{8\int_0^t \|\varphi(s)\|_\infty^2 ds}\|x\|_2^2 \\
& + 2\int_0^t e^{8\int_r^t \|\varphi(s)\|_\infty^2 ds}\|\varphi(r)\|_\infty^4\,dr,
\end{aligned} \quad (14.2.12)$$

and

$$\begin{aligned}
\int_0^T \|Y_\xi(t)\|^2 dt \;\leq\; & 8\int_0^T \|\varphi(t)\|_\infty^2\|Y(t)\|_2^4 dt \\
& + \int_0^T \|\varphi(t)\|_\infty^4 dt.
\end{aligned} \quad (14.2.13)$$

Since the trajectories of $W_A(\cdot)$ can be uniformly approximated, on any finite interval $[0,T]$, by functions φ from $C([0,T];H_0^1(0,1))$ and since we have that $D(A)$ is dense in $C([0,1])$ the estimates (14.2.12) and (14.2.13) hold true for any local solution Y of the equation with φ replaced by W_A. This shows that local solutions cannot explode in finite time and the proof of global existence of solutions to (14.2.1) is complete.

Moreover, again by the estimates (14.2.12) and (14.2.13) one sees that

$$Y \in C([0,T];L^2(0,1)) \cap L^2(0,T;C([0,1])),$$

which implies, in view of (14.2.9), that

$$X \in C([0,T]; L^2(0,1)) \cap L^2(0,T; C([0,1])),$$

as required. ∎

14.3   Strong Feller property

Let $X(\cdot, x)$ be the mild solution to problem (14.2.1) whose existence is proved in Theorem 14.2.4. In this section we will show that the corresponding transition semigroup P_t, $t \geq 0$, has the strong Feller property on H. For this purpose we consider a modified Burgers equation. For any $R > 0$ let ψ_R be a C^1 function on $[0, +\infty[$ taking values in $[0,1]$ and such that

$$\psi(r) = \begin{cases} 1 & \text{for } r \in [0, R], \\ 0 & \text{for } r \in [R+1, +\infty[, \end{cases}$$

and define a mollifier

$$M_R(x) = x\psi(x), \ x \in H. \qquad (14.3.1)$$

Note that $M_R \in C_b^1(H)$ and

$$D_x M_R(x) = \psi(\|x\|)I + \frac{\psi'_R(\|x\|)}{\|x\|} x \otimes x, \ x \in H.$$

Proposition 14.3.1 *If M_R is a mollifier given by (14.3.1) then the modified Burgers equation*

$$dX = \left[X_{\xi\xi} + \tfrac{1}{2} D_\xi \left(M_R(X) \right)^2 \right] dt + dW(t), \left.\begin{array}{c} \\ \\ \end{array}\right\} \qquad (14.3.2)$$
$$X(0) = x,$$

has a unique mild solution on a time interval $[0,T]$. Moreover the corresponding transition semigroup P_t^R, $t \geq 0$ has the strong Feller property on H.

Proof — Step 1—Existence and uniqueness

The equation (14.3.2) can be rewritten in the form

$$Y(t) = S(t)x + \frac{1}{2}F\left(M_R(Y + W_A)\right)^2(t),\ t \in [0,T], \qquad (14.3.3)$$

with $X(t) = Y(t) + W_A(t)$.

The proof of existence and uniqueness of solutions to (14.3.3) is similar to that of Theorem 14.2.4 provided one can derive an a–priori bound of the form (14.2.12). For this purpose we assume that Y satisfies the equation

$$\left.\begin{array}{l} Y'(t) = AY(t) + \frac{1}{2}D_\xi\left(M_R(Y + \varphi)\right)^2,\ t \in [0,T], \\[2mm] Y(0) = x, \end{array}\right\} \qquad (14.3.4)$$

with $\varphi \in C([0,T]; H_0^1(0,1))$, $x \in D(A)$, and proceed as in step 2 of the proof of Theorem 14.2.4. In particular we obtain that

$$\frac{1}{2}\frac{d}{dt}\|Y(t)\|_2^2 + \|Y_\xi(t)\|_2^2$$

$$= -\frac{1}{2}\psi_R(\|Y(t)+\varphi(t)\|)\int_0^1 (Y(t)+\varphi(t))^2\, Y_\xi(t)d\xi$$

$$= -\psi_R(\|Y(t)+\varphi(t)\|)\left[\int_0^1 \varphi(t)Y(t)Y_\xi(t)d\xi + \frac{1}{2}\int_0^1 \varphi^2(t)Y_\xi(t)d\xi\right]$$

$$\leq \|\varphi(t)\|_\infty\, \|Y_\xi(t)\|_2\|Y(t)\|_2 + \frac{1}{2}\,\|\varphi(t)\|_\infty^2\,\|Y_\xi(t)\|_2.$$

Therefore the estimate (14.2.12) also holds in the modified case.

Step 2—Strong Feller property

We proceed as in the proof of Theorem 7.1.1. We set

$$\langle X_x(t,x), h\rangle = \eta^h(t,x),\ t \geq 0,\ x \in H.$$

It is easy to check that the solution $X(t,x)$ of (14.3.2) is differentiable with respect to x and that $\eta^h(t,x)$ is a mild solution of the problem

$$\left.\begin{array}{l} \eta' = \eta_{\xi\xi} + D_\xi\left(M_R(X)D_x M_R(X)\eta\right),\ t \in [0,T], \\[2mm] \eta(0) = h,\ \mathbb{P}\text{–a.s.} \end{array}\right\} \qquad (14.3.5)$$

Note that by Hölder's inequality, for arbitary $x, z \in H$

$$\|M_R(x)D_x M_R(x)z\|_1 \leq \|M_R(x)\|_2 \|D_x M_R(x)z\|_2 \leq C\|z\|_2,$$
$$(14.3.6)$$

where C is a constant depending on the choice of the function ψ_R. Moreover equation (14.3.5) can be rewritten as

$$\eta(t) = S(t)h + \frac{1}{2}F\left(M_R(x)D_x M_R(x)\eta\right)(t), \; t \in [0,T].$$

Taking into account Lemma 14.2.1 and estimate (14.3.6) we conclude that the solution to (14.3.5) exists in $C([0,T];H)$ provided T is sufficiently small. Moreover, for a non–random constant C_T,

$$\sup_{t\in[0,T]} \|\eta^h(t,x)\| \leq C_T\|h\|^2, \; h,x \in H.$$

This implies, as in the proof of Theorem 7.1.1, the required strong Feller property on a small time interval, and then, by the semigroup property, on $[0,+\infty[$. ∎

Theorem 14.3.2 *The transition semigroup P_t, $t \geq 0$, corresponding to solutions of equation (14.2.1), has the strong Feller property on H.*

Proof — For any $R > 0$ let P_t^R, $t \geq 0$, be the transition semigroup corresponding to solutions X^R of equation (14.3.2). Let

$$\tau_x^R = \inf\{t \geq 0 : \|X^R(t,x)\| \geq R\}.$$

It is clear that

$$X(t,x) = X^R(t,x) \text{ for all } t \leq \tau_x^R, \; x \in H.$$

Let now $\varphi \in B_b(H)$, then

$$\left|P_t^R\varphi(x) - P_t\varphi(x)\right| = \left|\mathbb{E}\left(\varphi(X^R(t,x)) - \varphi(X(t,x))\right)\right|$$

$$= \left|\mathbb{E}\left(\varphi(X^R(t,x)) - \varphi(X(t,x))\right)\chi_{\tau_x^R \leq t}\right|$$

$$\leq 2\sup_{z\in H}|\varphi(z)|\mathbb{P}(\tau_x^R \leq t).$$

Since the functions $P_t^R \varphi$ are continuous for each $R > 0$ and $t > 0$ it is enough to show that for arbitrary $M > 0$ and $t > 0$

$$\lim_{R \to +\infty} \sup_{\|x\| \leq M} \mathbb{P}(\tau_x^R \leq t) = 0. \tag{14.3.7}$$

This follows, however, from the estimate (14.2.12) valid also for processes X_R. In fact if $\|x\| \leq M$ then

$$\sup_{s \leq t} \|X_R(s,x)\|_2^2 \leq 2 \sup_{s \leq t} \|Y_R(s,x)\|_2^2 + 2 \sup_{s \leq t} \|W_A(s,x)\|_2^2$$

$$\leq 2M^2 e^{8 \int_0^t \|W_A(s)\|_\infty^2 ds} + 2 \int_0^t e^{8 \int_r^t \|W_A(s)\|_\infty^2 ds} \|W_A(r)\|_\infty^4 dr$$

$$+ 2 \sup_{s \leq t} \|W_A(s,x)\|_2^2.$$

Since the right hand side of the estimate is finite and independent of x, (14.3.7) holds.   ∎

14.4   Invariant measure

14.4.1   Existence

To prove existence of an invariant measure for equation (14.2.1) it is convenient to write the problem in a slightly different form. For any $\alpha > 0$ we set

$$S_\alpha(t) = e^{-\alpha t} S(t), \ t \geq 0.$$

Then the mild form of (14.2.1), with $x = 0$, is equivalent to the integral equation

$$X(t) = \frac{1}{2} \int_0^t S_\alpha(t-s) D_\xi \left(X^2(s) \right) ds + \alpha \int_0^t S_\alpha(t-s) X(s) ds + W_A^\alpha(t),$$

where

$$W_A^\alpha(t) = \int_0^t S_\alpha(t-s) dW(s),$$

and α will be chosen later. Finally, setting

$$Y(t) = X(t) - W_A^\alpha(t),$$

we consider the initial value problem

$$Y'(t) = AY(t) + \frac{1}{2}D_\xi\left((Y + W_A^\alpha)^2\right) + \alpha W_A^\alpha, \left.\begin{array}{l}\\ \\ \end{array}\right\}$$

$$Y(0) = 0.$$

(14.4.1)

Note that equation (14.4.1) defines a transition semigroup P_t, $t \geq 0$, on $B_b(L^2(0,1))$ which is a Feller one (because the solutions to (14.4.1) depend continuously on the initial conditions). Therefore, by the general theory, see Chapter 6, to show existence of an invariant measure for (14.4.1) it is enough to prove that the family of measures

$$\left\{\frac{1}{T}\int_0^T P_s(0,\cdot)ds\right\}$$

is tight on $L^2(0,1)$. For this we need a lemma.

Lemma 14.4.1 *For any $\varepsilon > 0$ there exists $M_\varepsilon > 0$ such that, for all $T > 0$,*

$$\frac{1}{T}\int_0^T \mathbb{P}\left(\|Y(s)\|_2^2 > M_\varepsilon\right) ds < \varepsilon.$$

(14.4.2)

Proof — Without any loss of generality we can assume that Y is a strict solution of (14.4.1). Multiplying both sides of this equation by $Y(t)$ and integrating by parts in $[0,1]$, we find

$$\frac{1}{2}\frac{d}{dt}\|Y\|_2^2 + \int_0^1 Y_\xi^2\,d\xi = -\frac{1}{2}\int_0^1 (Y + W_A^\alpha)^2 Y_\xi\,d\xi + \alpha\int_0^1 YW_A^\alpha\,d\xi$$

$$= -\int_0^1 W_A^\alpha Y Y_\xi\,d\xi - \frac{1}{2}\int_0^1 (W_A^\alpha)^2 Y_\xi\,d\xi + \alpha\int_0^1 YW_A^\alpha\,d\xi,$$

since

$$\frac{1}{2}\int_0^1 Y^2 Y_\xi\,d\xi = 0.$$

By the Hölder inequality we find

$$\frac{1}{2}\frac{d}{dt}\|Y\|_2^2 + \int_0^1 Y_\xi^2\,d\xi \leq \|W_A^\alpha\|_4\,\|Y\|_4\left[\int_0^1 Y_\xi^2 d\xi\right]^{1/2}$$

$$+ \frac{1}{2}\|W_A^\alpha\|_4^2\left[\int_0^1 Y_\xi^2 d\xi\right]^{1/2} + \alpha\|W_A^\alpha\|_2\,\|Y\|_2.$$

(14.4.3)

We want now to express $\|Y\|_4$ in terms of $\|Y\|_2$ and $\|Y\|_{H^2(0,1)}$. By the Sobolev embedding theorem we have

$$L^4(0,1) \subset H^{1/4}(0,1),$$

so that there exists a constant $C > 0$ such that

$$\|Y\|_4 \leq C \|Y\|_{H^{1/4}(0,1)}.$$

It follows, from a well known interpolatory inequality, see e.g. J. L. Lions and E. Magenes [107], that

$$\|Y\|_4 \leq C_1 \|Y\|_2^{3/4} \|Y\|_{H^1(0,1)}^{1/4} = C_1 \|Y\|_2^{3/4} \left[\int_0^1 Y_\xi^2 d\xi \right]^{1/8}.$$

From (14.4.3) it follows that

$$\frac{1}{2} \frac{d}{dt} \|Y\|_2^2 + \int_0^1 Y_\xi^2 \, d\xi \leq C_1 \|W_A^\alpha\|_4 \|Y\|_2^{3/4} \left[\int_0^1 Y_\xi^2 d\xi \right]^{5/8}$$

$$+ \frac{1}{2} \|W_A^\alpha\|_4^2 \left[\int_0^1 Y_\xi^2 d\xi \right]^{1/2} + \alpha \|W_A^\alpha\|_2 \|Y\|_2.$$
$$(14.4.4)$$

Now recalling the inequality

$$xy \leq \frac{x^p}{p} + \frac{y^q}{q}, \qquad \frac{1}{p} + \frac{1}{q} = 1, \ x, y \geq 0,$$

and setting $p = \frac{8}{3}$, $q = \frac{8}{5}$, we see that there exists a constant $C_2 > 0$ such that

$$C_1 \|W_A^\alpha\|_4 \|Y\|_2^{3/4} \left[\int_0^1 Y_\xi^2 d\xi \right]^{5/8} \leq C_2 \|W_A^\alpha\|_4^{8/3} \|Y\|_2^2 + \frac{1}{4} \int_0^1 Y_\xi^2 \, d\xi.$$
$$(14.4.5)$$

Moreover

$$\frac{1}{2} \|W_A^\alpha\|_4^2 \left[\int_0^1 Y_\xi^2 d\xi \right]^{1/2} \leq 2 \|W_A^\alpha\|_2^4 + \frac{1}{4} \int_0^1 Y_\xi^2 d\xi. \qquad (14.4.6)$$

By substituting (14.4.5) and (14.4.6) in (14.4.4) we find

$$\frac{1}{2} \frac{d}{dt} \|Y\|_2^2 + \frac{1}{2} \int_0^1 Y_\xi^2 d\xi \leq C_2 \|W_A^\alpha\|_4^{8/3} \|Y\|_2^2 + 2 \|W_A^\alpha\|_4^2$$

$$+ \alpha \|W_A^\alpha\|_2 \|Y\|_2.$$
$$(14.4.7)$$

By using the Poincaré inequality

$$\int_0^1 Y_\xi^2 d\xi \geq \pi^2 \int_0^1 Y^2 d\xi,$$

and the following one

$$\alpha \|W_A^\alpha\|_2 \|Y\|_2 \leq \frac{4\alpha^2}{\pi^2} \|W_A^\alpha\|_2^2 + \frac{\pi^2}{4} \|Y\|_2^2,$$

we finally get

$$\frac{d}{dt} \|Y(t)\|_2^2 + \frac{\pi^2}{2} \|Y(t)\|_2^2 \leq 2C_2 \|W_A^\alpha(t)\|_4^{8/3} \|Y(t)\|_2^2 + 4 \|W_A^\alpha(t)\|_4^2$$

$$+ \frac{8\alpha^2}{\pi^2} \|W_A^\alpha(t)\|_2^2, \quad t \geq 0.$$

(14.4.8)

We now prove that there exists $\alpha > 0$ such that

$$\sup_{t \geq 0} \mathbb{E}\left(\|W_A^\alpha(t)\|_4^{8/3} \right) \leq \frac{\pi^2 \varepsilon}{8C_2}.$$

(14.4.9)

We have in fact

$$\mathbb{E}|W_A^\alpha(t,\xi)|^2 = \sum_{n=1}^\infty \mathbb{E}\left[\int_0^t e^{-(\alpha+\pi^2 n^2)(t-s)} d\beta_n(s) \right]^2 e_n^2(\xi)$$

$$= \sum_{n=1}^\infty \int_0^t e^{-2(\alpha+\pi^2 n^2)s} ds\, e_n^2(\xi)$$

$$\leq \frac{1}{\pi} \sum_{n=1}^\infty \frac{1}{\alpha + \pi^2 n^2}, \quad \xi \in [0,1].$$

Since $W_A^\alpha(t,\xi)$ is a Gaussian random variable, we have, for arbitrary $p \geq 1$ and a finite constant $c_p > 0$,

$$\mathbb{E}|W_A^\alpha(t,\xi)|^p = c_p \left(\mathbb{E}|W_A^\alpha(t,\xi)|^2 \right)^{p/2}$$

$$\leq c_p \left(\frac{1}{\pi} \sum_{n=1}^\infty \frac{1}{\alpha + \pi^2 n^2} \right)^{p/2}, \quad \xi \in [0,1].$$

Taking $p = 4$ it follows that

$$\mathbb{E}\left[\|W_A^\alpha(t)\|_4^4\right] \leq c_4 \left(\frac{1}{\pi} \sum_{n=1}^{\infty} \frac{1}{\alpha + \pi^2 n^2}\right)^2.$$

This implies (14.4.9) by the Hölder estimate.

Finally we use an argument due to D. Gątarek, see G. Da Prato and D. Gątarek [38]. We fix $M > 1$ and set

$$\zeta(t) = \log\left(\|Y(t)\|_2^2 \vee M\right).$$

We have

$$\zeta'(t) = \frac{1}{\|Y(t)\|_2^2} \chi_{\{\|Y(t)\|_2^2 \geq M\}} \frac{d}{dt} \|Y(t)\|_2^2.$$

By multiplying both sides of (14.4.8) by

$$\frac{1}{\|Y(t)\|_2^2} \chi_{\{\|Y(t)\|_2^2 \geq M\}},$$

we get

$$\zeta'(t) + \frac{\pi^2}{2} \chi_{\{\|Y(t)\|_2^2 \geq M\}} \leq 2C_2 \|W_A^\alpha(t)\|_4^{8/3}$$

$$+ \frac{4}{M} \|W_A^\alpha(t)\|_4^2 + \frac{8\alpha^2}{\pi^2 M} \|W_A^\alpha(t)\|_2^2.$$

Integrating from 0 to t and taking expectation, we find

$$\mathbb{E}(\zeta(t) - \zeta(0)) + \frac{\pi^2}{2} \int_0^t \mathbb{P}\left(\|Y(s)\|_2^2 \geq M\right) ds$$

$$\leq 2C_2 \int_0^t \left[\mathbb{E}\|W_A^\alpha(s)\|_4^{8/3} + \frac{4}{M}\mathbb{E}\|W_A^\alpha(s)\|_4^2 + \frac{8\alpha^2}{M\pi^2}\mathbb{E}\|W_A^\alpha(s)\|_2^2\right] ds.$$

But $\zeta(t) - \zeta(0) = \zeta(t) - M \geq 0$, and so, taking into account (14.4.9),

$$\frac{1}{t} \int_0^t \mathbb{P}\left(\|Y(s)\|_2^2 \geq M\right) ds \leq \frac{\varepsilon}{2} + \frac{8}{M t \pi^2} \int_0^t \mathbb{E}\|W_A^\alpha(s)\|_4^4 ds$$

$$+ \frac{16\alpha^2}{M t \pi^4} \mathbb{E}\|W_A^\alpha(s)\|_2^2 ds,$$

and the conclusion follows by choosing M sufficiently large.   ∎

We can now prove the result

Theorem 14.4.2 *There exists an invariant measure for the transition semigroup P_t, $t \geq 0$, corresponding to solutions of equation (14.2.1).*

Proof — We fix $\alpha > 0$ such that the statement of Lemma 14.4.1 holds. Note that for arbitrary $\sigma > 0$ the embedding

$$D((-A)^{\sigma}) \subset L^2(0,1)$$

is compact and therefore to prove the required tightness property it is enough to show that

For any $\varepsilon > 0$ there exists $M > 0$ such that for all $T > 1$

$$\frac{1}{T} \int_0^T \mathbb{P}\left(\|(-A)^{\sigma} X(t)\|_2^2 > M \right) dt < \varepsilon. \qquad (14.4.10)$$

From now on we fix not only α but also $\sigma \in]0, 1/4[$. Repeating the calculations which led to the estimate (14.4.9) we easily deduce that

$$K = \sup_{t \geq 1} \mathbb{E}\|(-A)^{\sigma} W_A^{\alpha}(t)\|_2^2 < +\infty.$$

Therefore by Chebyshev's inequality

$$
\begin{aligned}
I_T &= \frac{1}{T} \int_1^T \mathbb{P}\left(\|(-A)^{\sigma} W_A^{\alpha}(t)\|_2^2 > M \right) dt \\
&\leq \frac{1}{TM} \int_1^T \mathbb{E}\left(\|(-A)^{\sigma} W_A^{\alpha}(t)\|_2^2 \right) dt \\
&\leq \left(1 - \frac{1}{T} \right) \frac{K}{M}.
\end{aligned}
$$

So I_T can be made small uniformly in $T \geq 1$ provided M is chosen sufficiently large. This way, since $X = Y + W_A$, to show that (14.4.10) holds it is enough to prove it with the process X replaced by Y. This will be done in the same spirit as in the proof of Theorem 6.1.2 exploiting the regularizing effect of (14.1.3).

For the mild solution Y of (14.4.1) and arbitrary $t > 0$ we have that

$$
\begin{aligned}
(-A)^\sigma Y(t+1) \;=\;& (-A)^\sigma S(1)Y(t) \\[2mm]
&+\; (-A)^\sigma \int_t^{t+1} S(t+1-r)D_\xi \left(Y(r) + W_A^\alpha(r)\right)^2 dr \\[2mm]
&+\; \alpha(-A)^\sigma \int_t^{t+1} S(t+1-r)W_A^\alpha(r)dr.
\end{aligned}
$$

Taking into account Lemma 14.4.1 and Remark 14.2.3 we have (for any $p > 1$ such that $\frac{1}{p} + \sigma < \frac{1}{4}$)

$$
\begin{aligned}
\|(-A)^\sigma Y(t+1)\|_2 \;\leq\;& M_1\|Y(t)\|_2 + M_2 \sup_{0 \leq r \leq 1} \|Y(t+r)\|_2^2 \\[2mm]
&+\; M_3 \left(\int_t^{t+1} \|W_A^\alpha(r)\|_2^p dr \right)^{1/p},
\end{aligned}
$$

for some constants M_1, M_2, M_3 and for all $t > 0$.

Moreover from the estimate (14.4.8)

$$
\begin{aligned}
\|Y(t+r)\|_2^2 \leq\;& e^{2C_2 \int_t^{t+r} \|W_A^\alpha(s)\|_4^{8/3} ds}\|Y(t)\|_2^2 \\[2mm]
&+ \int_t^{t+r} e^{2C_2 \int_\rho^{t+r} \|W_A^\alpha(s)\|_4^{8/3} ds} \left[4\|W_A^\alpha(\rho)\|_4^2 + \frac{8\alpha^2}{\pi^2}\|W_A^\alpha(\rho)\|_2^2 \right] d\rho,
\end{aligned}
$$

and so

$$
\begin{aligned}
\sup_{0 \leq r \leq 1} \|Y(t+r)\|_2^2 \leq\;& e^{2C_2 \int_t^{t+1} \|W_A^\alpha(s)\|_4^{8/3} ds}\|Y(t)\|_2^2 \\[2mm]
&+ \int_t^{t+1} e^{2C_2 \int_\rho^{t+r} \|W_A^\alpha(s)\|_4^{8/3} ds} \left[4\|W_A^\alpha(\rho)\|_4^2 + \frac{8\alpha^2}{\pi^2}\|W_A^\alpha(\rho)\|_2^2 \right] d\rho.
\end{aligned}
$$

Consequently

$$\mathbb{P}(\|(-A)^\sigma Y(t+1)\|_2 \geq M) \leq \mathbb{P}\left(M_1\|Y(t)\|_2 \geq \frac{M}{3}\right)$$

$$+\mathbb{P}\left(M_2 \sup_{0 \leq r \leq 1} \|Y(t+r)\|_2^2 \geq \frac{M}{3}\right)$$

$$+\mathbb{P}\left(M_3 \left(\int_t^{t+1} \|W_A^\alpha(r)\|_p^2 dr\right)^{1/p} \geq \frac{M}{3}\right)$$

$$\leq I_1(t) + I_2(t) + I_3(t).$$

Moreover

$$I_2(t) \leq \mathbb{P}\left(M_2 e^{2C_2 \int_t^{t+1} \|W_A^\alpha(s)\|_4^{8/3} ds} \|Y(t)\|_2^2 \geq \frac{M}{6}\right)$$

$$+\mathbb{P}\left(M_2 e^{2C_2 \int_t^{t+1} \|W_A^\alpha(s)\|_4^{8/3} ds}\right.$$

$$\left.\times \int_t^{t+1} \left[4\|W_A^\alpha(s)\|_4^2 + \frac{8\alpha^2}{\pi^2}\|W_A^\alpha(s)\|_2^2\right] ds \geq \frac{M}{6}\right)$$

$$\leq I_{21}(t) + I_{22}(t).$$

In addition

$$I_{21}(t) \leq \mathbb{P}\left(M_2 e^{2C_2 \int_t^{t+1} \|W_A^\alpha(s)\|_4^{8/3} ds} \geq \sqrt{M/6}\right) + \mathbb{P}\left(\|Y(t)\|_2^2 \geq \sqrt{M/6}\right),$$

$$I_{21}(t) \leq \mathbb{P}\left(M_2 e^{2C_2 \int_t^{t+1} \|W_A^\alpha(s)\|_4^{8/3} ds} \geq \sqrt{M/6}\right)$$

$$+\mathbb{P}\left(\int_t^{t+1} \left[4\|W_A^\alpha(s)\|_4^2 + \frac{8\alpha^2}{\pi^2}\|W_A^\alpha(s)\|_2^2\right] ds \geq \sqrt{M/6}\right).$$

Finally we see that

$$\frac{1}{T}\int_0^T \mathbb{P}(\|(-A)^\sigma Y(t+1)\|_2 \geq M)\,dt \leq \frac{1}{T}\int_0^T \mathbb{P}\left(M_1\|Y(t)\|_2 \geq \frac{M}{3}\right) dt$$

$$+\frac{2}{T}\int_0^T \mathbb{P}\left(e^{2C_2\int_t^{t+1}\|W_A^\alpha(s)\|_4^{8/3}ds} \geq \sqrt{M/6}\right) dt$$

$$\leq \frac{1}{T}\int_0^T \mathbb{P}\left(\|Y(t)\|_2 \geq \sqrt{M/6}\right) dt$$

$$+\frac{1}{T}\int_0^T \mathbb{P}\left(\int_t^{t+1}\left[4\|W_A^\alpha(s)\|_4^2 + \frac{8\alpha^2}{\pi^2}\|W_A^\alpha(s)\|_2^2\right]ds \geq \sqrt{M/6}\right) dt$$

$$+\frac{1}{T}\int_0^T \mathbb{P}\left(M_3\left(\int_t^{t+1}\|W_A^\alpha(s)\|_p^2 ds\right)^{1/p} \geq \frac{M}{3}\right) dt.$$

The first two terms on the right hand side can be made arbitrarily small uniformly in T if M is chosen sufficiently large (by Lemma 14.4.1). The same is true by Chebyshev's inequality with the remaining terms, because, as is easy to check,

$$\sup_{s>0} \mathbb{E}\left[\|W_A^\alpha(s)\|_4^{8/3} + \|W_A^\alpha(s)\|_4^2 + \|W_A^\alpha(s)\|_2^2 + \|W_A^\alpha(s)\|_2^p\right] < +\infty,$$

compare the derivation of the estimate (14.4.9). ∎

14.4.2   Uniqueness

Finally we prove that the invariant measure for the equation (14.2.1) is unique. According to the general theory it is enough, see Chapter 7, to show that the corresponding transition semigroup is irreducible and has the strong Feller property. Since the latter was proved in Theorem 14.3.2, it remains to show irreducibility.

Proposition 14.4.3 *The transition semigroup P_t, $t \geq 0$, on the space $B_b(L^2(0,1))$, corresponding to (14.2.1) is irreducible.*

Proof —   Let us fix $T > 0, a \in L^2(0,1)$ and $b \in H_0^1(0,1)$. We show first that there exists $v \in L^2(0,T;L^2(0,1))$ such that for the solution

$y(t)$, $t \in [0,T]$ of the problem

$$y_t(t,\xi) = D_\xi^2 y(t,\xi) + \frac{1}{2} D_\xi \left(y^2(t,\xi) \right) + v(t,\xi),$$

$$t > 0, \ \xi \in [0,1],$$

$$y(t,0) = y(t,1) = 0, \ t > 0,$$

$$y(0,\xi) = a(\xi), \ \xi \in [0,1],$$

$$(14.4.11)$$

one has

$$y(T,\xi) = b(\xi), \ \xi \in [0,1].$$

Assume in addition that $a \in H_0^1(0,1)$. Then there exists $u \in L^2(0,T;L^2(0,1))$ such that for the solution $z(t)$, $t \in [0,T]$, of the linear problem

$$z_t(t,\xi) = D_\xi^2 z(t,\xi) + u(t,\xi), t > 0, \ \xi \in [0,1],$$

$$z(t,0) = z(t,1) = 0, \ t > 0,$$

$$z(0,\xi) = a(\xi), \ \xi \in [0,1],$$

$$(14.4.12)$$

one has

$$z(T,\xi) = b(\xi), \ \xi \in [0,1],$$

and one can easily check that $z \in C([0,T]; H_0^1(0,1))$.

Define

$$v(t,\xi) = -\frac{1}{2} D_\xi^2 z(t,\xi) + u(t,\xi), \ t > 0, \ \xi \in [0,1].$$

Then $v \in L^2(0,T;L^2(0,1))$ and by direct substitution one sees that z is the solution to problem (14.4.11).

Now, from the well known identity for solutions of (14.4.11),

$$\frac{1}{2}\frac{d}{dt}\|y(t)\|_2^2 + \|D_\xi y(t)\|_2^2 = (y,v),$$

it follows that, if $v = 0$, then for almost all $t \in [0,T]$, $y(t) \in H_0^1(0,1)$. It is therefore enough to define $v(s) = 0$ in $[0,\bar{t}]$ where $\bar{t} \in]0,T[$ is a moment such that $y(\bar{t}) \in H_0^1(0,1)$ and for the remaining interval $[\bar{t}, T]$ to use the first part of the proof.

To prove the proposition we will estimate the L^2 distance between the solution X to equation (14.2.1), with $x_0 = a$, and the function y. Using our previous notation,

$$X(t) - y(t) = \frac{1}{2}F(X^2 - y^2)(t) + W_A(t) - V_A(t),$$

where

$$V_A(t) = \int_0^t S(t-s)v(s)ds, \quad t \in [0,T].$$

From the proof of Lemma 14.2.1, see formula (14.2.6), it follows that

$$\|F(X^2 - y^2)(t)\|_2 \le C \int_0^t (t-s)^{-\frac{3}{4}} \|X^2(s) - y^2(s)\|_1 ds.$$

But, for arbitrary $s \in [0,T]$,

$$\|X^2(s) - y^2(s)\|_1 \le \|X(s) - y(s)\|_2 \cdot \|X(s) + y(s)\|_2,$$

and we see that

$$\|X(t) - y(t)\|_2 \le \frac{1}{2}C \int_0^t (t-s)^{-\frac{3}{4}}(\|X(s) - y(s)\|_2 \tag{14.4.13}$$

$$\times \|X(s) + y(s)\|_2)ds + \|W_A(t) - V_A(t)\|_2.$$

Defining

$$Y(t) = X(t) - W_A(t), \quad t \in [0,T],$$

we have that

$$\|X(t)\|_2 \le \|Y(t)\|_2 + \|W_A(t)\|_2.$$

Taking into account the estimate (14.2.11),

$$\frac{d}{dt}\|Y(t)\|_2^2 \le \|W_A(t)\|_\infty^4 + 4\|W_A(t)\|_\infty^2,$$

derived at the end of the proof of Theorem 14.2.4, we have that if

$$\sup_{t \le T} \|W_A(t)\|_\infty \le \gamma$$

then

$$\|Y(t)\|_2^2 \le e^{c_1 t} + c_2 \int_0^t e^{c_1 s} ds, \ t \in [0, T],$$

where

$$c_1 = \gamma^4,$$

$$c_2 = 4\gamma^2.$$

Consequently

$$
\begin{aligned}
\|X(s) + y(s)\|_2 &\le \|y(s)\|_2 + \|Y(s)\|_2 + \|W_A(s)\|_2 \\
&\le \|y(s)\|_2 + e^{c_1 s} + c_2 \int_0^s e^{c_1 r} dr + \|W_A(s)\|_2 \\
&\le \sup_{s \le T} \left[\|y(s)\|_2 + e^{c_1 s} + c_2 \int_0^s e^{c_1 r} dr + \gamma \right] \le c.
\end{aligned}
$$

where the constant c depends only on γ.

From (14.4.13) we have that

$$
\begin{aligned}
\|X(t) - y(t)\|_2 &\le \frac{1}{2} cC \int_0^t (t-s)^{-\frac{3}{4}} \|X(s) - y(s)\|_2 ds \\
&+ \|W_A(t) - V_A(t)\|_2.
\end{aligned}
$$

Still assuming

$$\sup_{t \le T} \|W_A(t)\|_2 \le \gamma,$$

by a generalization of Gronwall's lemma

$$\sup_{t \le T} \|X(t) - y(t)\|_2 \le \tilde{c} \sup_{t \le T} \|W_A(t) - V_A(t)\|_2, \qquad (14.4.14)$$

where the constant $\tilde{c}$ depends only on T.

Since the support of the distribution of the process $W_A(\cdot)$ in $C([0, T]; L^2([0, 1]))$ is exactly the closure of the set of all functions

$$\int_0^t S(t-s)w(s)ds, \ t \in [0, T], \ w \in L^2(0, T; L^2(0, 1)),$$

we have for arbitrary $\varepsilon > 0$

$$\mathbb{P}\left(\sup_{t \le T} \|W_A(t) - V_A(t)\|_2 < \varepsilon\right) > 0.$$

Let us fix

$$\gamma = \varepsilon + \sup_{t \le T} \|V_A(t)\|_2.$$

Then we have

$$\mathbb{P}\left(\sup_{t \le T} \|W_A(t) - V_A(t)\|_2 < \varepsilon\right)$$

$$= \mathbb{P}\left(\sup_{t \le T} \|W_A(t) - V_A(t)\|_2 < \varepsilon \ \& \ \sup_{t \le T} \|W_A(t)\|_2 \le \gamma\right) > 0.$$

Therefore taking into account the estimate (14.4.14)

$$\mathbb{P}\left(\sup_{t \le T} \|X(t) - y(t)\|_p < \varepsilon \tilde{c}\right)$$

$$\ge \mathbb{P}\left(\sup_{t \le T} \|X(t) - y(t)\|_p < \varepsilon \tilde{c} \ \& \ \sup_{t \le T} \|W_A(t)\|_2 \le \gamma\right)$$

$$\ge \mathbb{P}\left(\sup_{t \le T} \|W_A(t) - V_A(t)\|_2 < \varepsilon \ \& \ \sup_{t \le T} \|W_A(t)\|_2 \le \gamma\right) > 0.$$

In particular

$$\mathbb{P}\left(\sup_{t \le T} \|X(t) - b)\|_p < \varepsilon \tilde{c}\right) > 0.$$

Since $\varepsilon > 0$ can be arbitrarily chosen the result follows.   ∎

We can now state our final result.

Theorem 14.4.4 *There exists a unique invariant measure for the transition semigroup P_t, $t \ge 0$, corresponding to solutions of equation (14.2.1). Moreover μ is ergodic and strongly mixing.*

Proof — By Theorem 14.4.2 there exists an invariant measure μ. Since the semigroup P_t, $t \ge 0$, is a strong Feller one by Theorem 14.3.2 and irreducible by Proposition 14.4.3, the conclusion follows from Doob's theorem, 4.2.1.   ∎

Chapter 15

Navier–Stokes equations

In this chapter we first establish existence of a solution to the stochastic Navier–Stokes equations on a bounded domain $D \subset \mathbb{R}^2$ by a direct method. Then existence of an invariant measure is proved by a modification of the dissipativity method.

15.1 Preliminaries

Let D be an open subset of $\mathbb{R}^2$ with regular boundary ∂D. We are concerned with the problem

$$
\left.
\begin{aligned}
&d_t Z(t,\xi) = (\nu \Delta Z(t,\xi) + Z_\xi(t,\xi) \cdot Z(t,\xi) + \nabla p(t,\xi))dt + dW_Q(t), \\[4pt]
&\qquad \text{on } [0,+\infty[\times D, \\[8pt]
&Z(t,\xi) = 0, \quad \text{on } [0,+\infty[\times \partial D, \\[8pt]
&\text{div } Z(t,\xi) = 0, \quad \text{on } [0,+\infty[\times D, \\[8pt]
&Z(0,\xi) = x(\xi) \quad \text{on } D.
\end{aligned}
\right\}
$$

$$(15.1.1)$$

Here $Z(t,\xi)$ is the velocity of an incompressible fluid, ν is the viscosity parameter, and $p(t,\xi)$ the pressure of the fluid. Moreover W_Q is a Q–Wiener process in a probability space $(\Omega, \mathcal{F}, \mathbb{P})$. The distributional derivative of $W_Q(t)$, $t \geq 0$, represents external force acting on the fluid.

In this chapter we first prove the existence of a solution to the stochastic Navier–Stokes equation on an open subset of $\mathbb{R}^2$. We use a simple and new fixed point argument instead of the more common Galerkin–Faedo approximation scheme. Then we show the existence of an invariant measure by adopting a method of F. Flandoli [66]. For the proof of uniqueness we refer to F. Flandoli and B. Maslowski [67]. These results can be generalized to the Bénard problem, that is a Navier–Stokes equation perturbed by white noise coupled with the heat equation, see B. Ferrario [65].

We shall denote by $\mathcal{V}$ the set of all mappings $u : D \to \mathbb{R}^2$ of C^∞ class and having compact support in D, such that

$$\text{div } u(\xi) = \text{ Tr } [D_\xi u(\xi)] = 0, \ \ \xi \in D.$$

For all $u = (u_1, u_2)$, $v = (v_1, v_2) \in \mathcal{V}$ we set

$$|u(\xi)|^2 = u_1^2(\xi) + u_2^2(\xi), \ \xi \in D,$$

$$\langle u(\xi), v(\xi) \rangle = u_1(\xi)v_1(\xi) + u_2(\xi)v_2(\xi), \ \xi \in D.$$

For any $p \geq 2$, we shall denote by L^p the closure of $\mathcal{V}$ with respect to the norm

$$\|u\|_p = \left(\int_D |u(\xi)|^2 d\xi \right)^{1/p},$$

and by V the closure of $\mathcal{V}$ with respect to the norm

$$\|u\|_V^2 = \int_D (|D_\xi u(\xi)|^2) \, d\xi.$$

We set $H = \mathrm{L}^2$ and $|\cdot| = |\cdot|_2$.

Obviously H is a closed subspace of $L^2(D) \times L^2(D) = L^2(D; \mathbb{R}^2)$, we shall denote by P the orthogonal projection of $L^2(D) \times L^2(D)$ onto H.

We introduce finally the linear operator

$$\left. \begin{array}{l} D(A) = [H^2(D) \times H^2(D)] \cap V, \\[2mm] Au = P\Delta u, \text{ forall } u \in D(A). \end{array} \right\} \tag{15.1.2}$$

A is a self–adjoint strictly negative operator on H, and $V = D((-A)^{1/2})$ see e.g. R. Temam [150]. We shall denote by $\{e_k\}$ an orthonormal complete system of eigenvectors of A and by $\{\lambda_k\}$ a sequence of positive numbers such that

$$Ae_k = -\lambda_k e_k, \ k \in \mathbf{N}.$$

We assume that $\{\lambda_k\}$ is not decreasing.

A central role in our considerations will be played by the space

$$E = L^4(0, T; \mathbf{L}^4).$$

Since $\mathbf{L}^4$ is the closure of $\mathcal{V}$ with respect to

$$\|(x_1, x_2)\|_4 = \left(\int_D \left(|x_1(\xi)|^4 + |x_2(\xi)|^4 \right) d\xi \right)^{1/4}, \ (x_1, x_2) \in \mathcal{V},$$

we have

$$\|u\|_E = \left[\int_0^T \int_D \left(|u_1(t,\xi)|^4 + |u_2(t,\xi)|^4 \right) d\xi \, dt \right]^{1/4}$$

$$= \left(\int_0^T \|u(t)\|_4^4 dt \right)^{1/4}.$$

We will also need the following inclusion result.

Proposition 15.1.1 *For any $T > 0$ we have*

$$L^\infty(0, T; H) \cap L^2(0, T; V) \subset E,$$

and there exists a constant K, independent of $T > 0$, such that

$$\|u\|_E \le K(\|u\|_{L^\infty(0,T;H)} + \|u\|_{L^2(0,T;V)}), \ u \in E. \tag{15.1.3}$$

Proof — We recall that due to the Sobolev embedding theorem

$$H^{1/2}(D) \subset L^4(D)$$

and there exists $C_1 > 0$ such that

$$\|v\|_{L^4(D)} \le C_1 \|v\|_{H^{1/2}(D)}, \ v \in H^{1/2}(D). \tag{15.1.4}$$

Define
$$V^{1/2} = \left(H^{1/2}(D) \times H^{1/2}(D) \right) \cap V.$$

Then $V^{1/2} = D((-A)^{1/4})$. By a well known interpolatory inequality, see J. L. Lions and E. Magenes [107, pag.22], we have for some constant $C_2 > 0$ and all $t \in [0, T]$

$$\|u(t)\|_{V^{1/2}} \le C_2 \|u(t)\|_H^{1/2} \, \|u(t)\|_V^{1/2}.$$

It therefore follows from (15.1.3) that

$$\|u(t)\|_4^4 \le 2C_1 \, \|u(t)\|_{V^{1/2}}^4 \le 2C_1 C_2^4 \, \|u(t)\|_H^2 \, \|u(t)\|_V^2.$$

Now integrating this inequality in $[0, T]$ we easily get (15.1.4). ∎

15.2 Local existence and uniqueness results

By applying the projection operator P to the first equation in (15.1.1) and setting $X = PZ$ we find the problem

$$\left. \begin{array}{l} dX(t) = [AX(t) + D_\xi(X(t)) \cdot X(t)] \, dt + dW(t), \ T > 0, \\[12pt] X(0) = x, \end{array} \right\}$$

$$(15.2.1)$$

where $T > 0$ is fixed and W is defined by

$$W(t) = \sum_{h=1}^{\infty} \sigma_h \beta_h g_h. \qquad (15.2.2)$$

Here $\{\beta_h\}$ is a sequence of mutually independent real Brownian motions in a fixed probability space $(\Omega, \mathcal{F}, \mathbb{P})$ adapted to a filtration $\{\mathcal{F}_t\}_{t \ge 0}$. Moreover $\{\sigma_h\}$ is a sequence of positive numbers and $\{g_h\}$ is an orthonormal basis in H.

We reduce problem (15.2.1) as usual to a deterministic one by setting

$$Y(t) = X(t) - W_A(t), \ \ t \in [0, T], \ \mathbb{P}\text{--a.s.}$$

We find

$$Y'(t) = AY(t) + D_\xi(Y(t) + W_A(t)) \cdot (Y(t) + W_A(t)) + f(t), \ t \geq 0, \left.\vphantom{\begin{array}{c}1\\1\\1\end{array}}\right\}$$

$$Y(0) = x,$$

$$(15.2.3)$$

where $W_A(t)$ is the stochastic convolution

$$W_A(t) = \int_0^t S(t-s)dW(s), \qquad (15.2.4)$$

and $S(t) = e^{tA}$, $t \geq 0$.

We now make the following basic assumption

Hypothesis 15.1 *Process $W_A(\cdot)$ has a version in L^4 with trajectories integrable in the fourth power.*

Notice that an equivalent formulation of the above hypothesis is that $W_A(\cdot)$ has an E–valued version.

We will interpret equation (15.2.3) as an integral equation

$$\begin{aligned}
Y(t) &= S(t)x + \int_0^t S(t-s)D_\xi(Y(s) + W_A(s)) \cdot (Y(s) + W_A(s))ds \\
&= S(t)x + F(Y + W_A)(t), \ t \in [0, T],
\end{aligned}$$

where $F : E \to E$ is a continuous extension of the operator

$$F_0 : C^1([0, T]; V) \to E$$

given by the formula

$$(F_0 u)(t) = \int_0^t S(t-s)(G_0 u)(s)ds, \ t \in [0, T],$$

where

$$G_0 : C^1([0, T]; V) \to E, \ (G_0 u)(t) = D_\xi u(t) \cdot u(t), \ t \in [0, T].$$

To show that the operators G and F are well defined it is useful to introduce a trilinear form on $V \times V \times V$ by setting

$$\begin{aligned}
b(u, v, z) &= \int_D \langle D_\xi v(\xi) \cdot u(\xi), z(\xi) \rangle d\xi \\
&= \sum_{h,k=1}^2 \int_D \frac{\partial v_h}{\partial \xi_k} u_k z_h \, d\xi.
\end{aligned}$$

By the Hölder inequality we have

$$|b(u,v,z)| \leq 4\|v\|_V \|u\|_4 \|z\|_4, \qquad (15.2.5)$$

so, by the inclusion (15.1.3), the definition of b is meaningful.

We now list some useful properties of b.

Lemma 15.2.1 *The following identities hold:*

$$b(u,v,z) = -b(u,z,v), \ \forall \, u,v,z \in V, \qquad (15.2.6)$$

and

$$b(u,v,v) = 0, \ \forall \, u,v \in V. \qquad (15.2.7)$$

Proof of (15.2.6) — We have

$$
\begin{aligned}
b(u,v,z) &= \sum_{h,k=1}^{2} \int_D \frac{\partial v_h}{\partial \xi_k} u_k \, z_h \, d\xi \\
&= -\sum_{h,k=1}^{2} \int_D v_h \frac{\partial u_k}{\partial \xi_k} z_h \, d\xi \\
&\quad - \sum_{h,k=1}^{2} \int_D v_h \, u_k \frac{\partial z_h}{\partial \xi_k} d\xi \\
&= \int_D \operatorname{div} u \, \langle v, z \rangle d\xi - b(u,z,v),
\end{aligned}
$$

and the conclusion follows since div $u = 0$.

Proof of (15.2.7) — We have

$$
\begin{aligned}
b(u, v, v) &= \sum_{h,k=1}^{2} \int_D \frac{\partial v_h}{\partial \xi_k} u_k \, v_h \, d\xi \\
&= \frac{1}{2} \sum_{h,k=1}^{2} \int_D u_k \frac{\partial}{\partial \xi_k} \left(v_h^2 \right) \, d\xi \\
&= -\frac{1}{2} \sum_{h,k=1}^{2} \int_D \frac{\partial u_k}{\partial \xi_k} v_h^2 d\xi \\
&= -\frac{1}{2} \int_D \operatorname{div} u \, |v|^2 d\xi = 0.
\end{aligned}
$$

■

Lemma 15.2.2 *The mapping G_0 can be continuously extended to*

$$
G : E \to L^2(0, T; V').
$$

Moreover

$$
\|G(u) - G(v)\|_{L^2(0,T;V')} \le 4(\|u\|_E + \|v\|_E) \, \|u - v\|_E, \quad u, v \in E.
$$

Proof — Let $u \in C^2([0,T]; V)$, $\varphi \in L^2(0, T; V)$. Then by Lemma 15.2.1, denoting by $\ll \cdot, \cdot \gg$ the duality mapping between $L^2(0, T; V')$ and $L^2(0, T; V)$,

$$
\ll G_0 u, \varphi \gg = \int_0^T \int_D \langle D_\xi u(t, \xi) \cdot u(t, \xi), \varphi(t, \xi) \rangle dt \, d\xi
$$

$$
= \int_0^T b(u(t), u(t), \varphi(t)) dt = -\int_0^T b(u(t), \varphi(t), u(t)) dt.
$$

In addition if $u, v \in C^2([0, T]; V)$ and $\varphi \in L^2(0, T; V)$ we have

$$\ll G_0 u - G_0 v, \varphi \gg$$

$$= \int_0^T \int_D \langle D_\xi u_{t,\xi} \cdot u(t, \xi) - D_\xi v(t, \xi) \cdot v(t, \xi) \rangle \varphi dt \, d\xi$$

$$= -\int_0^T [b(u(t), u(t), \varphi(t)) - b(v(t), v(t), \varphi(t))]dt$$

$$= -\int_0^T [b(u(t), \varphi(t), u(t)) - b(v(t), \varphi(t), v(t))]dt$$

$$= \int_0^T [b(u(t), \varphi(t), -u(t)) + b(v(t), \varphi(t), v(t))]dt$$

$$= \int_0^T [b(u(t), \varphi(t), v(t) - u(t)) + b(v(t) - u(t), \varphi(t), v(t))]dt.$$

It follows from (15.2.5) that

$$| \ll G_0 u - G_0 v, \varphi \gg | \le 4 \int_0^T \Big[\|u(t) - v(t)\|_4 \, \|u(t)\|_4 \, \|\varphi(t)\|_V$$

$$+ \|u(t) - v(t)\|_4 \, \|v(t)\|_4 \, \|\varphi(t)\|_V \Big] dt$$

$$\le 4 \left(\int_0^T \|u(t) - v(t)\|_4^4 dt \right)^{1/4} \left(\int_0^T \left(\|u(t)\|^4 + \|v(t)\|^4 dt \right) \right)^{1/4}$$

$$\times \left(\int_0^T \|\varphi(t)\|_V^2 dt \right)^{1/2}$$

$$\le 4\|u - v\|_E \left(\|u\|_E + \|v\|_E \right) \|\varphi\|_{L^2(0,T;V)}.$$

Consequently

$$\|G_0 u - G_0 v\|_{L^2(0,T;V)} \le 4\|u - v\|_E (\|u\|_E + \|v\|_E),$$

and we see that the extension G exists and satisfies the required inequality. ∎

The following lemma is classical, see e.g. J.L. Lions and E. Magenes [107].

Lemma 15.2.3 *Assume that A is a negative self-adjoint operator on H and*

$$V = D((-A)^{1/2}) \subset H \subset V'.$$

Then A and $S(t) = e^{tA}$ have continuous extensions from V to V'. If

$$y(t) = y(t; g) = \int_0^t e^{(t-s)A} g(s) ds, \ t \in [0, T], \ g \in L^2(0, T; V'),$$

then

$$y \in L^\infty(0, T; H) \cap L^2(0, T; V),$$

and for a constant $L > 0$, independent of $T > 0$,

$$\|y\|_{L^\infty(0,T;H)} + \|y\|_{L^2(0,T;V)} \le L\|g\|_{L^2(0,T;V')}.$$

Lemma 15.2.4 *The transformation F_0 can be continuously extended to $F : E \to E$. Moreover there exists a constant $M > 0$, independent of $T > 0$, such that*

$$\|F(u) - F(v)\|_E \le M (\|u\|_E + \|v\|_E) \|u - v\|_E, \ u, v \in E.$$

Proof — Using the notation of the previous lemma we define the extension F by the formula

$$F(u)(t) = y(t, G(u)), \ t \in [0, T], \ u \in E.$$

By Lemma 15.2.2 $F \in L^\infty(0, T; H) \cap L^2(0, T; V)$. Moreover, for $u, v \in E$,

$$\|F(u) - F(v)\|_E \le K\|y(\cdot, G(u)) - y(\cdot, G(v)))\|_{L^\infty(0,T;H)}$$
$$+ \|y(\cdot, G(u) - G(v))\|_{L^2(0,T;V)}.$$

Taking into account Lemma 15.2.3 again we get

$$\|F(u) - F(v)\|_E \le KL \|G(u) - G(v)\|_{L^2(0,T;V')}, \ u, v \in E.$$

Finally, by Lemma 15.2.2,

$$\|F(u) - F(v)\|_E \le 4KL \|u - v\|_E (\|u\|_E + \|v\|_E), \ u, v \in E,$$

and it is enough to define $M = 4KL$. ∎

Theorem 15.2.5 *Assume that Hypothesis 15.1 holds. Then for arbitrary $x \in H$ there exists a random variable τ taking values $\mathbb{P}$-a.s. in $]0, T]$ such that equation (15.2.1) has a unique solution on the interval $[0, \tau]$.*

Proof —   We first remark that the function $\psi(s) = S(s)x$, $s \in [0, T]$, belongs to

$$L^2(0, T; V) \cap C([0, T]; H).$$

By Proposition 15.1.1 $\psi \in E$. Moreover $W_A(\cdot) \in E$, $\mathbb{P}$-a.s.

Let us fix $\omega \in \Omega$ such that $W_A(\cdot, \omega) \in E$ and write $\varphi(s) = W_A(s, \omega)$, $s \in [0, T]$. To prove the theorem it is enough to show that the fixed point problem

$$u = \psi + F(u + \varphi), \ u \in E,$$

has a unique solution in the space $E = L^4(0, T; \mathbb{L}^4)$ for $T > 0$ sufficiently small. We apply the following version of the contraction mapping theorem.

Lemma 15.2.6 *Let $\mathcal{F}$ be a transformation from a Banach space $\mathcal{E}$ into $\mathcal{E}$, a an element of $\mathcal{E}$ and $\alpha > 0$ a positive number. If $\mathcal{F}(0) = 0$, $\|a\|_{\mathcal{E}} \leq \frac{1}{2}\alpha$ and*

$$\|\mathcal{F}(z_1) - \mathcal{F}(z_2)\|_{\mathcal{E}} \leq \frac{1}{2} \|z_1 - z_2\|_{\mathcal{E}} \ for \ \|z_1\|_{\mathcal{E}} \leq \alpha, \ \|z_2\|_{\mathcal{E}} \leq \alpha,$$

then the equation

$$z = a + \mathcal{F}(z), \ z \in \mathcal{E},$$

has a unique solution $z \in \mathcal{E}$ satisfying $\|z\|_{\mathcal{E}} \leq \alpha$.

Proof —   Let $\tilde{\mathcal{F}}(z) = \mathcal{F}(z) + a$, $z \in \mathcal{E}$. By an induction argument

$$\|\tilde{\mathcal{F}}^k(0)\| \leq \alpha \left(1 - 2^{-k}\right), \ k \in \mathbb{N}.$$

From the Lipschitz condition it follows that the sequence $\{\tilde{\mathcal{F}}^k(0)\}$ converges to a fixed point of $\tilde{\mathcal{F}}$.   ∎

We now continue the proof of the theorem. Define $\mathcal{E} = E$, $a = \psi + F(\varphi)$,

$$\mathcal{F}(u) = F(u + \varphi) - F(\varphi), \ u \in \mathcal{E}.$$

Then $\mathcal{F}(0) = 0$ and for $u_1, u_2 \in E$ and $\alpha = \frac{1}{4M}$,

$$\|\mathcal{F}(u_1) - \mathcal{F}(u_2)\|_E \leq 2M\alpha\|u_1 - u_2\| \leq \frac{1}{2}\|u_1 - u_2\|.$$

It is clear that we can find $T > 0$ such that in the space $E = L^4(0, T; L^4)$

$$\|\psi + F(\varphi)\|_E \leq \|\psi\|_E + \|F(\varphi)\|_E$$

$$\leq \left(\int_0^T \|S(s)x\|_4^4 ds\right)^{1/4} + \left(\int_0^T \|F(\varphi)(s)\|_4^4 ds\right)^{1/4} \leq \alpha.$$

The desired result follows from Lemma 15.2.6. ∎

15.3 A priori estimates and global existence

We now prove the main result of the present section.

Theorem 15.3.1 *Under Hypothesis 15.1 for arbitrary $x \in H$ there exists $\mathbb{P}$-a.s. a unique solution $X(\cdot, x) \in E$ of the equation (15.2.5). The solution depends continuously on $x \in E$.*

Proof— The first part of the theorem is an immediate consequence of Theorem 15.2.5, and the a priori estimate of Proposition 15.3.2 below. Continuous dependence on initial data is clear for the local solutions, see the proof of Theorem 15.2.5, and therefore it holds also for global solutions.

Proposition 15.3.2 *Assume that $\varphi \in E$, $x \in E$ and that $y \in E$ is a solution to the equation*

$$y(t) = S(t)x + F(y + \varphi)(t), \ t \in [0, T].$$

Then we have

$$\sup_{t \in [0,T]} \|y(t)\|_H^2 \leq e^{K_1 \int_0^T \|\varphi(r)\|_4^4 dr} \|x\|_H^2$$

$$+ K_2 \int_0^T e^{K_1 \int_s^T \|\varphi(r)\|_4^4 dr} \|\varphi(s)\|_4^4 ds, \tag{15.3.1}$$

$$\int_0^T \|y(t)\|_V^2 dt \;\le\; \|x\|_H^2 + K_1 \sup_{s\in[0,T]} \|y(s)\|_H^2 \int_0^T \|\varphi(s)\|_4^4 ds$$

$$+ \; K_2 \int_0^T \|\varphi(s)\|_4^4 ds,$$

(15.3.2)

where

$$K_1 = 4(24)^3 C_1^4, \quad K_2 = 32 C_1^2,$$

(15.3.3)

and C_1 is the constant from (15.1.4).

Proof —   It is enough to show (15.3.1) and (15.3.2) for $x \in D(A)$, regular φ and for the (strong) solution to the differential equation

$$\left.\begin{array}{l} \dfrac{dy(t)}{dt} = Ay(t) + D_\xi(y(t) + \varphi(t))\,(y(t) + \varphi(t)),\; t \in [0,T], \\[4mm] y(0) = x. \end{array}\right\}$$

(15.3.4)

We will show first that

$$\frac{d}{dt}\|y(t)\|_H^2 + \|y(t)\|_V^2 \le K_1 \|y(t)\|_H^2\,\|\varphi(t)\|_4^4 + K_2 \|\varphi(t)\|_4^4, \quad (15.3.5)$$

where K_1, K_2 are given by (15.3.3).

Multiplying (15.3.4) by $y(t)$ and integrating in D we obtain by Lemma 15.2.2

$$\begin{aligned} \frac{1}{2}\frac{d}{dt}\,\|y(t)\|_H^2 + \|y(t)\|_V^2 &= b(y(t) + \varphi(t), y(t) + \varphi(t), y(t)) \\[2mm] &= b(y(t) + \varphi(t), \varphi(t), y(t)) \\[2mm] &= -b(y(t), y(t), \varphi(t)) - b(\varphi(t), y(t), \varphi(t)). \end{aligned}$$

We first estimate $b(y(t), y(t), \varphi(t))$. Since

$$b(y(t), y(t), \varphi(t)) = \sum_{k,l=1}^2 \int_D \frac{\partial y_l(t,\xi)}{\partial \xi_k}\, y_k(t,\xi)\,\varphi_l(t,\xi),$$

we have

$$|b(y(t), y(t), \varphi(t))| \le \sum_{k,l=1}^2 \|y_l(t)\|_{H^1(D)}\,\|y_k(t)\|\,\|\varphi_l(t)\|_4.$$

By (15.1.4) it follows that

$$|b(y(t), y(t), \varphi(t))| \le C_1 \sum_{k,l=1}^{2} \|y_l(t)\|_{H^1(D)} \|y_k(t)\|_{H^{1/2}(D)} \|\varphi_l(t)\|_4.$$

Using the interpolatory estimate

$$\|y_k(t)\|_{H^{1/2}(D)} \le \|y_k(t)\|_{H^1(D)}^{1/2} \|y_k(t)\|_2^{1/2}$$

we get

$$|b(y(t), y(t), \varphi(t))|$$

$$\le C_1 \sum_{k,l=1}^{2} \|y_l(t)\|_{H^1(D)} \|y_k(t)\|_{H^1(D)}^{1/2} \|y_k(t)\|_2^{1/2} \|\varphi_l(t)\|_4$$

$$\le C_1 \left[\sum_{k=1}^{2} \|y_k(t)\|_{H^1(D)}\right] \left[\sum_{l=1}^{2} \|y_l(t)\|_{H^1(D)}^{1/2}\right]$$

$$\times \left[\sum_{m=1}^{2} \|y_m(t)\|_2^{1/2}\right] \left[\sum_{n=1}^{2} \|\varphi_n(t)\|_4\right],$$

and so

$$|b(y(t), y(t), \varphi(t))|$$

$$\le C_1 \left[2^{1/2} \left(\sum_{k=1}^{2} \|y_k(t)\|_{H^1(D)}^2\right)^{1/2}\right] \left[2^{3/4} \left(\sum_{l=1}^{2} \|y_l(t)\|_{H^1(D)}^2\right)^{1/4}\right]$$

$$\times \left[2^{3/4} \left(\sum_{m=1}^{2} \|y_m(t)\|_2^2\right)^{1/4}\right] \left[2^{3/4} \left(\sum_{n=1}^{2} \|\varphi_n(t)\|_4^2\right)^{1/4}\right]$$

$$\le 8 C_1 \|y(t)\|_V^{3/2} \|y(t)\|_H^{1/2} \|\varphi(t)\|_4.$$

Finally this implies that

$$|b(y(t), y(t), \varphi(t))| \le \frac{1}{4}\|y(t)\|_V^2 + 2 \cdot 24^3\, C_1^4\, \|y(t)\|_H^2\, \|\varphi(t)\|_4^4, \ t \in [0, T].$$

$$\tag{15.3.6}$$

To estimate $b(\varphi(t), y(t), \varphi(t))$ notice that

$$|b(y(t), y(t), \varphi(t))| = \left| \sum_{h,k=1}^{2} \int_D \frac{\partial y_h(t,\xi)}{\partial \xi_k} \, \varphi_k(t,\xi) \, \varphi_h(t,\xi) d\xi \right|,$$

$$\leq C_1 \sum_{h,k=1}^{2} \|y_h(t)\|_{H^1(D)} \, \|\varphi_k(t)\|_4 \, \|\varphi_h(t)\|_4$$

$$\leq C_1 \sum_{h=1}^{2} \|y_h(t)\|_{H^1(D)} \left(\sum_{k=1}^{2} \|\varphi_k(t)\|_4 \right)^2$$

$$\leq C_1 2^{1/2} \left(\sum_{h=1}^{2} \|y_h(t)\|_{H^1(D)}^2 \right)^{1/2} \left[2^{3/4} \left(\sum_{k=1}^{2} \|\varphi_k(t)\|_4^4 \right)^{1/4} \right]^2$$

$$\leq 4C_1 \, \|y(t)\|_V \|\varphi(t)\|_4^2.$$

Therefore

$$|b(\varphi(t), y(t), \varphi(t))| \leq \frac{1}{4}\|y(t)\|_V^2 + 16 \, C_1^2 \, \|\varphi(t)\|_4^4, \; t \in [0,T]. \quad (15.3.7)$$

Adding the inequalities (15.3.6) and (15.3.7) we get (15.3.5).

Taking into account that (15.3.5) is a first order differential inequality for $\|y(t)\|_H^2$, $t \in [0,T]$, we arrive at (15.3.1). Moreover, integrating (15.3.5) from 0 to T we obtain (15.3.2). ∎

With obvious changes in the proof we can deduce the following a priori estimate needed in the next section and generalizing (15.3.5).

Proposition 15.3.3 *If for an $\alpha > 0$, y is a mild solution of the equation*

$$\frac{dy}{dt} = Ay(t) + D_\xi(y(t) + \varphi(t)) \cdot (y(t) + \varphi(t)) + \alpha\varphi(t), \quad (15.3.8)$$

then for arbitrary $\varepsilon \geq 0$

$$\frac{d}{dt} \|y(t)\|_H^2 + \|y(t)\|_V^2 \leq (\varepsilon + K_1 \|\varphi(t)\|_4^4) \|y(t)\|_H^2$$

$$+ K_2 \|\varphi(t)\|_4^4 + \frac{4\alpha^2}{\varepsilon^2} \|\varphi(t)\|_2^2, \ t \geq 0.$$
$$(15.3.9)$$

15.4 Existence of an invariant measure

We pass now to the existence of an invariant measure for equation (15.2.1) under the condition

Hypothesis 15.2 *Process $W_A(\cdot)$ has a continuous version in* $D((-A)^{\frac{1+2\theta}{4}})$ *for some $\theta \in]0, \frac{1}{2}[$.*

We remark that since

$$D((-A)^{\frac{1+2\theta}{4}}) \subset D((-A)^{\frac{1}{4}}) \subset L^4,$$

Hypothesis 15.2 is stronger than Hypothesis 15.1.

The main result of this section is the following theorem. We will adapt, following F. Flandoli, see [66], the technique which proved to be useful for dissipative systems in §6.3. For different ways of investigating asymptotic properties of the stochastic Navier–Stokes equation, we refer to A. B. Cruzeiro [31] and to S. Albeverio and A. B. Cruzeiro [3].

Theorem 15.4.1 *Under Hypothesis 15.2 there exists at least one invariant measure for equation (15.2.1).*

Let $V(t)$, $t \geq 0$, be a Wiener process independent of V but having the same law as $W(t)$, $t \geq 0$. We denote by $\overline{W}$ a Wiener process defined on the whole real line by the formula

$$\overline{W}(t) = \begin{cases} W(t), & \text{if } t \geq 0, \\ \\ V(-t), & \text{if } t \geq 0, \end{cases}$$

and by $\overline{\mathcal{F}}_t$ the filtration

$$\overline{\mathcal{F}}_t = \sigma\left(\overline{W}(s),\ s \le t\right),\ t \in \mathbb{R}.$$

For an arbitrary real number s, $X(t,s)$, $t \ge s$, is the unique solution of the stochastic Navier–Stokes equation

$$\left.\begin{array}{l} dX(t) = AX(t) + D_\xi X(t) \cdot X(t) + d\overline{W}(t),\ t \ge s, \\[3mm] X(s) = 0. \end{array}\right\} \qquad (15.4.1)$$

Let moreover $Z(t)$, $t \in \mathbb{R}$, be a $D((A)^{-\frac{1+2\theta}{4}})$-continuous, stationary solution of the linear equation,

$$dZ(t) = AZ(t) + d\overline{W}(t),\ t \in \mathbb{R}.$$

Since the semigroup $S(t)$, $t \ge 0$, is exponentially stable on H we have

$$Z(t) = \int_{-\infty}^{t} S(t-s)d\overline{W}(s),\ t \in \mathbb{R}.$$

However, the explicit form of Z will not be needed in the proof.

Remark 15.4.2 Since

$$D((-A)^{\frac{1+2\theta}{4}}) \subset D((-A)^{\theta}),$$

Hypothesis 15.2 implies that the process Z is also $D((-A)^{\theta})$-continuous.

Note that the space $D((-A)^{\theta})$ is compactly embedded into H. Consequently if one proves that

the process $X(t,0)$, $t \ge 1$, is bounded in probability as a process

with values on $D((-A)^{\theta})$,

$$(15.4.2)$$

one gets immediately that the laws $\mathcal{L}(X(t,0))$, $t \ge 1$, are tight on H. This is certainly sufficient for the existence of an invariant measure.

The proof of Theorem 15.4.1 will consist of two steps.

Step 1– For any $\alpha \ge 0$ denote by Z_α the stationary solution of the equation

$$dZ_\alpha = (A - \alpha)Z_\alpha dt + dW(t).$$

Since

$$Z_\alpha(t) = Z(t) + e^{(A-\alpha)(t-s)}(Z_\alpha(s) - Z(s))$$

$$- \alpha \int_s^t e^{(A-\alpha)(t-\sigma)} Z(\sigma) d\sigma, \ t \geq s,$$

one can assume, see Remark 15.4.2, that $Z_\alpha(\cdot)$ is also a $D((-A)^\theta)$-continuous process.

Moreover let

$$Y_\alpha(t, s) = X(t, s) - Z_\alpha(t), \ t \geq s.$$

Then

$$Y_\alpha(t) = Y_\alpha(t, s), \ t \geq s,$$

is the mild solution of the equation

$$\left. \begin{array}{rl}
\dfrac{d}{dt} Y_\alpha(t) = & AY_\alpha(t) + D_\xi(Y_\alpha(t) + Z_\alpha(t)) \cdot (Y_\alpha(t) + Z_\alpha(t)) \\[2mm]
 & + \ \alpha Z_\alpha(t), \ t \geq s, \\[2mm]
Y(s) \ \ = & -Z_\alpha(s).
\end{array} \right\}$$

The main aim of Step 1 is to show the following proposition.

Proposition 15.4.3 *There exist $\alpha > 0$ and a random variable ξ such that $\mathbb{P}$-a.s.*

$$\|Y_\alpha(t, s)\|_2 \leq \xi \ \text{for all} \ t \in [-1, 0] \ \text{and all} \ s \leq -1, \qquad (15.4.3)$$

$$\int_{-1}^0 \|Y_\alpha(t, s)\|_{D((-A)^{1/2})}^2 \ ds \leq \xi \ \text{for all} \ t \in [-1, 0] \ \text{and all} \ s \leq -1. \qquad (15.4.4)$$

Proof — Taking into account Proposition 15.3.3 and the obvious estimate

$$\lambda_1 \|Z\|_H^2 \leq \|Z\|_{D((-A)^{1/2})}, \ z \in D((-A)^{1/2}),$$

one obtains that

$$\|Y_\alpha(t, s)\|_H^2 \leq e^{\int_s^t [-\lambda_1 + \varepsilon + K_1 \|Z_\alpha(\sigma)\|_4^4] d\sigma} \|Z_\alpha(s)\|_2^2$$

$$+ \int_s^t e^{\int_\sigma^t [-\lambda_1 + \varepsilon + K_1 \|Z_\alpha(\zeta)\|_4^4] d\zeta} \left[K_2 \|Z_\alpha(\sigma)\|_4^4 + \dfrac{4\alpha^2}{\varepsilon^2} \right] d\sigma. \qquad (15.4.5)$$

Let μ_α be the unique invariant (and Gaussian) measure for $Z_\alpha(\cdot)$. It is supported by $D((-A)^{\frac{1+2\theta}{4}}) \subset L_4(D)$ and therefore, by the strong law of large numbers, see Remark 3.3.2, $\mathbb{P}$–a.s

$$\lim_{s \to -\infty} \frac{1}{t-s} \int_s^t \|Z_\alpha(\sigma)\|_4^4 \, d\sigma = \int_V \|z\|_4^4 \, \mu_\alpha(dz).$$

It is easy to see that

$$\lim_{\alpha \to +\infty} \int_V \|z\|_4^4 \mu_\alpha(dz) = 0.$$

Consequently there exist $\alpha > 0$ and $\varepsilon > 0$ such that

$$-\lambda_1 + \varepsilon + K_1 \int_V \|z\|_4^4 \mu_\alpha(dz) = -\gamma < 0.$$

From now on we assume that $\alpha > 0$ and $\varepsilon > 0$ are fixed and that (15.4.5) holds. Then for a sufficiently large random $s_0 > 0$ and all $s \leq -s_0$,

$$e^{\int_s^t [-\lambda_1 + \varepsilon + K_1 \|Z_\alpha(\sigma)\|_4^4] d\sigma} \leq e^{-\frac{\gamma}{2}(t-s)}. \tag{15.4.6}$$

We will need the following elementary lemma.

Lemma 15.4.4 *Assume that Z is a continuous and stationary Gaussian process on a separable Banach space L. Then for arbitrary $\delta > 0$ there exist random variables ξ_1 and ξ_2 such that $\mathbb{P}$–a.s.*

$$\|Z(t)\|_L \leq \xi_1 + \xi_2 \, |t|^\delta,$$

for all $t \leq 0$.

Proof — Write

$$\eta_n = \sup_{-n \leq s \leq -n+1} \|Z(s)\|_L, \ n \in \mathbb{N}.$$

Then, by Gaussianity, for arbitrary $p > 0$,

$$\mathbb{E}\eta_n^p = \mathbb{E} \sup_{-n \leq s \leq -n+1} \|Z(s)\|_L^p. \tag{15.4.7}$$

Consequently

$$\mathbb{P}\left(\eta_n \geq n^\delta\right) \leq \frac{\mathbb{E}\eta_n^p}{n^{\delta p}}.$$

If $\delta p > 1$ then

$$\sum_{n=1}^{\infty} \mathbb{P}\left(\eta_n \geq n^\delta\right) < +\infty$$

and by the Borel–Cantelli lemma, $\mathbb{P}$–a.s., for all sufficiently large n

$$\eta_n \leq n^\delta.$$

This proves the lemma. ∎

To finish the proof of Proposition 15.4.3 note that (15.4.3) is a direct consequence of estimates (15.4.5), (15.4.6) and the lemma. The estimate (15.4.4) follows similarly using in addition the a priori estimate (15.3.2).

Step 2— We show that condition (15.4.3) holds using Proposition 15.4.3 and a regularizing property of the equation (15.3.8).

The following result is a generalization of Proposition 15.3.3.

Proposition 15.4.5 *For any* $\theta \in [0, 1/2]$ *there exists a constant* $C = C(\theta)$ *such that for an arbitrary mild solution* $y(\cdot)$ *of* (15.3.8)

$$\|(-A)^\theta y(t)\|_2^2 \leq e^{c \int_0^t \|y(s)\|_2^2 \|(-A)^{1/2} y(s)\|_2^2 ds} \, \|(-A)^\theta y(0)\|_2^2$$

$$+ c \int_0^t e^{c \int_\sigma^t \|y(s)\|_2^2 \|(-A)^{1/2} y(s)\|_2^2 ds} \left[\|\varphi(\sigma)\|_2^2 + \|(-A)^{\frac{1+2\theta}{4}} \varphi(\sigma)\|_2^4 \right] d\sigma.$$

Proof— Multiplying equation (15.3.8) by $(-A)^\theta y$ and integrating over D we find that

$$\frac{1}{2} \frac{d}{dt} \|(-A)^\theta y(t)\|_2^2 + \|(-A)^{\frac{1}{2}+\theta} y(t)\|_2^2$$

$$= -b(y(t) + \varphi(t), y(t) + \varphi(t), (-A)^{2\theta} y(t)) \qquad (15.4.8)$$

$$= \alpha \langle (-A)^\theta y(t), (-A)^\theta \varphi(t) \rangle, \ t \geq 0.$$

We estimate first, in two steps,

$$b(y(t) + \varphi(t), y(t) + \varphi(t), (-A)^{2\theta} y(t)).$$

Step 1— We show that there exists a constant $C_1 > 0$ such that

$$|b(u, u, (-A)^{2\theta} v)| \leq C_1 \left\|(-A)^{\frac{\theta}{2}} v\right\|_2 \left\|(-A)^{\frac{1+2\theta}{2}} u\right\|_2. \qquad (15.4.9)$$

We have in fact, setting $b(u, v, z) = \langle B(u, v), z \rangle$, that

$$
\begin{aligned}
|b(u, u, (-A)^{2\theta} v)| &\leq C_2 \|B(u, u)\|_{H^{2\theta-1}(D)} \|(-A)^{2\theta} v\|_{H^{1-2\theta}(D)} \\
&\leq C_2 \|B(u, u)\|_{H^{2\theta-1}(D)} \|(-A)^{\theta+\frac{1}{2}} v\|_2.
\end{aligned}
$$
$$(15.4.10)$$

Since $\operatorname{div} u = 0$ we have

$$(B(u, u))_h = \sum_{k=1}^{2} \frac{\partial u_h}{\partial \xi_k} u_k = \sum_{k=1}^{2} \frac{\partial}{\partial \xi_k}(u_h u_k)$$

and so

$$\|B(u, u)\|_{H^{2\theta-1}(D)} \leq C_2 \sum_{h,k=1}^{2} \|u_h u_k\|_{H^{2\theta}(D)}.$$

By a classical result on multipliers, see W. Sickel [140],

$$\|vz\|_{H^{2\theta}(D)} \leq C_3 \|v\|_{H^{\frac{1}{2}+\theta}(D)} \|z\|_{H^{\frac{1}{2}+\theta}(D)},$$

and therefore

$$\|B(u, u)\|_{H^{2\theta-1}(D)} \leq C_4 \|u^2\|_{H^{\frac{1}{2}}(D)}. \qquad (15.4.11)$$

Taking into account (15.4.9) we find

$$
\begin{aligned}
|b(u, u, (-A)^{2\theta} v)| &\leq C_2 C_4 \|u^2\|_{H^{\frac{1}{2}+\theta}(D)} \|(-A)^{\theta+\frac{1}{2}} v\|_2 \\
&\leq C_2 C_4 \|(-A)^{\frac{1+2\theta}{2}} u\|_2 \|(-A)^{\theta+\frac{1}{2}} v\|_2,
\end{aligned}
$$

which proves (15.4.9).

Step 2— We prove that for arbitrary $\varepsilon > 0$ there exists a constant $K(\varepsilon)$ such that

$$|b(v + u, v + u, (-A)^{2\theta} v)| \leq 2\varepsilon \|(-A)^{\theta+\frac{1}{2}} v\|_2^2$$
$$(15.4.12)$$
$$+ K(\varepsilon) \left[\|v\|_2^2 \|(-A)^{\frac{1}{2}} v\|_2^2 \|(-A)^{\theta} v\|_2^2 + \|(-A)^{\frac{1+2\theta}{4}} u\|_2^4 \right].$$

From (15.4.9)

$$|b(v+u,v+u,(-A)^{2\theta}v)| \le C_1\|(-A)^{\theta+\frac{1}{2}}v\|_2^2\|(-A)^{\frac{1+2\theta}{4}}(v+u)\|_2^2$$

$$\le 2C_1\left[\|(-A)^{\theta+\frac{1}{2}}v\|_2\,\|(-A)^{\frac{1+2\theta}{4}}v\|_2^2 + \|(-A)^{\theta+\frac{1}{2}}v\|_2\,\|(-A)^{\frac{1+2\theta}{4}}u\|_2^2\right]$$

$$\le \frac{\varepsilon}{2}\|(-A)^{\theta+\frac{1}{2}}v\|_2^2 + 2\frac{C_1^2}{\varepsilon}\|(-A)^{\frac{1+2\theta}{4}}v\|_2^4$$

$$+\frac{\varepsilon}{2}\|(-A)^{\theta+\frac{1}{2}}v\|_2^2 + 2\frac{C_1^2}{\varepsilon}\|(-A)^{\frac{1+2\theta}{4}}u\|_2^4$$

$$\le \varepsilon\|(-A)^{\theta+\frac{1}{2}}v\|_2^2 + 2\frac{C_1^2}{\varepsilon}\|(-A)^{\frac{1+2\theta}{4}}v\|_2^4 + +2\frac{C_1^2}{\varepsilon}\|(-A)^{\frac{1+2\theta}{4}}u\|_2^4.$$
$$(15.4.13)$$

By interpolatory inequalities

$$\|(-A)^{\frac{1+2\theta}{4}}v\|_2^4 \;\le\; C_5\|(-A)^{\frac{1}{4}}v\|_2^2\|(-A)^{\frac{1}{4}+\theta}v\|_2^2$$

$$\le\; C_6\|v\|_2\|(-A)^{\frac{1}{2}}v\|_2\|(-A)^{\theta}v\|_2\|(-A)^{\frac{1}{2}+\theta}v\|_2.$$

Therefore

$$\frac{2C_1^2}{\varepsilon}\|(-A)^{\frac{1+2\theta}{4}}v\|_2^4 \;\le\; \frac{\varepsilon}{2}\|(-A)^{\frac{1}{2}+\theta}v\|_2^2$$

$$+\; \frac{C_1^4 C_6^2}{\varepsilon^3}\|v\|_2^2\,\|(-A)^{\frac{1}{2}}v\|_2^2\,\|(-A)^{\theta}v\|_2^2.$$
$$(15.4.14)$$

By substituting (15.4.14) into (15.4.13) we arrive at (15.4.12).

From (15.4.8) and (15.4.12),

$$\frac{1}{2}\frac{d}{dt}\|(-A)^{\theta}y(t)\|_2^2 + \|(-A)^{\frac{1}{2}+\theta}y(t)\|_2^2 \le 2\varepsilon\|(-A)^{\frac{1}{2}+\theta}y(t)\|_2^2$$

$$+K(\varepsilon)\left[\|y(t)\|_2^2\|(-A)^{1/2}y(t)\|_2^2\|(-A)^{\theta}y(t)\|_2^2 + \|(-A)^{\frac{1+2\theta}{4}}\varphi(t)\|_4^4\right]$$

$$+\alpha\|(-A)^{2\theta}y(t)\|_2\|\varphi(t)\|_2.$$
$$(15.4.15)$$

Since $\theta \le \frac{1}{2}$, $\|(-A)^{2\theta} z)\|_2 \le \|(-A)^{\theta+\frac{1}{2}} z\|_2$. Consequently

$$\alpha \|(-A)^{2\theta} y(t)\|_2 \|\varphi(t)\|_2 \le 2\sqrt{\varepsilon}\, \|(-A)^{\theta+\frac{1}{2}} y(t)\|_2 \frac{\alpha}{2\sqrt{\varepsilon}} \|\varphi(t)\|_2$$

$$\le \varepsilon\, \|(-A)^{\theta+\frac{1}{2}} y(t)\|_2^2 + \frac{\alpha^2}{4\varepsilon} \|\varphi(t)\|_2^2.$$

$$(15.4.16)$$

Combining (15.4.15) with (15.4.16) we see that

$$\frac{1}{2} \frac{d}{dt} \|(-A)^{\theta} y(t)\|_2^2 + (1 - 3\varepsilon)\|(-A)^{\frac{1}{2}+\theta} y(t)\|_2^2$$

$$\le \left[K(\varepsilon)\|y(t)\|_2^2 \|(-A)^{1/2} y(t)\|_2^2 \right] \|(-A)^{\theta} y(t)\|_2^2$$

$$+ K(\varepsilon)\|(-A)^{\frac{1+2\theta}{4}} \varphi(t)\|_4^4 + \frac{\alpha^2}{4\varepsilon} \|\varphi(t)\|_2.$$

Thus the proof of the proposition is complete.

We are ready now to complete the proof of the theorem. It follows from the proposition that for arbitrary $t \le -1 \le r \le 0$

$$\|(-A)^{\theta} Y_\alpha(0,t)\|_2^2 \le e^{c\int_r^0 \|Y_\alpha(s,t)\|_2^2 \|(-A)^{1/2} Y_\alpha(s,t)\|_2^2 ds} \|(-A)^{\theta} Y_\alpha(r,t)\|_2^2$$

$$+ c\int_r^0 e^{c\int_\sigma^0 \|Y_\alpha(s,t)\|_2^2 \|(-A)^{1/2} Y_\alpha(s,t)\|_2^2 ds} \left(\|Z_\alpha(\sigma)\|_2^2 + \|(-A)^{\frac{1+2\theta}{4}} Z_\alpha(\sigma)\|_2^4 \right)$$

$$\le e^{c[\sup_{-1 \le s \le 0} \|Y_\alpha(s,t)\|_2^2] \int_{-1}^0 \|(-A)^{1/2} Y_\alpha(s,t)\|_2^2 ds}$$

$$\times \left[\|(-A)^{1/2} Y_\alpha(r,t)\|_2^2 + c\int_{-1}^0 \left(\|Z_\alpha(\sigma)\|_2^2 + \|(-A)^{\frac{1+2\theta}{4}} Z_\alpha(\sigma)\|_2^4 \right) d\sigma \right].$$

Consequently, integrating the above inequality over the interval $[-1,0]$ we get for $t \le -1$ that

$$\|(-A)^{\theta} Y_\alpha(0,t)\|_2^2 \le e^{c[\sup_{-1 \le s \le 0} \|Y_\alpha(s,t)\|_2^2] \int_{-1}^0 \|(-A)^{1/2} Y_\alpha(s,t)\|_2^2 ds}$$

$$\times \left[\|(-A)^{1/2} Y_\alpha(r,t)\|_2^2 + c\int_{-1}^0 \left(\|Z_\alpha(\sigma)\|_2^2 + \|(-A)^{\frac{1+2\theta}{4}} Z_\alpha(\sigma)\|_2^4 \right) d\sigma \right].$$

By Proposition 15.4.3 there exists a random variable ξ_2 such that $\mathbb{P}$–a.s.

$$\|(-A)^\theta Y_\alpha(0,t)\|_2 \le \xi_2, \quad \text{for all } t \le -1.$$

Moreover

$$\|(-A)^\theta X(0,t)\|_2 \le \|(-A)^\theta Y_\alpha(0,t)\|_2 + \|(-A)^\theta Z\alpha(0)\|_2.$$

Since $Z\alpha(0)$ takes values in $D((-A)^\theta)$ there exists a random variable ξ_3 such that $\mathbb{P}$–a.s.

$$\|(-A)^\theta X(0,t)\|_2 \le \xi_3 \quad \text{for all } t \le -1.$$

This proves the tightness of $\mathcal{L}(X(0,t))$, $t \le -1$. ∎

Part IV

Appendices

Appendix A

Smoothing properties of convolutions

A.1

Let $A : D(A) \subset H \to H$ be the infinitesimal generator of an analytic semigroup $S(t)$, $t \geq 0$, in the Hilbert space H. We assume that, for some $M > 0$ and $\omega > 0$, we have

$$\|S(t)\| \leq M e^{-\omega t}, \ t \geq 0.$$

For any $\alpha, \gamma \in]0,1[$ and any $p > 1$, we define a linear bounded operator in $L^p(0,T;H)$ by setting

$$R_{\alpha,\gamma}\psi(t) = \int_0^t (t - \sigma)^{\alpha-1}(-A)^\gamma S(t - \sigma)\varphi(\sigma)d\sigma, \ \psi \in L^p(0,T;H).$$

$$(A.1.1)$$

Proposition A.1.1 *(i) If $\gamma > 0$ and $\alpha > \gamma + \frac{1}{p}$ then $R_{\alpha,\gamma}$ is a bounded operator from $L^p(0,T;H)$ into $C^{\alpha-\gamma-\frac{1}{p}}([0,T]; D((-A)^\gamma))$.*

(ii) If $\gamma = 0$ and $\alpha > \frac{1}{p}$ then for arbitrary $\delta \in]0, \alpha - \frac{1}{p}[$, $R_{\alpha,\gamma}$ is a bounded operator from $L^p(0,T;H)$ into $C^\delta([0,T];H)$.

Proof —   Let $\varphi \in L^p(0, T; H)$ and set $\psi = R_{\alpha,\gamma}\varphi$. Then

$$\psi(t) - \psi(s) = \int_s^t (t - \sigma)^{\alpha-1}(-A)^\gamma S(t - \sigma)\varphi(\sigma)d\sigma$$

$$+ \int_0^s [(t - \sigma)^{\alpha-1} - (s - \sigma)^{\alpha-1}](-A)^\gamma S(t - \sigma)\varphi(\sigma)d\sigma$$

$$+ \int_0^s (s - \sigma)^{\alpha-1}[(-A)^\gamma S(t - \sigma) - (-A)^\gamma S(s - \sigma)]\varphi(\sigma)d\sigma$$

$$= I_1 + I_2 + I_3.$$

We will estimate each term I_1, I_2, I_3 separately.

$$|I_1| \leq \left(\int_s^t ((t - \sigma)^{\alpha-1} \|(-A)^\gamma S(t - \sigma)\|)^q d\sigma \right)^{1/q} \left(\int_s^t |\varphi(s)|^p ds \right)^{1/p},$$

where $\frac{1}{p} + \frac{1}{q} = 1$. Since the semigroup $S(t)$, $t \geq 0$, is analytic there exists a constant $M > 0$ such that

$$\|(-A)^\gamma S(\sigma)\| \leq \frac{M}{\sigma^\gamma}, \ \sigma \in [0, T].$$

Consequently

$$|I_1| \leq M \left(\int_0^{t-s} \sigma^{(\alpha-1)q} \sigma^{-\gamma q} d\sigma \right)^{1/q} \|\varphi\|_p,$$

where $\|\varphi\|_p$ denotes the $L^p(0, T; H)$ norm. Thus for a constant M_1

$$|I_1| \leq M_1 (t - s)^{\alpha-\gamma-\frac{1}{p}}.$$

In a similar way

$$|I_2| \leq M \left(\int_0^s [(s - \sigma)^{\alpha-1} - (t - \sigma)^{\alpha-1}]^q (t - \sigma)^{-\gamma q} d\sigma \right)^{1/q} \|\varphi\|_p.$$

Since for arbitrary positive numbers $a < b$ and $q \geq 1$

$$(b - a)^q \leq b^q - a^q,$$

we have

$$|I_2|^p \leq M^p \|\varphi\|_p^p \left[\int_0^s \frac{(s-\sigma)^{(\alpha-1)q}}{(t-\sigma)^{\gamma q}} d\sigma - \int_0^s (t-\sigma)^{(\alpha-1)q-\gamma q} d\sigma \right]$$

$$\leq M^p \|\varphi\|_p^p \left[\int_0^s (s-\sigma)^{(\alpha-1)q-\gamma q} d\sigma - \int_0^s (t-\sigma)^{(\alpha-1)q-\gamma q} d\sigma \right]$$

$$\leq \frac{M^p \|\varphi\|_p^p}{(\alpha-1)q - \gamma q + 1} \Big[s^{(\alpha-1)q-\gamma q+1} - t^{(\alpha-1)q-\gamma q+1}$$

$$+ (t-s)^{(\alpha-1)q-\gamma q+1} \Big]$$

$$\leq \frac{M^p \|\varphi\|_p^p}{(\alpha-1)q - \gamma q + 1} (t-s)^{(\alpha-1)q-\gamma q+1}.$$

Therefore, for a constant M_2,

$$|I_2| \leq M_2 (t-s)^{\alpha-\gamma-\frac{1}{p}}.$$

It remains to estimate $|I_3|$. Note that

$$(-A)^\gamma S(t-\sigma) - (-A)^\gamma S(s-\sigma) = \int_{s-\sigma}^{t-\sigma} (-A)^{\gamma+1} S(\rho) d\rho.$$

Therefore for a constant $M_3 > 0$

$$|I_3| \leq M_3 \int_0^s (s-\sigma)^{\alpha-1} \int_{s-\sigma}^{t-\sigma} \frac{1}{\rho^{1+\gamma}} d\rho |\varphi(\sigma)| d\sigma.$$

If $\gamma > 0$ we have

$$|I_3| \leq \frac{M_3}{\gamma} \int_0^s (s-\sigma)^{\alpha-1} \left[(s-\sigma)^{-\gamma} - (t-\sigma)^{-\gamma} \right] |\varphi(\sigma)| d\sigma$$

$$\leq \frac{M_3}{\gamma} \left(\int_0^s (s-\sigma)^{(\alpha-1)q} \left[(s-\sigma)^{-\gamma q} - (t-\sigma)^{-\gamma q} \right] d\sigma \right)^{1/q} \|\varphi\|_p$$

$$\leq \frac{M_3}{\gamma} \left(\int_0^s (s-\sigma)^{(\alpha-1)q-\gamma q} d\sigma - \int_0^s (s-\sigma)^{(\alpha-1)q}(t-\sigma)^{-\gamma q} d\sigma \right)^{1/q} \|\varphi\|_p$$

$$\leq \frac{M_3}{\gamma} \left(\int_0^s (s-\sigma)^{(\alpha-1)q-\gamma q} d\sigma - \int_0^s (t-\sigma)^{(\alpha-1)q}(t-\sigma)^{-\gamma q} d\sigma \right)^{1/q} \|\varphi\|_p.$$

By a similar argument as in the estimation of $|I_2|$ we arrive at the inequality

$$|I_3| \leq M_4(t - s)^{\alpha - \gamma - \frac{1}{p}}.$$

If $\gamma = 0$ then, for arbitrary $\delta \in \,]0, 1[$,

$$|I_3| \leq M_3 \int_0^s (s - \sigma)^{\alpha - 1} \int_{s - \sigma}^{t - \sigma} \frac{1}{\rho} d\rho |\varphi(\sigma)| d\sigma$$

$$\leq M_3 \int_0^s (s - \sigma)^{\alpha - 1} \int_{s - \sigma}^{t - \sigma} \frac{1}{\rho^\delta \rho^{1 - \delta}} d\rho |\varphi(\sigma)| d\sigma$$

$$\leq M_3 \int_0^s (s - \sigma)^{\alpha - 1} (s - \sigma)^{-\delta} \int_{s - \sigma}^{t - \sigma} \frac{1}{\rho^{1 - \delta}} d\rho |\varphi(\sigma)| d\sigma$$

$$\leq \frac{M_3}{\delta} \int_0^s (s - \sigma)^{\alpha - 1 - \delta} \left[(t - \sigma)^\rho - (s - \sigma)^\rho \right] \varphi(\sigma) | d\sigma$$

$$\leq \frac{M_3}{\delta} (t - s)^\delta \int_0^s (s - \sigma)^{\alpha - 1 - \delta} \varphi(\sigma) | d\sigma$$

$$\leq \frac{M_3}{\delta} (t - s)^\delta \left(\int_0^s \sigma^{(\alpha - 1 - \delta)q} \, d\sigma \right)^{1/q} \|\varphi\|_p.$$

Therefore if $\delta \in \,]0, \alpha - \frac{1}{p}[$ then

$$|I_3| \leq M_5 (t - s)^\delta,$$

and the required result follows.   ■

Appendix B

An estimate on modulus of continuity

B.1

The following result, which goes back to A. M. Garsia, E. RademIch and H. Rumsey Jr. [76], is very useful in proving regularity of stochastic processes. It can be used to prove the Kolmogorov continuity theorem and also Sobolev embeddings theorem for spaces with fractional norms, see R. A. Adams [1]. The formulation below is a generalization of that given in D.W. Stroock and S. R. S. Varhadan [148, pp. 60–61]. The proof differs in a few minor points from that in [148].

Let $\mathcal{O}$ be an open bounded subset of $\mathbb{R}^d$ equipped with a norm $|\cdot|$. Let

$$D = \sup_{x,y \in \mathcal{O}} |x - y|$$

be the diameter of $\mathcal{O}$. By $\lambda(\cdot)$ we will denote the Lebesgue measure on $\mathbb{R}^d$, although different choices are possible. Moreover we set

$$\kappa = \kappa(\mathcal{O}) = \inf_{0 < \rho < D} \inf_{x \in \mathcal{O}} \frac{\lambda(\mathcal{O} \cap B(x, \rho))}{\rho^d}, \qquad (\text{B.1.1})$$

where

$$B(x, \rho) = \{y \in \mathbb{R}^d : |x - y| < \rho\}.$$

311

Theorem B.1.1 *Let ψ and p be strongly increasing functions from $[0, +\infty[$ into $]0, +\infty[$ such that $\psi(0) = p(0) = 0$ and $\lim_{r \to \infty} \psi(r) = +\infty$. For an arbitrary continuous function $f : \overline{\mathcal{O}} \to E$ such that*

$$\int_{\mathcal{O}} \int_{\mathcal{O}} \psi\left(\frac{|f(\xi) - f(\eta)|}{p(|\xi - \eta|)}\right) \lambda(d\xi)\lambda(d\eta) \leq B,$$

we have

$$|f(x) - f(y)| \leq 8 \int_0^{2|x-y|} \psi^{-1}\left(\frac{4^{d+1}B}{\kappa^2 u^{2d}}\right) p(du),$$

provided $x, y \in \mathcal{O}$ and $\frac{x+y}{2} \in \mathcal{O}$.

Proof — We start from two lemmas.

Lemma B.1.2 *If $\varphi \geq 0$ on $\mathcal{O}$ then for arbitrary $\alpha > 0$*

$$\lambda\left\{\eta \in \mathcal{O} : \int_{\mathcal{O}} \varphi(\xi)\lambda(d\xi) < \alpha\varphi(\eta)\right\} < \alpha.$$

Proof — Let

$$\tilde{\mathcal{O}} = \left\{\eta \in \mathcal{O} : \int_{\mathcal{O}} \varphi(\xi)\lambda(d\xi) < \alpha\varphi(\eta)\right\}.$$

If $\lambda(\tilde{\mathcal{O}}) = 0$ then the result is true. If $\lambda(\tilde{\mathcal{O}}) > 0$ then

$$\lambda(\tilde{\mathcal{O}})\int_{\mathcal{O}} \varphi(\xi)\lambda(d\xi) \quad < \quad \alpha \int_{\tilde{\mathcal{O}}} \varphi(\eta)\lambda(d\eta)$$

$$< \quad \alpha \int_{\mathcal{O}} \varphi(\eta)\lambda(d\eta).$$

Dividing both sides of the inequality by the positive number $\int_{\mathcal{O}} \varphi(\xi)\lambda(d\xi)$ we get the result. ∎

Lemma B.1.3 *Let*

$$I(\xi) = \int_{\mathcal{O}} \psi\left(\frac{|f(\xi) - f(\eta)|}{p(|\xi - \eta|)}\right) \lambda(d\eta), \; \xi \in \mathcal{O}.$$

For arbitrary $a, b \in \mathcal{O}$ and for an arbitrary number $r \in \,]0, D[$ there exists $c \in \mathcal{O} \cap B(a, r)$ such that

$$I(c) \leq 2\frac{B}{\kappa + d}, \quad \psi\left(\frac{|f(b) - f(c)|}{p(b - c)}\right) \leq 2\frac{I(b)}{\kappa r^d}.$$

Proof — Let

$$\mathcal{O}_1 = \left\{ \zeta \in \mathcal{O} \cap B(a,r) : I(\zeta) > 2\frac{B}{\kappa r^d} \right\},$$

$$\mathcal{O}_2 = \left\{ \zeta \in \mathcal{O} \cap B(a,r) : \frac{\kappa r^d}{2}\varphi(\xi) > I(b) \right\},$$

where

$$\varphi(\zeta) = \psi\left(\frac{|f(b) - f(\zeta)|}{p(|b - \zeta|)}\right), \quad \zeta \in \mathcal{O}\backslash\{b\}.$$

It is enough to show that

$$\lambda(\mathcal{O}_1 \cup \mathcal{O}_2) < \lambda(\mathcal{O} \cap B(a,r)). \tag{B.1.2}$$

Since $I(b) = \int_{\mathcal{O}} \varphi(\zeta)d\zeta$ we have, by Lemma B.1.2,

$$\lambda(\mathcal{O}_2) \leq \lambda\left\{\zeta \in \mathcal{O} : \frac{\kappa r^d}{2}\varphi(\zeta) > I(b)\right\} < \frac{\kappa r^d}{2}.$$

Moreover

$$\lambda(\mathcal{O} \cap B(a,r)) \geq \kappa r^d.$$

So

$$\lambda(\mathcal{O}_2) < \frac{1}{2}\lambda(\mathcal{O} \cap B(a,r)). \tag{B.1.3}$$

On the other hand if $\lambda(\mathcal{O}_1) > 0$ then

$$2\frac{B}{\kappa r^d}\lambda(\mathcal{O}_1) < \int_{\mathcal{O}_1} I(\zeta)\lambda(d\zeta) \leq \int_{\mathcal{O}} I(\zeta)\lambda(d\zeta) \leq B,$$

and consequently

$$\lambda(\mathcal{O}_1) < \frac{\kappa r^d}{2} \leq \frac{1}{2}\lambda(\mathcal{O}_1 \cap B(a,r)). \tag{B.1.4}$$

The same inequality holds, of course, if $\lambda(\mathcal{O}_1) = 0$. Taking into account (B.1.3) and (B.1.4) one gets (B.1.2).

Proof of Theorem B.1.1 — Let $x, y \in \mathcal{O}$, $x \neq y$ and $\frac{x+y}{2} \in \mathcal{O}$. Write $\rho = |x - y|$. From Lemma B.1.2 there exists $x_0 \in \mathcal{O}$ such that

$x_0 \in B\left(\frac{x+y}{2}, \frac{\rho}{2}\right)$ and $I(x_0) \le 2^{d+1}\frac{B}{\kappa \rho^d}$, (one simply takes $a = \frac{x+y}{2}$, $r = \frac{\rho}{2}$). Let d_n, $n \in \{-1,0\} \cup \mathbb{N}$, be a decreasing sequence such that

$$d_{-1} = 2\rho \text{ and } p(d_n) = \frac{1}{2}p\left(\frac{1}{2}(d_n + d_{n+1})\right), \; n \in \{0\} \cup \mathbb{N}.$$

Such a sequence does exist by the continuity of the function p, since

$$p(0) < \frac{1}{2}\,p\left(\frac{\alpha}{2}\right), \; p(\alpha) > \frac{1}{2}\,p\left(\frac{1}{2}2\alpha\right).$$

From Lemma B.1.3 there exists a sequence $\{x_n\}$, $n \in \mathbb{N} \cup \{0\}$, such that

$$x_n \in B\left(x, \frac{d_{n-1}}{2}\right) \cap \mathcal{O}, \; I(x_n) \le \frac{2B}{\kappa\left(\frac{d_{n-1}}{2}\right)^d}, \; n \in \mathbb{N}, \qquad \text{(B.1.5)}$$

$$\psi\left(\frac{|f(x_{n-1}) - f(x_n)|}{p(|x_{n-1} - x_n|)}\right) \le \frac{2I(x_{n-1})}{\kappa\left(\frac{d_{n-1}}{2}\right)^d}, \; n \in \mathbb{N}, \qquad \text{(B.1.6)}$$

(one takes $a = x$, $b = x_{n-1}$, $n \in \mathbb{N}$). In a similar way one can costruct a sequence $\{y_n\}$, $n \in \mathbb{N} \cup \{0\}$, such that $y_0 = x_0$ and

$$y_n \in B\left(y, \frac{d_{n-1}}{2}\right) \cap \mathcal{O}, \; I(y_n) \le \frac{2B}{\kappa\left(\frac{d_{n-1}}{2}\right)^d}, \; n \in \mathbb{N}.$$

Since $d_n \to 0$, $x_n \to x$ and $y_n \to y$. From the inequalities (B.1.5) and (B.1.6)

$$\psi\left(\frac{|f(x_0) - f(x_1)|}{p(|x_0 - x_1|)}\right) \;\le\; \frac{2^{d+1}}{\kappa d_0^d}\frac{B2^{d+1}}{\kappa d_{-1}^d}$$

$$\le\; 4^{d+1}\frac{B}{\kappa^2 d_{-1}^d d_0^d},$$

and consequently

$$\psi\left(\frac{|f(x_n) - f(x_{n-1})|}{p(|x_n - x_{n-1}|)}\right) \le 2^{d+1}\frac{I(x_{n-1})}{\kappa d_{n-1}^d} \le 4^{d+1}\frac{B}{d_{n-1}^d d_{n-2}^d},$$

and

$$|f(x_n) - f(x_{n-1})| \le \psi^{-1}\left(4^{d+1}\frac{B}{d_{n-1}^d d_{n-2}^d}\right)p(|x_n - x_{n-1}|).$$

Continuity of f implies that

$$|f(x) - f(x_0)| \leq \sum_{n=1}^{\infty} |f(x_n) - f(x_{n-1})|$$

$$\leq \psi^{-1}\left(4^{d+1}\frac{B}{\kappa^2 d_{n-1}^d d_{n-2}^d}\right) p(|x_n - x_{n-1}|) = I_1.$$

To estimate I_1 we need the following elementary lemma.

Lemma B.1.4 *For $n \in N$*

$$p(|x_n - x_{n-1}|) \leq 4\left(p(d_{n-1}) - p(d_n)\right).$$

Proof — From (B.1.5)

$$(|x_n - x_{n-1}|) \leq p(|x_{n-1} - x| + |x_n - x|)$$

$$\leq p\left(\tfrac{1}{2}(d_{n-2} + d_{n-1})\right), \; n \in \mathbb{N},$$

and from the definition of $\{d_n\}$

$$p\left(\frac{1}{2}(d_{n-2} + d_{n-1})\right) = 2p(d_{n-1}).$$

It is therefore enough to show that

$$2p(d_n) \leq p(d_{n-1}).$$

However,

$$p(d_n) = \frac{1}{2}p\left(\frac{1}{2}(d_n + d_{n-1})\right) \leq \frac{1}{2}p(d_{n-1})$$

and the result follows. ∎

From Lemma B.1.4

$$I_1 \leq 4\sum_{n=1}^{\infty} \psi^{-1}\left(4^{d+1}\frac{B}{\kappa^2 d_{n-1}^d d_{n-2}^d}\right)(p(d_{n-1}) - p(d_n))),$$

and therefore

$$I_1 \leq 4\int_0^{d_0} \psi^{-1}\left(4^{d+1}\frac{B}{\kappa^2 d_{n-1}^{2d}}\right)p(du)$$

$$\leq 4\int_0^{2|x-y|} \psi^{-1}\left(4^{d+1}\frac{B}{\kappa^2 u^{2d}}\right)p(du).$$

Since
$$|f(x) - f(y)| \le |f(x) - f(x_0)| + |f(x_0) - f(y)|$$
the result follows.   ∎

Applying the above theorem to the functions

$$\psi(u) = u^\alpha, \ p(u) = u^{\beta/\alpha}, \ \alpha > 0, \ \beta > 0,$$

one gets the following corollary.

Theorem B.1.5 *Assume that* $\kappa(\mathcal{O}) > 0$. *For arbitrary* $\alpha > 0$ *and* $\beta > 2d$ *there exists a constant* $c > 0$ *such that for an arbitrary continuous function* $f : \overline{\mathcal{O}} \to E$ *one has*

$$|f(x) - f(y)|^\alpha \le c|x - y|^{\beta - 2d} \int_\mathcal{O} \int_\mathcal{O} \frac{|f(\xi) - f(\eta)|^\alpha}{|\xi - \eta|^\beta} \lambda(d\xi)\lambda(d\eta).$$
$$(B.1.7)$$

Proof — We can assume that

$$B = \int_\mathcal{O} \int_\mathcal{O} \frac{|f(\xi) - f(\eta)|^\alpha}{|\xi - \eta|^\beta} \lambda(d\xi)\lambda(d\eta) < +\infty.$$

From Theorem B.1.1

$$|f(x) - f(y)| \le 8 \int_0^{2|x-y|} \frac{4^{\frac{d+1}{\alpha}} B^{1/\alpha}}{\kappa^{\frac{2}{\alpha}} u^{\frac{2d}{\alpha}}} \frac{\beta}{\alpha} u^{\frac{\beta}{\alpha} - 1} du.$$

However, if $\beta > 2d$ we have

$$\int_0^{2|x-y|} u^{\frac{\beta}{\alpha} - 1 - \frac{2d}{\alpha}} du = |x - y|^{\frac{\beta - 2d}{\alpha}} \frac{2^{\frac{\beta - 2d}{\alpha}}}{\frac{\beta - 2d}{\alpha}}.$$

This completes the proof.   ∎

Appendix C

A result on implicit functions

C.1

We are given two Banach spaces Λ and E and a mapping

$$\mathcal{F} : \Lambda \times E \to E, \ (\lambda, x) \to \mathcal{F}(\lambda, x).$$

We assume that the following holds.

Hypothesis C.1 *(i)* $\mathcal{F}$ *is continuous.*
(ii) There exists $\alpha \in {]}0, 1{[}$ *such that*

$$\|\mathcal{F}(\lambda, x_1) - \mathcal{F}(\lambda, x_2)\|_E \le \alpha \|x_1 - x_2\|_E, \ x_1, x_2 \in E.$$

The following result is an immediate consequence of the contraction principle.

Proposition C.1.1 *Under Hypothesis C.1 there exists a unique continuous mapping* $\varphi \in C(\Lambda; E)$ *such that*

$$\mathcal{F}(\lambda, \varphi(\lambda)) = \varphi(\lambda), \ \lambda \in \Lambda. \tag{C.1.1}$$

To study the differentiability of the mapping φ we need the following hypothesis.

Hypothesis C.2 *(i) For all $x \in E$, $\mathcal{F}(\cdot, x)$ is differentiable and the derivative $D_\lambda \mathcal{F}$ is continuous.*

(ii) The mapping $\mathbb{R} \to E$, $h \to \mathcal{F}(\lambda, x_1 + hx_2)$, is continuously differentiable for all $x_1, x_2 \in E$, $\lambda \in \Lambda$. Moreover there exists a linear operator $D_x \mathcal{F}(\lambda, x_1 + hx_2)$ in $L(E)$ such that

$$\frac{d}{dh} \mathcal{F}(\lambda, x_1 + hx_2) = D_x \mathcal{F}(\lambda, x_1 + hx_2) \cdot x_2.$$

We can now prove the following result.

Proposition C.1.2 *Under Hypotheses C.1 and C.2, φ is differentiable at any point $\lambda \in \Lambda$ in any direction $\mu \in \Lambda$, and*

$$D\varphi(\lambda) \cdot \mu = [I - D_x \mathcal{F}(\lambda, \varphi(\lambda))]^{-1} \cdot D_\lambda \mathcal{F}(\lambda, \varphi(\lambda)) \cdot \mu. \qquad (\text{C.1.2})$$

Proof — We note that by C.1(ii) it follows

$$\|D_x \mathcal{F}(\lambda, x)\| \leq \alpha, \; x \in E, \; \lambda \in \Lambda. \qquad (\text{C.1.3})$$

For $\lambda, \mu \in \Lambda$ and $h \in \mathbb{R}$, we have

$$\varphi(\lambda + h\mu) - \varphi(\lambda) = \mathcal{F}(\lambda + h\mu, \varphi(\lambda + h\mu)) - \mathcal{F}(\lambda, \varphi(\lambda))$$

$$= \int_0^1 D_\lambda \mathcal{F}(\lambda + \xi h\mu, \varphi(\lambda + h\mu)) \cdot h\mu d\xi$$

$$+ \int_0^1 D_x \mathcal{F}(\lambda, \xi\varphi(\lambda + h\mu) + (1 - \xi)\varphi(\lambda)) \cdot (\varphi(\lambda + h\mu) - \varphi(\lambda))d\xi.$$

Setting

$$G(\lambda, h, \mu) = \int_0^1 D_x \mathcal{F}(\lambda, \xi\varphi(\lambda + h\mu) + (1 - \xi)\varphi(\lambda))d\xi$$

we have $\|G(\lambda, h, \mu)\|_{\mathcal{L}(E)} \leq \alpha < 1$, and so

$$\varphi(\lambda + h\mu) - \varphi(\lambda) = [I - G(\lambda, h, \mu)]^{-1} \int_0^1 D_\lambda \mathcal{F}(\lambda + \xi h\mu, \varphi(\lambda + h\mu)) \cdot h\mu d\xi,$$

and the conclusion follows by dividing by h and letting h tend to 0.
■

To obtain the existence of the second derivative of $\varphi(\lambda)$ we need an additional assumption. We assume

Hypothesis C.3 *(i) There exists another Banach space G continuously and densely included in E such that the restriction of $\mathcal{F}$ to $\Lambda \times G$ fulfils (C.1.1) with E replaced by G.*

If (i) holds, we have, again by the Contraction Principle, that $D_\lambda \varphi(\lambda) \cdot \mu \in G$ for all $\lambda, \mu \in \Lambda$.

(ii) For all $x \in X$, $\mathcal{F}(\cdot, x)$ is twice differentiable and the derivative $D_\lambda^2 \mathcal{F}$ is continuous.

(iii) The mapping $\mathbb{R} \to E$, $h \to D_\lambda \mathcal{F}(\lambda, x_1 + hx_2)\, \mu$, is continuously differentiable for all $x_1, x_2 \in E$, $\lambda, \mu \in \Lambda$. Moreover there exists a linear operator $D_x D_\lambda \mathcal{F}(\lambda, x_1 + hx_2) \cdot \mu$ in $L(E)$ such that

$$\frac{d}{dh} D_\lambda \mathcal{F}(\lambda, x_1 + hx_2) \cdot \mu = D_x(D_\lambda \mathcal{F}(\lambda, x_1 + hx_2) \cdot x_2 \cdot \mu) \cdot x_2.$$

(iv) The mapping $\mathbb{R} \to E$, $h \to \mathcal{F}(\lambda, y_1 + hy_2)\, \mu$, is twice continuously differentiable for all $y_1, y_2 \in G$, $\lambda, \mu \in \Lambda$. Moreover there exists a linear operator $D_x^2 \mathcal{F}(\lambda, y_1 + hy_2)$ in $L(E; L(E))$ such that

$$\frac{d^2}{dh^2} \mathcal{F}(\lambda, y_1 + hy_2) = D_x^2 \mathcal{F}(\lambda, y_1 + hy_2)(y_1, y_2)\ \lambda \in \Lambda, y_1, y_2 \in G.$$

We can now prove the following result.

Proposition C.1.3 *Under Hypotheses $C.1, C.2, C.3$, φ is twice differentiable at any point $\lambda \in \Lambda$ in any directions $\mu, \xi \in \Lambda$, and we have*

$$D^2 \varphi(\lambda)(\mu, \xi) = [I - D_x \mathcal{F}(\lambda, \varphi(\lambda))]^{-1}$$

$$\left\{ D_\lambda^2 \mathcal{F}(\lambda, \varphi(\lambda))(\mu, \xi) + 2 D_\lambda D_x \mathcal{F}(\lambda, \varphi(\lambda))(D\varphi(\lambda)\mu, \xi) \right. \qquad \text{(C.1.4)}$$

$$\left. + D_x^2 \mathcal{F}(\lambda, \varphi(\lambda))(D\varphi(\lambda) \cdot \mu, D\varphi(\lambda) \cdot \xi) \right\}.$$

Proof — For any $\lambda, \mu, \xi \in \Lambda$ and any $h \in \mathbb{R}$ we have

$$D\varphi(\lambda + h\xi) \cdot \mu - D\varphi(\lambda) \cdot \mu = J_1(h, \lambda) + J_2(h, \lambda)$$

$$+ D_x \mathcal{F}(\lambda, \varphi(\lambda)) \left[D\varphi(\lambda + h\xi) \cdot \mu - D\varphi(\lambda) \cdot \mu \right],$$

where

$$J_1(h, \lambda) = D_\lambda \mathcal{F}(\lambda + h\xi, \varphi(\lambda + h\xi)) \cdot \mu - D_\lambda \mathcal{F}(\lambda, \varphi(\lambda)) \cdot \mu,$$

$$J_2(h,\lambda) = [D_x\mathcal{F}(\lambda+h\xi,\varphi(\lambda+h\xi)) - D_x\mathcal{F}(\lambda,\varphi(\lambda))]\cdot(D\varphi(\lambda+h\xi)\cdot\mu).$$

It follows that

$$D\varphi(\lambda+h\xi)\cdot\mu - D\varphi(\lambda)\cdot\mu = [I - D_x\mathcal{F}(\lambda,\varphi(\lambda))]^{-1}\,(J_1(\lambda,x)+J_2(\lambda,x)).$$

Now the conclusion follows by dividing by h and by letting h tend to 0. ∎

Bibliography

[1] ADAMS R. A. (1975) SOBOLEV SPACES, Academic Press.

[2] AHMED N. U., FUHRMAN M. & ZABCZYK J. (1995) *Regular solutions of filtering equations in infinite dimensions,* Polish Academy of Sciences, Preprint 531.

[3] ALBEVERIO S. & CRUZEIRO A. B. (1990) *Global flows with invariant (Gibbs) measures for Euler and Navier-Stokes two dimensional fluids,* Commun. Math. Phys. **129**, 431–444.

[4] ALBEVERIO S., KONDRATIEV Y.G. & TSY-CALENKO T. V. (1993) *Stochastic dynamics for quantum lattice systems and stochastic quantization I*: Ergodicity. BiBos, Universität Bielefeld, Nr. 601/11/93.

[5] BALAKRISHNAN A. V. (1976) APPLIED FUNCTIONAL ANALYSIS, Springer–Verlag.

[6] BENSOUSSAN A., DA PRATO G., DELFOUR M. & MITTER S.K. (1992) REPRESENTATION AND CONTROL OF INFINITE DIMENSIONAL SYSTEMS, Birkhäuser., Vol. 1.

[7] BERTINI L. & CANCRINI N. (1995) *The stochastic heat equation: Feynman–Kač formula and intermittence,* J. Stat. Phys. **78**, 1377-1401.

[8] BERTINI L., CANCRINI N. & JONA–LASINIO G. (1994) *The stochastic Burgers equation,* Commun. Math. Phys. **165**, 211–232.

[9] BHATT A. G. & KARANDIKAR R. L. (1993) *Invariant measures and evolution equations for Markov processes characterized via Martingale problems,* Annals of Probability, **21**, 2246–2268.

[10] BLUMENTHAL R. M. & GETOOR R. K. (1968) MARKOV PROCESSES AND POTENTIAL THEORY, Academic Press.

[11] BOGACHEV V. I. & RÖCKNER M. *Regularity of invariant measures on finite and infinite dimensional spaces and applications,* Preprint 1994. To appear in J. Funct. Anal.

[12] BOGACHEV V. I. , RÖCKNER M. & SCHMULAND B. (1994) *Generalized Mehler semigroups and applications,* Preprint 94–088, Universität Bielefeld.

[13] BRACE A. & GĄTAREK D. (1995) *The marked model of interest rate dynamics,* UNSW Department of Statistics. Report No. S95-2.

[14] BRACE A. & MUSIELA M. (1994) *A multifactor Gauss Markov implementation of Health, Jarrow and Morton,* Mathematical Finance, 4, 3, 259–283.

[15] BURGERS J. M. (1970) THE NONLINEAR DIFFUSION EQUATION. ASYMPTOTIC SOLUTIONS AND STATISTICAL PROBLEMS, Reidel.

[16] CAHN J. W. & HILLIARD J. E. (1958) *Free energy for a non–uniform system I. Interfacial free energy,* J. Chem. Phys, **2**, 258–267.

[17] CERRAI S. (1994) *A Hille-Yosida theorem for weakly continuous semigroups,* Semigroup Forum, **49**, 349–367.

[18] CERRAI S. (to appear) *Elliptic and parabolic equations in $\mathbb{R}^n$ with coefficients having polynomial growth,* Comm. Partial Diff. Eq.

[19] CERRAI S. (1995) *Invariant measures for a class of SDEs with drift term having polynomial growth,* Preprint Scuola Normale Superiore, Pisa.

[20] CERRAI S. & GOZZI F. (1994) *Strong solutions of Cauchy problems associated to weakly continuous semigroups,* Differential and Integral Equations, **8**, 3, 465–486.

[21] CHAMBERS D. H., ADRIAN R. J., MOIN P., STEWART D.S. & SUNG H.J. (1988) *Karhunen–Loève expansion of Burgers' model of turbulence,* Phys. Fluids **31**, (9), 2573–2582.

[22] CHOI H., TEMAM R., MOIN P. & KIM J. (1992) *Feedback control for unsteady flow and its application to Burgers equation,* Center for Turbulence Research, Stanford University, CTR Manuscript 131. To appear in J. Fluid Mechanics.

[23] CHOJNOWSKA-MICHALIK A. (1978) *A semigroup approach to boundary problems for stochastic hyperbolic systems,* preprint.

[24] CHOJNOWSKA-MICHALIK A. (1978) *Representation theorem for general stochastic delay equations,* Bull. Acad. Pol. Sci. Ser. Sci. Math., **26**, 7, 634–41.

[25] CHOJNOWSKA-MICHALIK A. (1987) *On processes of Ornstein–Uhlenbeck in Hilbert spaces,* Stochastics, **21**, 251–286.

[26] CHOJNOWSKA-MICHALIK A. & GOLDYS B. (1995) *Existence, uniqueness and invariant measures for stochastic semilinear equations on Hilbert spaces,* Probability Theory and Relat. Fields, **102**, 331–356.

[27] CHOW P. L. (1987) *Expectation functionals associated with some stochastic evolution equations*, Lecture Notes in Mathematics No. 1236, eds. G. Da Prato & L. Tubaro, Springer–Verlag, 40–56.

[28] CHOW P.L. & MENALDI J.L. (1990) *Exponential estimates in exit probability for some diffusion processes in Hilbert spaces*. Stochastics, **23**, 377-393.

[29] CRAUEL H. & FLANDOLI F. (1994) *Attractors for Random Dynamical Systems*, Probab. Theory Relat. Fields **100**, 365–393.

[30] CRAUEL H., DEBUSSCHE A. & FLANDOLI F. (1995) *Random attractors* , Preprint No. 27, Université de Paris Sud.

[31] CRUZEIRO A. B. (1979) *Solutions et mesures invariantes pour des équations d'évolution stochastiques du type Navier–Stokes*, Expositiones Mathematicae, **7**, 73–82.

[32] CURTAIN R.F. & ZWART H. J. (1995) AN INTRODUCTION TO INFINITE–DIMENSIONAL LINEAR SYSTEMS, Springer–Verlag.

[33] DAH–TENG JENG (1969) *Forced model equation for turbulence*, The Physics of Fluids, **12**, 10, 2006–2010.

[34] DA PRATO G. (1976) APPLICATIONS CROISSANTES ET ÉQUATIONS D'ÉVOLUTIONS DANS LES ESPACES DE BANACH, Academic Press.

[35] DA PRATO G. (1995) *Null controllability and strong Feller property of Markov transition semigroups*, Nonlinear Analysis TMA, **25**, 9–10, 941–949.

[36] DA PRATO G., DEBUSSCHE A. & TEMAM R. (1994) *Stochastic Burgers equation*, NoDEA, 389–402.

[37] DA PRATO G., ELWORTHY D. & ZABCZYK J. (1995) *Strong Feller property for stochastic semilinear equations*, Stochastic Analysis and Applications, **13**, n.1, 35–45.

[38] DA PRATO G., & GĄTAREK (1995) *Stochastic Burgers equation with correlated noise,* Stochastics and Stochastic Reports, **52**, 29–41.

[39] DA PRATO G., GĄTAREK D. & ZABCZYK J. (1992) *Invariant measures for semilinear stochastic equations,* Stochastic Analysis and Applications. **10**, n.4, 387-408.

[40] DA PRATO G., MALLIAVIN P. & NUALART D. (1992) *Compact families of Wiener functionals,* C. R. Acad. Sci. Paris, 315, 1287–1291.

[41] DA PRATO G. & PARDOUX E. (1995) *Invariant measures for white noise driven stochastic partial differential equations,* Stochastic Analysis and Applications, **13**, (3), 295–305.

[42] DA PRATO G. & ZABCZYK J. (1991) *Smoothing properties of transition semigroups in Hilbert spaces,* Stochastics and Stochastic Reports, **35**, 63-77.

[43] DA PRATO G. & ZABCZYK J. (1992) *Non–explosion, boundedness and ergodicity for stochastic semilinear equations,* J. Differential E. **98**, 1, 181–195.

[44] DA PRATO G. & ZABCZYK J. (1992) STOCHASTIC EQUATIONS IN INFINITE DIMENSIONS. Encyclopedia of Mathematics and its Applications, Cambridge University Press.

[45] DA PRATO G. & ZABCZYK J. (1993) *Evolution equations with white-noise boundary conditions,* Stochastics and Stochastics Reports, **42**, 167–182.

[46] DA PRATO G. & ZABCZYK J. (1994) *Convergence to equilibrium for spin systems.* Scuola Normale Superiore, Pisa, Preprint No.1.

[47] DA PRATO G. & ZABCZYK J. (1995) *Regular densities of invariant measures for nonlinear stochastic equations.* J. Functional Analysis, **130**, 427–449.

[48] DA PRATO G. & ZABCZYK J. (to appear) *Convergence to equilibrium for classical and quantum spin systems*, Probability Theory and Relat. Fields.

[49] DAVIES E. B. (1980) ONE PARAMETER SEMIGROUPS, Academic Press.

[50] DAWSON D. A. (1975) *Stochastic evolution equations and related measure processes*, J. Multivar. Anal. **5**, 1–52.

[51] DAWSON D. A. (1977) *The critical measure diffusion process*, Z. Wahr. verw. Gebiete, **40**, 125–145.

[52] DAWSON D. A. & PAPANICOLAU G. C. (1984) *Random wave process*, Appl. Math. Optimiz. **12**, 97–114.

[53] DAWSON D. A. & SALEHI H. (1980) *Spatially homogeneous random evolutions*, J. Multivar. Anal. **10**, 141–180.

[54] DERMOUNE A. (to appear) *Around the stochastic Burgers equation*, Stochastic Analysis and Applications.

[55] DONATI–MARTIN C . & PARDOUX E. (1993) *White noise driven SPDEs with reflection*, Probab. Theory Relat. Fields 95, 413–425.

[56] DOOB J. L. (1948) *Asymptotic properties of Markoff transition probabilities*. Trans. Amer. Math. Soc. **63**, 394–421.

[57] DOSS H. & ROYER G. (1978) *Processus de diffusions associées aux mesures de Gibbs sur* $\mathbb{R}^{\mathbb{Z}^d}$, Z. Wahr. verw. Gebiete, **46**, 107–124.

[58] DYM H. (1966) *Stationary measures for the flow of a linear differential equation driven by white noise*, Trans. Amer. Math. Soc. **123**, 130–167.

[59] DYNKIN E. B. (1965) MARKOV PROCESSES I, II, Springer–Verlag.

[60] ECHEVERIA P. E. (1982) *A criterion for invariant measures of Markov processes,* Z. Wahr. verw. Gebiete, **61**, 1–16.

[61] ELWORTHY K. D. (1992) *Stochastic flows on Riemannian manifolds.* In M.A. Pinsky and V. Vihstutz, editors, Diffusion processes and related problems in analysis, Vol. II, 33–72, Birkhäuser.

[62] ERICKSON R. (1971) *Constant coefficient linear differential equations driven by white noise,* Ann. Math. Statistics, **42**, 820–823.

[63] ETHIER S. N. & KURTZ T. G. (1986) MARKOV PROCESSES: CHARACTERIZATION AND CONVERGENCE, Wiley.

[64] FELLER W. (1966) AN INTRODUCTION TO PROBABILITY THEORY AND ITS APPLICATIONS, Wiley.

[65] FERRARIO B. (to appear) *The Bénard problem with random perturbations: dissipativity and invariant measures,* NoDEA.

[66] FLANDOLI F. (1994) *Dissipativity and invariant measures for stochastic Navier–Stokes equations,* NoDEA, **1**, 403–423.

[67] FLANDOLI F. & MASLOWSKI B. (1995) *Ergodicity of the 2–D Navier–Stokes equation under random perturbations,* Commun. Math. Phys. **171**, 119–141.

[68] FORTET R. & MOURIER B. (1953) *Convergence de la répartition empirique vers la répartition théoretique,* Ann. Sc. École Normale, Sup., **70** , 267–285.

[69] FREIDLIN M. I (1988) *Random perturbations of reaction–diffusion equations: The quasi–deterministic approximation,* Trans. Amer. Math. Soc. **305**, 2, 665–697.

[70] FREIDLIN M. I & WENTZELL A. (1993) *Reaction-diffusions equations with randomly perturbed boundary conditions*, Annals of Probability, **20**, 963–986.

[71] FRITZ J. (1982) *Infinite lattice systems of interacting diffusion processes, existence and regularity properties*, Z. Wahr. verw. Gebiete, **59**, 291–309.

[72] FRITZ J. (1982) *Stationary measures of stochastic gradient systems. Infinite lattice models*, Z. Wahr. verw. Gebiete, **59**, 479–490.

[73] FUHRMAN M. (1995) *Analyticity of transition semigroups and closability of bilinear forms in Hilber spaces*, Studia Mathematica, **115**, 53–71.

[74] FUKUSHIMA M. (1980) DIRICHLET FORMS AND MARKOV PROCESSES, North Holland.

[75] FUNAKI T. (1983) *Random motions of string and related stochastic evolution equations*, Nagoya Math. J. **89**, 129–93.

[76] GARSIA A. M. RADEMICH E. & RUMSEY H. Jr. (1970) *A real variable lemma and the continuity of paths of some gaussian process*, Indiana University Mathematical Journal, **20**, 565–578.

[77] GĄTAREK D. & GOLDYS B. (1994) *On invariant measures for diffusions on Banach spaces*, University of New South Wales, Department of Statistics, Preprint S94-3, to appear in Potential Analysis.

[78] GAVEAU B. & MOULINIER J. M. (1985) *Régularité des mesures et perturbations stochastiques de champs de dimension infinie*, Publications of the Research Institute for Mathematical Sciences, Kyoto University, **21**, 3, 593–16.

[79] GOLDYS B. & MUSIELA M. (to appear) *Lognormality of rates and term structure models*, Mathematical Finnce.

[80] GYÖNGY I. (1995) *On non–degenerate quasi–linear stochastic parabolic partial differential equations*, Potential Analysis, **4**, 157–171.

[81] GYÖNGY I. & PARDOUX E. (1993) *On the regularization effect of space–time white noise on quasi–linear parabolic partial differential equations*, Probab. Theory Relat. Fields, **97**, 211–229.

[82] HALE J. (1977) THEORY OF FUNCTIONAL DIFFERENTIAL EQUATIONS, Springer–Verlag.

[83] HALMOS P. R. (1967) A HILBERT SPACE PROBLEM BOOK, Van Nostrand.

[84] HAYES N. D. (1950) *Roots of the transcendental equation associated with a certain differential difference equation*, J. London Math. Soc. **25**, 226–232.

[85] HENRY D. (1981) GEOMETRIC THEORY OF SEMILINEAR PARABOLIC EQUATIONS, Springer–Verlag.

[86] HILLE E. & PHILLIPS R. S. (1965) FUNCTIONAL ANALYSIS AND SEMIGROUPS, Amer. Math. Soc. Coll. Pub. Vol. XXXI.

[87] HOLDEN H, LINDSTRØM T., ØKSENDAL B., UBØE J., & ZHANG T. S. (1994) *The Burgers equation with a noisy force and the stochastic heat equation*, Comm. Patial Diff. Eq. **19**, 119–141.

[88] HOLLEY R. A. & STROOK D. W. (1978) *Generalized Ornstein–Uhlenbeck processes and infinite particle branching Brownian motions*, Publ. RIMS, Kyoto Univeristy, **14**, 741–788.

[89] HOLLEY R. A. & STROOK D. W. (1980) *The D. L. R. conditions for translation invariant Gaussian measures in* $S'(\mathbb{R}^d)$, Z. Wahr. verw. Gebiete, **53**, 293–304.

[90] HOLLEY R. A. & STROOK D. (1981) *Diffusions in infinite dimensional torus*, J. Functional Analysis, **42**, 29–53.

[91] ICHIKAWA A.(1985) *A semigroup model for parabolic equations with boundary and pointwise noise*, in Stochastic Space-Time Models and Limit Theorems, L.Arnold and P. Kotelenez Editors, Reidel Publishing Company, (1985), 1–94.

[92] ITO K. & NISIO M. (1964) *On stationary solutions of a stochastic differential equation*, J. Math. Kyoto Univ. 4, 1, 1–75.

[93] JACQUOT S. & ROYER G.(1995) *Ergodicité d'une classe d'équations aux dérivées partielles stochastiques*, C. R. Acad. Sci. Paris, 320, 231–236.

[94] KARDAR M., PARISI M. & ZHANG J.C. (1986) *Dynamical scaling of growing interfaces*, Phys. Rev. Lett. **56**, 889–892.

[95] KHAS'MINSKII R. Z. (1960) *Ergodic properties of recurrent diffusion processes and stabilization of the solution to the Cauchy problem for parabolic equations*, Theory of Probability and its applications, Vol. 5, No 2, 179–196.

[96] KHAS'MINSKII R. Z. (1980) STOCHASTIC STABILITY OF DIFFERENTIAL EQUATIONS, Sijthoff and Noordhoff.

[97] KIFER Yu. (1995) *The Burgers equation with a random force and a general model for directed polymers in random environments*, preprint.

[98] KUSUOKA S. & STROOCK D. W. (1985) *Properties of Certain Stationary Diffusion Semigroups*, J. Funct. Anal. **60**, 243–264.

[99] LASIECKA I. & TRIGGIANI R. (1981) *A cosine operator approach to modelling $L_2(0, T; L_2(\Gamma))$–boundary imput hyperbolic equations*, Applied Math. Optimization, 7, 35–93.

[100] LASOTA A. (to appear) *From fractals to stochastic differential equations* , Proceedings of the XXI Winter

School of Theoretical Physics, Karpacz 1995, Editors P. Garbaczewski and A. Weron.

[101] LASOTA A. & YORKE J. A. (1994) *Lower bound technique for Markov operators and iterated functions systems*, Random and Computational Dynamics, **2**, 41–77.

[102] LEBOWITZ J. & PRESUTTI E. (1976) *Statistical mechanics of systems of unbounded spins*, Comm. Math. Phys. **50**, 125.

[103] LEHA G. & RITTER G. (1984) *On diffusion processes and their semigroups in Hilbert spaces with an application to interacting stochastic systems*, Annals of Probability, **12**, n. 4, 1077– 1112.

[104] LEHA G. & RITTER G. (1985) *On solutions to stochastic differential equations with discontinuous drift in Hilbert spaces*, Matematische Annalen, **270**, 109–123.

[105] LEHA G. & RITTER G. (1994) *Lyapunov type conditions for stationary distributions of diffusion processes on Hilbert spaces*, Stochastics and Stochastic Reports, **48**, 195–225.

[106] LEHA G. & RITTER G. (to appear) *Stationary distributions of diffusion processes with singular drift on Hilbert spaces*, preprint.

[107] LIONS J.L & MAGENES E. (1968) PROBLEMES AUX LIMITES NON HOMOGENES ET APPLICATIONS, Dunod.

[108] LUNARDI A. (1995) ANALYTIC SEMIGROUPS AND OPTIMAL REGULARITY IN PARABOLIC PROBLEMS, Birkhäuser.

[109] MA Z. M. & ROCKNER M. (1992) INTRODUCTION TO THE THEORY OF (NON–SYMMETRIC) DIRICHLET FORMS, Springer–Verlag.

[110] MANITIUS A. (1981) *Necessary and sufficient conditions of approximate controllability to general linear retarded systems*, SIAM J. Control Optimiz. **19**, 516–532.

[111] MAO X. & MARCUS L. (1990) *Wave equation with stochastic boundary values*, Mathematical Institute, University of Warwick, preprint.

[112] MARCUS R. (1974) *Parabolic Ito equations* , Trans. Amer. Math. Soc. **198**, 177–190.

[113] MARCUS R. (1978) *Parabolic Ito equations with monotone nonlinearities*, J. Functional Analysis, **29**, 275–286.

[114] MARCUS R. (1979) *Stochastic diffusions on an unbounded domain*, Pacific J. Mathematics, **84**, n.1, 143–153.

[115] MARTIN R. (1970) *A global existence theorem for autonomous differential equations in Banach spaces*, Proc. Amer. Math. Soc. **26**,307–314.

[116] MASLOWSKI B. (1989) *Strong Feller property for semilinear stochastic evolution equations and applications*, in Proceedings of IFIP Conference on Stochastic Systems and Optimization, Warsaw 1989, Lecture Notes in Control and Information Sciences No. 136, ed. J. Zabczyk, Springer–Verlag, 210–235.

[117] MASLOWSKI B. (1992) *On probability distributions of solutions of a semilinear stochastic PDE*, Akademie věd České Republiky, Mathematický Ústav.

[118] MASLOWSKI B. (to appear) *Stability of semilinear equations with boundary and pointwise noise*, Annali Scuola Normale Superiore di Pisa.

[119] MOHAMMED S. A. & SCHEUTZOW M. (1990) *Lyapunov exponents and stationary solutions for affine stochastic delay equations*, Stochastics and Stochastic Reports, **29**, 259–283.

[120] MUELLER C. (1993) *Coupling and invariant measures for the heat equation with noise*, Ann. Prob. **21**, 2189–2199.

[121] MUSIELA M. (1993) *Stochastic PDEs and term structure models*, International Conference in Finance, La Baule.

[122] OLBROT A. W. & PANDOLFI L. (1988) *Null controllability of a class of functional differential systems*, Int. J. Control, **47**, 193–208.

[123] PESZAT S. (1993) *On a Sobolev space of function of infinite numbers of variables*, Preprint 510 of Institute of Mathematics, Polish Academy of Sciences, 1993, to appear in Bull. Pol. Acad. Sci.

[124] PESZAT S. & ZABCZYK J. (to appear) *Strong Feller property and irreducibility for diffusions on Hilbert spaces*, Annals of Probability.

[125] PETERSEN K. (1983) ERGODIC THEORY, Cambridge University Press.

[126] PRITCHARD A. & ZABCZYK J. (1981) *Stability and stabilizability of infinite dimensional systems*, SIAM Review, **23**, 25–52

[127] REED M. & SIMON B. (1972) METHODS OF MODERN MATHEMATICAL PHYSICS, Academic Press, New York and London, Vol. I.

[128] ROYER G. (1979) *Processus de diffusions associées à certains modèls d'Ising à spin continu*, Z. Wahr. verw. Gebiete, **59**, 479–490.

[129] ROZANOV A. Yu. (1991) On stochastic evolution equations with nonhomogeneous boundary conditions, Preprint.

[130] RUELLE D. (1976) *Probability estimates for continuous spin systems*, Comm. Math. Phys. **50**, 189.

[131] SCHEUTZOV M. (1982) *Qualitatives Verschatten der Lösungen von eindimensionalen nichtlinearen stochastischen Differentialgleichungen mit Gedächtnis*, Ph. D. thesis, Keiserslantern.

[132] SCHEUTZOV M. (1984) *Qualitative behaviour of stochastic delay equations with a bounded memory,* Stochastics, **12**, 41–80.

[133] SCHEUTZOV M. (1988) *Stationary and periodic stochastic differential systems,* Habilitationsschrift, Keiserslantern.

[134] SCHMIDT G. (1988) *Some introductory lectures on boundary controllability of partial differential equations,* Departement of Mathematics and Statistics Mc–Gill University.

[135] SCHWARTZ L. (1966) MATHEMATICS FOR THE PHYSICAL SCIENCES, Hermann.

[136] SEIDLER J. (1994) *Ergodic behaviour of stochastic parabolic equations,* Preprint 91, Akademie vĕd České Republiky, Mathematický Ústav.

[137] SEIDMAN T. (1988) *How violent are fast controls?,* Math. Control Signals Systems **1**, 89–95.

[138] SHIGEKAWA I. (1980) *Derivatives of Wiener functionals and absolute continuity of induced measures,* J. Math. Kyoto Univ. **20**, n.2, 263–289.

[139] SHIGEKAWA I. (1987)*Existence of invariant measures of diffusions on an abstract Wiener space,* Osaka J. Math., **24**, 37–59.

[140] SICKEL W. (1987) *On pointwise multipliers in Besov-Triebel–Lizorkin spaces,* in Seminar Analysis of the Karl-Weierstrass–Institute 1985–86, Teubner–Texte Math. **96**, Teubner, 45–103.

[141] SIMAO I. (1993) *Regular transition densities for infinite dimensional diffusions,* Stochastic Analysis and Applications, **11**, 309–336.

[142] SKOROKHOD A. V. (1987) Asymptotic methods in the theory of stochastic differential equations, Naukova Dumka. (in Russian).

[143] SNYDERS J. & ZAKAI M. (1970) *Stationary probability measures for linear differential equations driven by white noise*, J. Diff. Eq. 8, 27–32.

[144] SOWERS R. (1991) *New asymptotic results for stochastic partial equations*, Ph. D. dissertation, Dept. of Math. University of Maryland.

[145] SOWERS R. (1992) *Large deviations for the invariant measure of a reaction–diffusion equation with non Gaussian perturbation*, Prob. Theory Related Fields, **92**, 393–491.

[146] SOWERS R. (1994) *Multi–dimensional reaction–diffusion equations with white noise boundary perturbations*, Annals of Probability, **22**, 2071–2121.

[147] STETTNER L. (to appear) *Remarks on ergodic conditions for Markov processes on Polish spaces*, Bull. Polish. Acad. Sci.

[148] STROOCK D. W. & VARHADAN S.R.S (1979) Multidimensional Diffusion Processes, Springer–Verlag.

[149] STROOK D. W. & ZEGARLINSKI B. (1992) *The logaritmic Sobolev inequality for discrete spin systems on a lattice*, Commun. Math. Phys., **149**, 175–193.

[150] TEMAM R. (1984) Navier–Stokes Equations, North–Holland.

[151] TESSITORE G. & ZABCZYK J. *Invariant measures for Burgers equation*, preprint.

[152] T. VARGIOLU (1995), Tesi Scuola Normale Superiore di Pisa.

[153] VINTSCHGER R. V. (1989) *The existence of invariant measures for $C([0,1])$-valued diffusions*, Prob. Theory and Rel. Fields, **82**, 307–313.

[154] VIOT M. (1974) *Solution en lois d'une équation aux dérivées partielles non linéaires: Methodes de compacité*, C. R. Acad. Sci. Paris Sér. A, Math. **278**, 1185–1188.

[155] VIOT M. (1974) *Solution en lois d'une équations aux dérivées partielles non linéaires: methodes de monotonie*, C. R. Acad. Sci. Paris Sér. A, Math. **278**, 1405–1408.

[156] VIOT M. (1976) *Solutions faibles d'équations aux dérivées partielles non linéaires*, Thèse, Université Pierre et Marie Curie, Paris.

[157] VRKOC I. (1993) *A dynamical system in a Hilbert space with a weakly attractive nonstationary point*, Mathematica Bohemica, **118**, n.4, 401–423.

[158] WALSH J. B. (1984) *An introduction to stochastic partial differential equations*, in École d'eté de Probabilité de Saint Flour, XIV, ed. P.L. Hennequin, Lecture Notes in Mathematics No. 1180, 265–439.

[159] WEBB G. F. (1972) *Continuous nonlinear perturbations of linear accretive operators*, J. Functional Analysis, **10**, 191–203.

[160] WEBB G. F. (1974) *Autonomous nonlinear functional differential equations and nonlinear semigroups*, J. Math. Anal. Appl. **46**, 1–12.

[161] WEBB G. F. (1976) *Functional differential equations and nonlinear semigroups in L^p spaces*, J. Diff. Equations **20**, 71–89.

[162] YOSIDA K. (1965) FUNCTIONAL ANALYSIS, Springer-Verlag.

[163] ZABCZYK J. (1977) *Infinite dimensional systems in control theory*, Bull. Int. Stat. Inst. XLVII (2).

[164] ZABCZYK J. (1977) *Linear stochastic systems in Hilbert spaces: spectral properties and limit behaviour*, Institute of Mathematics, Polish Academy of Sciences, report No. 236. Also in Banach Center Publications vol. 41, (1985), 591–609.

[165] ZABCZYK J. (1980) *On stochastic controllability*, FDS Report No. 34, Universität Bremen.

[166] ZABCZYK J. (1981) *Controllability of stochastic linear systems*, Systems and Control Letters, 1, 25–35.

[167] ZABCZYK J. (1983) *Stationary distributions for linear equations driven by general noise*, Bull. Pol. Acad. Sci. 31, 197–209.

[168] ZABCZYK J. (1989) *Symmetric solutions of semilinear stochastic equations*, Lecture Notes in Mathematics No. 1390, eds. G. Da Prato & L. Tubaro, 237-56, Springer–Verlag.

[169] ZABCZYK J. (1992) J. MATHEMATICAL CONTROL THEORY. AN INTRODUCTION, Birkhäuser.

[170] ZEGARLINSKI B. *The strong exponential decay to equilibrium for the stochastic dynamics associated to the unbounded spin systems on a lattice*, preprint.

Index

Angle variable, 9
Approximately controllable, 141

Birkhoff's ergodic theorem, 6
Boundary noise, 239
Burgers equation, 255
    existence of invariant measure, 266
    strong Feller property, 263
    uniqueness of invariant measure, 274

Configuration, 223

Damped wave equation, 182
Dirichlet form, 19, 165
Dissipative mapping, 73
Doob's theorem, 43
Dynamical system, 3
    angle variable, 9
    canonical, 16
    continuous, 4
    ergodic, 4
    infinitesimal generator, 5
    invariant set, 9
    strongly mixing, 5
    weakly mixing, 4

Factorization formula, 58
Feller semigroups, 19
Fortet–Mourier norm, 39

Genuinely dissipative systems, 114
Gibbs measure, 223
Global interactions, 223
Gradient systems, 160

Invariant measure, 12
    dissipative case, existence, 105, 117
    absolutely continuous, 148, 158, 159, 162
    density, 147
    ergodic, 22
    existence, 191
    existence from boundedness, 89
    existence in the linear case, 97
    exponentially mixing, 40
    recurrence, 102
    regularity of density, 165, 168, 170
    strongly mixing, 33
    uniqueness, 191
    weakly mixing, 33

Kolmogorov equation, 71, 135
    mild, 136
Koopman–von Neumann theorem, 5

Krylov–Bogoliubov theorem, 21

Local interactions, 223

Markovian semigroup, 12
    Feller, 20
    irreducible, 42
    recurrent, 37
    regular, 41
    stochastically continuous, 12
    strongly Feller, 42
    symmetric, 18
Markovian transition function, 12

Navier–Stokes equation, 279
    existence of invariant measure, 293
Null controllability, 130

Ornstein–Uhlenbeck process, 56, 177
    in chaotic environment, 186
    in Finance, 184
    wave type, 178

Random environment, 53
Reaction–diffusion equations, 210

Sobolev spaces, 149
    compact embedding, 152
Spectral measure, 54
Spin systems, 223
    classical, 225
    quantum lattice, 233
Stationary measure, 12
Stochastic convolution, 56
    smoothing properties, 305
Stochastic evolution equation, 65

    dependence on initial conditions, 69
    dissipative, 80, 87
    linear, 96
    mild solution, 66
    second order dissipative, 181
    wave type, 178
    weak solution, 56
    with delay, 197
Strong law of large numbers, 30
Strongly dissipative mapping, 73

Transition function, 12
Transition semigroup, 12, 56, 70, 97, 105, 122, 148
    irreducible, 137, 141
    strong Feller, 121, 129, 134, 153, 200

Wiener process, 53
    coloured, 53
    cylindrical, 53

Printed in the United States
By Bookmasters